Biosurfactants: A Boon to Healthcare, Agriculture & Environmental Sustainability

Edited by

Arun Kumar Pradhan

Centre for Biotechnology
Siksha 'O' Anusandhan (Deemed to be University)
Bhubaneswar, Odisha-751030
India

&

Manoranjan Arakha

Centre for Biotechnology
Siksha 'O' Anusandhan (Deemed to be University)
Bhubaneswar, Odisha-751030
India

Biosurfactants: A Boon to Healthcare, Agriculture & Environmental Sustainability

Editors: Arun Kumar Pradhan and Manoranjan Arakha

ISBN (Online): 978-981-5196-92-4

ISBN (Print): 978-981-5196-93-1

ISBN (Paperback): 978-981-5196-94-8

Published by Bentham Science Publishers Pte. Ltd. Singapore. All Rights Reserved.

First published in 2024.

need for a court order if at any point you breach any terms of this License Agreement. In no event will any delay or failure by Bentham Science Publishers in enforcing your compliance with this License Agreement constitute a waiver of any of its rights.

3. You acknowledge that you have read this License Agreement, and agree to be bound by its terms and conditions. To the extent that any other terms and conditions presented on any website of Bentham Science Publishers conflict with, or are inconsistent with, the terms and conditions set out in this License Agreement, you acknowledge that the terms and conditions set out in this License Agreement shall prevail.

Bentham Science Publishers Pte. Ltd.
80 Robinson Road #02-00
Singapore 068898
Singapore
Email: subscriptions@benthamscience.net

CONTENTS

PREFACE

Biosurfactants are surface-acting molecules isolated from microbes (bacteria and fungi) and plants. The amphipathic nature of biosurfactants makes their application broader. Different types of biosurfactants are isolated and characterized *i.e.* glycolipids, lipopeptides, phospholipids and polymerics. These molecules have emulsifying, wetting, and foaming properties. So biosurfactants are used in pharmaceuticals, cosmetics, food, mining, and petrochemical industries. Research confirms their antimicrobial, anticancer and anti-biofilm formation capacities. They can remove oil from the contaminated soil of oil mines. Biosurfactants can be used in oil spillage aquatic areas to remove oil. They can scavenge heavy metals in contaminated soil and make the environment free from xenobiotics. With the help of nanotechnology, biosurfactants can be used to detect heavy metals in the environment. Nowadays, biosurfactants are used as potential agents in agricultural post-harvest technology. They play an important role in agrochemical industries.

In view of the above discussion, the overarching aim of this book is to discuss major applications of biosurfactants in the fields of healthcare, agriculture and the environment.

The unique features of this book are as follows:

• The book is a contribution of experts from microbiology, cancer biology, pharmaceutical science, nanotechnology, plant biotechnology and environmental sciences.

• The book discusses some novel applications of biosurfactants in the fields of medicine, agriculture and the environment.

The book, in total, comprises 11 chapters written by experts working in the respective aspects of biosurfactants. For instance, Chapter 1 and 2 describe the screening of biosurfactant-producing organisms, the production of biosurfactants and their various applications. Chapter 3 and 4 interpret the application of biosurfactants in the healthcare sector, like various cancer treatments and neurological disorders.

Agricultural applications and food processing with the help of biosurfactants are well explained in chapters 5 and 6. Biosurfactants show vast applications in the environmental cleaning sector. Bioremediation, removal of heavy metal from soil and water by biosurfactants, and removal of oil from contaminated soil by biosurfactants are described in chapters 8, 9, 10 and 11.

We are very much thankful to all the contributing authors for their timely cooperation and support, without which this book would not have taken its final shape. We are also thankful to the series editors for their critical reviews and positive remarks to bring positive developments in the book.

Arun Kumar Pradhan
Centre for Biotechnology
Siksha 'O' Anusandhan (Deemed to be University)
Bhubaneswar, Odisha-751030
India

&

Manoranjan Arakha
Centre for Biotechnology
Siksha 'O' Anusandhan (Deemed to be University)
Bhubaneswar, Odisha-751030
India

List of Contributors

Ananya Kuanar — Centre for Biotechnology, Siksha 'O' Anusandhan (Deemed to be University), Bhubaneswar, Odisha-751030, India

Arun Kumar Pradhan — Centre for Biotechnology, Siksha 'O' Anusandhan (Deemed to be University), Bhubaneswar, Odisha-751030, India

Arabinda Jena — Fisheries, UNDP (Collaboration with Directorate Fisheries), Cuttack, Odisha-753001, India

Banishree Sahoo — Centre for Biotechnology, Siksha 'O' Anusandhan (Deemed to be University), Bhubaneswar, Odisha-751030, India

Bhabani Shankar Das — Centre for Biotechnology, School of Pharmaceutical Sciences, Siksha 'O' Anusandhan (Deemed to be University), Bhubaneswar, Odisha-751030, India

Elina Khatua — Environment and Sustainability Department, CSIR-Institute of Minerals and Materials Technology, Bhubaneswar, Odisha-751003, India
Academy of Scientific and Innovative Research, Ghaziabad, Uttar Pradesh, 201002, India

Gargi Balabantaray — Department of Immunology and Rheumatology, Institute of Medical Science, Sum Hospital, Siksha 'O' Anusandhan (Deemed to be University), Bhubaneswar, Odisha-751003, India

Manisha Mahapatra — Centre for Biotechnology, Siksha 'O' Anusandhan (Deemed to be University), Bhubaneswar, Odisha-751030, India

Manoranjan Arakha — Centre for Biotechnology, Siksha 'O' Anusandhan (Deemed to be University), Bhubaneswar, Odisha-751030, India

Meena Choudhary — Analytical and Environmental Science Division & Centralized Instrument Facility, CSIR-Central Salt & Marine Chemicals Research Institute, G.B. Marg, Bhavnagar-364002, India
Academy of Scientific and Innovative Research (AcSIR), Ghaziabad-201002, India

Monali Muduli — Analytical and Environmental Science Division & Centralized Instrument Facility, CSIR-Central Salt & Marine Chemicals Research Institute, G.B. Marg, Bhavnagar-364002, India
Academy of Scientific and Innovative Research (AcSIR), Ghaziabad-201002, India

Muchalika Satapathy — AIPH University, Bhubaneswar, India

Nilotpala Pradhan — Environment and Sustainability Department, CSIR-Institute of Minerals and Materials Technology, Bhubaneswar, Odisha-751003, India
Academy of Scientific and Innovative Research, Ghaziabad, Uttar Pradesh, 201002, India

Pratyasha Panda — AIPH University, Bhubaneswar, India

Pradeepta Sekhar Patro — Department of Immunology and Rheumatology, Institute of Medical Science, Sum hospital, Siksha 'O' Anusandhan (Deemed to be University), Bhubaneswar, Odisha-751003, India

Pyari Payal Beura Department of Biotechnology, Rama Devi Women's University, Vidya Vihar, Bhubaneswar, Odisha-751003, India

Sameer Ranjan Sahoo Centre for Biotechnology, Siksha 'O' Anusandhan (Deemed to be University), Bhubaneswar, Odisha-751030, India

Sanak Ray Analytical and Environmental Science Division & Centralized Instrument Facility, CSIR-Central Salt & Marine Chemicals Research Institute, G.B. Marg, Bhavnagar-364002, India
Academy of Scientific and Innovative Research (AcSIR), Ghaziabad-201002, India

Sanjay Kumar Raul Department of Biotechnology, Rama Devi Women's University, Vidya Vihar, Bhubaneswar, Odisha-751003, India

Sibani Sahoo Gangadhar Meher University, Sambalpur, India

Soumyashree Rout Department of Neurology, Siksha 'O' Anusandhan (Deemed to be University), Bhubaneswar, Odisha-751003, India

Srikanta Kumar Sahoo Department of Neurology, Siksha 'O' Anusandhan (Deemed to be University), Bhubaneswar, Odisha-751003, India

Swapnashree Satapathy Centre for Biotechnology, Siksha 'O' Anusandhan (Deemed to be University), Bhubaneswar, Odisha-751030, India

Swastika Mallick Environment and Sustainability Department, CSIR-Institute of Minerals and Materials Technology, Bhubaneswar, Odisha-751003, India
Academy of Scientific and Innovative Research, Ghaziabad, Uttar Pradesh, 201002, India

Truptirekha Das Rautara Govt. High School, Binjharpur, Jajapur,, Odisha-755004, India

Twinkle Rout Department of Surgical Oncology, Institute of Medical Sciences and Sum Hospital, Siksha 'O' Anusandhan (Deemed to be University), Bhubaneswar, Odisha-751003, India

CHAPTER 1

Biosurfactants: An Amazing Bioactive Compound

Sameer Ranjan Sahoo[1], **Manisha Mahapatra**[1] and **Arun Kumar Pradhan**[1,*]

[1] *Centre for Biotechnology, Siksha 'O' Anusandhan (Deemed to be University), Bhubaneswar, Odisha-751003, India*

Abstract: Biosurfactants are organic compounds synthesized by microorganisms such as bacteria, yeasts, or fungi. These compounds exhibit intricate chemical compositions and unique physical properties, often surpassing or rivaling synthetic surfactants. Furthermore, they typically have low toxicity towards freshwater, marine, and terrestrial ecosystems, making them environmentally favorable for various applications. To date, the primary focus of biosurfactant research has centered on enhancing the biodegradation and recovery of oil. These substances have demonstrated their utility in aiding the removal of hazardous contaminants from polluted areas due to their capacity to solubilize and emulsify harmful pesticides. Their versatility as chemical agents renders them suitable for deployment in both biotechnological and industrial applications. This review aims to provide readers with an extensive comprehension of recent advancements in utilizing biosurfactants and the microorganisms responsible for their production. This knowledge encompasses their medical applications, metal remediation technology, and hydrocarbon-related industries.

Keywords: Antimicrobial, Biosurfactants, Bioremediation, Emulsification, Glycolipid, Lipopeptide.

INTRODUCTION

Surfactants are a class of chemicals found in a wide range of fields, such as chemicals, fast-moving consumer goods, pharmaceuticals, and oil/water treatment. Biosurfactants are a family of chemical compounds synthesized by microorganisms, including hydrophilic and hydrophobic moieties. They tend to disperse interfaces between liquid phases (oil/water) of varying degrees of polarity by lowering interfacial and surface tension. Surfactants derived from microorganisms have numerous advantages over synthetic ones, including low toxicity, high biodegradability, multifunctionality, and mild manufacturing conditions.

---

[*] **Corresponding author Arun Kumar Pradhan:** Centre for Biotechnology, Siksha 'O' Anusandhan (Deemed to be University), Bhubaneswar, Odisha-751003, India; E-mail: arunpradhan@soa.ac.in

Arun Kumar Pradhan and Manoranjan Arakha (Eds.)

An *emulsifier* is a substance composed of polymers of polysaccharides, lipopolysaccharides, proteins, lipoproteins, and glycolipids with a low molecular weight, known as a biosurfactant [1 - 3]. The former's molecular components are typically more effective in maintaining oil-in-water emulsions, although they reduce surface tension less than the latter's. It comprises non-interfering amphiphilic and multi-affinity polymers [4].

These compounds, primarily biosynthetically derived as secondary metabolites, play crucial roles in the proliferation and localization of microorganisms. The hydrophobic segment of these molecules comprises long-chain fatty acids, hydroxy fatty acids, or alkyl hydroxy fatty acids. On the other hand, hydrophilic components may include carbohydrates, amino acids, cyclic peptides, phosphates, carboxylic acids, and alcohols. The commercial surfactant market is from petrochemicals, plants, animal fats, and microorganisms. Research indicates that a substantial majority of commercially available surfactant products are petrochemical-based. One of the environmental agencies' transformative challenges involves developing innovative technologies to replace fossil fuel-based products with sustainable, biodegradable, green energy sources.

Numerous microorganisms demonstrate the capacity to utilize diverse substrates, including sugars, oils, alkanes, and various waste materials, for the biosynthesis of biodegradable surfactants [5]. Biosurfactants typically exhibit critical micelle concentrations (CMC) within the 1-200 mg/l range, accompanied by molecular weights ranging from 500 to 1500 Da [6]. Notably, they can rival or surpass the effectiveness of synthetic surfactants due to their distinctive advantages, such as high specificity, biocompatibility, and inherent biodegradability [7]. Biosurfactants represent a category of biodegradable surfactants primarily produced by bacteria and yeast [8-10].

Lipopeptides are mainly produced by Bacillus spp. followed by *Brevibacterium aureum* and *Nocardiopsis alba*. Glycolipids by Pseudomonas, Burkholderia, Mycobacterium, Rhodococcus, Arthrobacter, Nocardia, Gordonia. Some yeasts and fungi such as Starmerella, Yarrowia and Pseudozyma, *Ustilago scitaminea* also found to produce glycolipids. Thiobacillus sp. are found associated with phospholipids production [11].

The adaptability and utility of the molecules have generated attention in industrial and ecological applications such as bioremediation, soil cleansing, better oil recovery, and processing [12]. Emerging commercial applications have been found in chemical, textiles, agricultural, pharmaceuticals, food, and manufacturing industries [10, 13].

Biosurfactants accumulate at the intersection of two immiscible phases or the contact of a liquid and a solid. Lowering surface and interfacial tension minimizes repulsive forces between two distinct phases, allowing for easier mixing and contact. Below are some examples of biosurfactants and their application refer to Table **1**.

Table 1. Some examples of biosurfactants and their applications.

Biosurfactants		Microorganism	Environmental Biotechnology Applications	References
Group	**Class**			
Glycolipids	Rhamnolipids	*Pseudomonas aeruginosa*, Pseudomonas sp.	Improvement of hydrocarbon degradation and dispersion; emulsification of hydrocarbons and vegetable oils; elimination of metals from soil	[17]
	Trehalolipids	*Mycobacterium tuberculosis*, *Rhodococcus erythropolis*, Arthrobacter sp., Nocardia sp., Corynebacterium sp., *Serratia marcescens*.	Increase of hydrocarbon bioavailability	[18]
	Sophorolipids	*Torulopsis bombicola*, *Torulopsis petrophilum*, *Torulopsis apicola*, *Candida apicola*, *Candida bombicola*, *Candida bogoriensis*, *Candida lipolytica*	Hydrocarbon recovery from dregs and muds; heavy metal reduction from sediments; oil recovery improvement	[19]

(Table 1) cont.....

Biosurfactants		Microorganism	Environmental Biotechnology Applications	References
Group	**Class**			
Fatty acids, phospholipids and neutral lipids	Corynomycolic acid	*Corynebacterium lepus*	Improved bitumen recovery	[20]
	Spiculisporic acid	*Penicillium spiculisporum*	Metal ion removal from aqueous solution; dispersion of hydrophilic pigments; creation of novel emulsion-type organogels, superfine microcapsules, heavy metal sequestrants	[21]
	Phosphati-dylethanolamine	*Acinetobacter* sp., *Rhodococcus erythropolis*	Enhancing bacteria's tolerance to heavy metals	
Lipopeptides	Surfactin	*Bacillus subtilis*	Improvement of hydrocarbon and chlorinated pesticide biodegradation; removal of heavy metals from polluted soil, sediment, and water improving the efficiency of phytoextraction	[22]
	Lichenysin	*Bacillus licheniformis*	Increased oil recovery	[23]
Polymeric biosurfactants	Emulsan	*Acinetobacter calcoaceticus RAG-1*	Hydrocarbon-in-water emulsion stabilisation	[24]
	Alasan	*Acinetobacter radioresistens KA-53*		
	Biodispersan	*Acinetobacter calcoaceticus A2*	Limestone dispersion in water	[25]
	Liposan	*Candida lipolytica*	Hydrocarbon-in-water stabilisation	[26]
	Mannoprotein	*Saccharomyces cerevisiae*		

A biosurfactant's ability to alter surface tension, maintain emulsions, and probe hydrophilic-lipophilic properties determines its potency and balance (HLB). The HLB number shows if the biosurfactant is a water-in-oil emulsion or an oil-in-water emulsion. This criterion may be used to assess the suitability of biosurfactants. Emulsifiers with low HLBs help stabilize water-in-oil emulsions, while emulsifiers with high HLBs promote water solubility [14, 15]. The hydrophobicity of the bacterial cell surface is also influenced by biosurfactants

(CSH). This capacity was described by Al Tahan *et al.* [16]—modifications in the bacterial cell's structure surface make carbohydrates more accessible to the microbial cell.

Biosurfactants: Real Prospects for Industrial Use

Increasing environmental concerns, biotechnology advancements, and increasingly severe laws have made biosurfactants a viable alternative to synthetic surfactants on the market [27]. The number of biosurfactant papers and patents has expanded dramatically in recent years [28]. Worldwide, patents on biosurfactants, emulsifiers, and other features have been issued. Petroleum-related sectors acquired the most patents (33%), followed by cosmetics (15%), pharmaceuticals (12%), and bioremediation (11%), with the bulk of patents covering sophorolipids, surfactins, and rhamnolipids.

Biosurfactants are increasingly being used in industry. Many difficulties, however, must be solved before broader adoption may be contemplated. These concerns are connected to yield and manufacturing costs, as well as the time and effort necessary to adapt the molecule of choice for a specific application [29, 30]. Transparency Market Research recently conducted a comprehensive assessment of the biosurfactants industry and market trends™.

Environmental Applications

The unintended or purposeful release of organic and inorganic substances into the environment is a common cause of pollution from industrial activity. These chemicals quickly bond to soil particles, making clean-up difficult. Is it possible to replicate the removal of pollutants by using biosurfactants to clean organic molecules such as hydrocarbons? The washing and solubilization procedure fluidizes and eliminates emulsification. The use of biosurfactants in the remediation of inorganic substances such as heavy metals, on the other hand, promotes chelation promotes the removal of such ions during cleaning phases, which is assisted by chemical interactions between amphiphiles and metal ions.

It is becoming increasingly common for industry to use biosurfactants. In order to increase the adoption rate, several issues must be addressed. These issues involve yields and production costs, including downstream processing and tailoring molecules for specific applications [29, 30]. A market research firm, Transparency Market Research, analyzed the biosurfactant industry and trends recently.

Biosurfactants and Bioremediation

Bioremediation is now famous as a potential low-cost, high-performance solution for tackling environmental contamination problems. Ranging from polyaromatic hydrocarbons, jet fuel, gasoline, diesel fuel, and the benzene, toluene, ethyl benzene, and xylene cluster, acid mine drainage pesticides, munitions compounds (*e.g.*, trinitrotoluene), and inorganic heavy metals to crude oil, biotechnology products objects are being created in commercial and government laboratories all over the world to find and develop biotechnological solutions that will help in the economically and effectively resolution of pollution concerns in an ecologically acceptable way [31, 32]. Amphiphiles can alter interface physicochemical conditions, influencing chemical distribution among phases. Biosurfactants introduced to this system can interact with abiotic particles and bacterial cells.

This changes the processes of organic contaminants' micellarization and emulsification, interaction with absorbed pollutants, and sorption to soil particles, resulting in changes in cell-envelope composition and hydrophobicity. Among the most critical changes in bacterial composition are interactions between micelles and cells [33]. These processes can be employed to boost the bioavailability of sparingly soluble pollutants and, hence, the biodegradation rate. It can also be used to prevent biodegradation.

Surfactants are developing as a technique for improving the accessibility and bioavailability of hydrophobic compounds, hence supplementing conventional (bio)remediation strategies. Unfortunately, most of the research on biosurfactants have been done in the laboratory, and large-scale plotting is uncommon. Additionally, most field service technicians must become more familiar with surfactant technology, especially biosurfactants. To date, the literature on the effects and efficacy of the molecules needs to be more extensive and conclusive.

Biosurfactants are mainly used in the biodegradation of crude oil [34] and improved oil recovery systems [35, 36]. Strikingly, a large-scale bioremediation study conducted after the 1989 Alaska oil spill showed that nutrient-stimulated biomass effectively increased the availability and biodegradability of oil-affected shorelines. A potential role for biosurfactants produced by microbial flora has been thwarted [37]. According to Chakrabarty [38], biosurfactants produced by *Pseudomonas aeruginosa* can effectively disperse oil into tiny droplets, improving the bioremediation of oil-contaminated shorelines.

Biosurfactants' Role in Biological Degradation Processes

Using biosurfactants to improve bioremediation efficiency in hydrocarbon-contaminated settings is a potential method. They have the potential to increase hydrocarbon bioremediation in two ways. The first is to increase the bioavailability of the microbial substrate, and these conditions interact with the cell surface, which makes the surface more hydrophobic, making it easier for the hydrophobic substrate to adhere to bacterial cells [39]. Biosurfactants increase the surface area of insoluble molecules and lower surface and interfacial tension, enhancing hydrocarbon mobility and bioavailability. As a result, biosurfactants promote biodegradation and hydrocarbon elimination. Biosurfactants, which mobilize, solubilize, or emulsify hydrocarbons, are predicted to promote hydrocarbon biodegradation.

The mobilization process takes place at concentrations lower than the biosurfactant CMC. Biosurfactants diminish surface and interfacial tension between air/water and soil/water systems at such concentrations. Contact between a biosurfactant and a soil/oil system increases the contact angle. It decreases the capillary forces that hold the oil and soil together as the interfacial forces decrease. The solubilization process takes place sequentially over the biosurfactant CMC. The biosurfactant molecules form micelles at these concentrations, considerably improving oil solubility. Outside the aqueous phase, the hydrophilic ends of the biosurfactant molecules are exposed, while the hydrophobic ends are bound together within the micelles. This creates a favorable environment for hydrophobic organic compounds within the micelles.

Emulsification produces an emulsion, which is made up of microscopic droplets of fat or oil floating in a liquid (usually water). Emulsifiers are also high molecular weight biosurfactants refer to Fig. (1). They are frequently utilized as adjuvants in bioremediation and removing oil molecules from ecosystems.

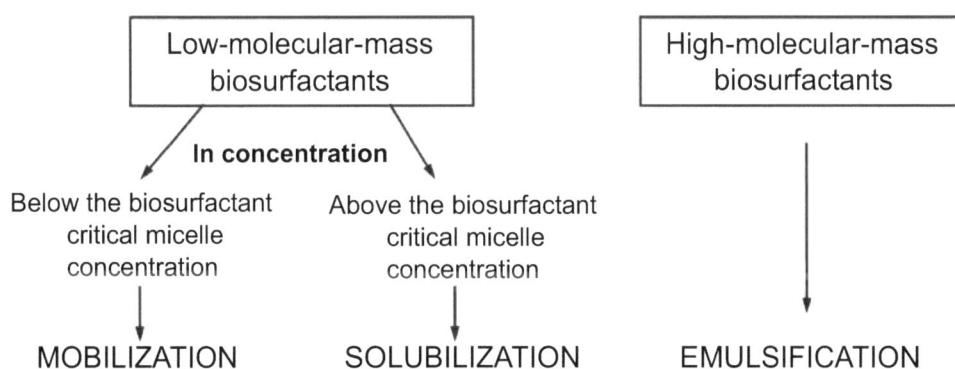

Fig. (1). Methods of hydrocarbon elimination by biosurfactants based on molecular mass and concentration.

Bacteria with low cell hydrophobicity can attach to micelles or emulsified oils, whereas bacteria with strong cell hydrophobicity can directly contact oil droplets and solid hydrocarbons [40]. Microorganisms and hydrocarbons interact in three ways: access to water-solubilized hydrocarbons, direct contact of cells with big oil droplets, and interaction with quasi-solubilized or emulsified oils. Biosurfactants are presumed to be abletoalter carbohydrate up take patterns during different stages of microbial growth. Growth of On hydrocarbons, Gordonia sp. BS 29 developed extracellular bio emulsifiers and cell-associated glycolipid biosurfactants that changed surface hydrophobicity during hexadecane formation [40, 41].

Recent studies have revealed the then-alkane uptake process of *Pseudomonas aeruginosa* and the importance of rhamnolipids in internalizing hydrocarbons for later degradation. Hexadecane was made more readily available to bacterial cells due to the formation of microdroplets produced by the action of the biosurfactant. Recordings of biosurfactant-coated hydrocarbon droplets were performed following experiments using electron-microscopy. Remarkably, the biosurfactant-coated hydrocarbon droplet absorption mechanism was similar to active pinocytosis.

Numerous researchers [42 - 44] have observed that biosurfactants and the bacterial strains responsible for their production can expedite the availability and biodegradation of organic pollutants. For instance, Obayori *et al.* [45] investigated the biodegradability of biosurfactants derived from the genus *Pseudomonas*. Their study focused on the LP1 strain's exposure to crude oil and diesel. Reddy *et al.* [46] reported that the bacteria Brevibacterium sp., particularly the PDM-3 strain, achieved a remarkable degradation rate of 93.92% for phenanthrene, anthracene, and fluorene. In another study, Kang *et al.* [47] characterized sophorolipid and its effectiveness in the biodegradation of aliphatic and aromatic hydrocarbons. Their work emphasized enhancing the bioavailability of microbial consortia for biodegradation, thereby facilitating the bioremediation of hydrocarbon-contaminated sites characterized by low water solubility.

Employing biosurfactant-producing bacteria to bioremediate hydrocarbon-contaminated locations without specifying the surface-active chemicals' nature is an effective microbiological method. Cell-free culture medium containing biosurfactants can either be administered directly to the polluted region or diluted well beforehand. Biosurfactants are highly stable and effective in the media utilized for their synthesis. This is another advantage of this strategy.

Das and Mukherjee [48] discovered the importance of biosurfactant-producing strains in the bioremediation of crude petroleum hydrocarbon-contaminated environments. It was demonstrated that three biosurfactant-producing strains, *Bacillus subtilis* DM-04, *Pseudomonas aeruginosa* M, and *Pseudomonas aeruginosa* NM, could remediate petroleum-contaminated soil samples by treating soil samples with aqueous solutions of the biosurfactant obtained from each bacterial strain. Joseph and Joseph [49] were able to extract oil from petroleum sludge, which promotes bacterial development of biosurfactants. When refining crude oil, refineries generate a large amount of petroleum sludge. Storage tanks are typically used to store crude oil—oily contaminants remain on the bottom of the tank. Sludge is collected during tank cleaning and treated as waste. The use of biosurfactants often enhances the bioavailability and biodegradability of hydrophobes, but little is known about how emulsifier formation affects the biodegradation of complex hydrocarbon mixtures.

Bilge waste is a hazardous waste made up of saltwater and hydrocarbons. Residues, the main components of which are a complex mixture of n-alkanes, total soluble hydrocarbons, and insoluble solvents. Non-solvent complex mixtures predominantly comprise branched, cycloaliphatic, and aromatic hydrocarbons that exhibit the highest resistance to biodegradation. A consortium of bacteria that produce emulsifiers investigated the biodegradation of oily bilge waste. They discovered that the degree of biodegradation was 85% for n-alkanes, 75% for total dissolved hydrocarbons, and 58% for undissolved complex combinations.

Barkey *et al.* [50] investigated the solubilization of polyaromatic hydrocarbons (PAH), phenanthrene (PHE), and fluoranthene by alasane generated by *Acinetobacter radioresistens* KA53 (FLA). They also looked at the impact of arasan on the mineralization of PHE and FLA by *Sphingomonas paucimobilis* EPA505. They found that increasing bio-emulsifier concentration(from 50 to 500 gml1) linearly increased the water solubility of phenanthrene and fluoranthene and increased the mineralization of PAHs. Including arasan at concentrations up to 300 g mL1 quadrupled the rate of fluoranthene degradation and greatly enhanced the rate of phenanthrene degradation. The solubilization curves revealed that the apparent solubility of these compounds increased linearly with the addition of alacane in this concentration range, but increasing the alacane concentration beyond 300 gmL1 did not result in PAHM ineralization was not stimulated further. It has also been discovered that *Enterobacter cloacae* strain TU secretes an emulsifier exopolysaccharide (EPS) [51]. EPS has a robust emulsifying activity (E24=75). EPS can make the bacterial cell surface more hydrophobic while also neutralizing the cell's surface charge.

Microbial Improved Oil Recovery (MOER)

Initial oil recovery from wells frequently relies on conventional primary and secondary technologies, with certain wells recovering only 20% to 30% of their total oil reserves. Once these conventional methods reach their limits, as much as two-thirds of the oil in the reservoir may remain untapped. In such instances, tertiary oil recovery techniques, collectively known as Enhanced Oil Recovery (EOR), extract 10% to 15% of the remaining oil. These techniques encompass both chemical and microbiological approaches [52].

The microbiological approach, specifically Microbial Enhanced Oil Recovery (MEOR), leverages microbial processes involving the partial degradation of large oil molecules, gas production, selective plugging, and biosurfactant synthesis. In the case of biosurfactant production, it leads to reduced oil/water interfacial tension and the generation of oil-in-water emulsions, thereby enhancing oil mobility through rock fracture. Biosurfactants can be generated off-site in digesters and subsequently injected into oil reservoirs. Alternatively, introducing allophytic microbes or stimulating local bacteria through nutrient injection can produce them *in situ* [12].

A significant challenge in developing *in situ* production techniques involves the isolation of microbial strains capable of thriving in harsh conditions characterized by high pressure, salinity, elevated temperatures reaching up to 85 °C, and extreme pH values. Furthermore, some operators have reported issues related to clogging and corrosion when introducing microorganisms into wells.

The earliest possibilities for such applications were rhamnolipids, although all prominent families of microbial surfactants have been proposed for his MEOR uses. Lipopeptides such as surfactin, lichenin, and emulsan have been proven to increase oil recovery. Temperatures, pH levels, and salinities identical to those seen in petroleum reserves are detected in *B. subtilis, P. aeruginosa*, and *Bacillus cereus*. Oil displacement studies in glass micromodels were also carried out with the most promising suggested *B. subtilis* biosurfactant [51].

Most of the MEOR laboratory studies with biosurfactants were carried out in core flood systems, which simulated the features of oil reservoirs [52]. The core flood test used lipopeptide biosurfactants generated by the *Bacillus mojavensis* strain to assess oil recovery from carbonate reserves. It has been observed that treatment can recover up to 60% of the original oil in such cores [53].

She *et al.* [54] found that introducing various Bacillus cultures increased oil output from 4.89% to 6.96%. Xia *et al.* [55] achieved a 9.02% efficiency while injecting *Pseudomonas aeruginosa* cells. Numerous efforts have been attempted to identify unconventional biosurfactant-producing bacteria from sources capable of creating particularly efficient compounds in oil mobilization. Castrena-Cortez *et al.* [56] identified a stable consortium with the leading species Thermoanaerobacter. When evaluated in a low-volume oil recovery experiment, a granular porous medium revealed a 12% increase in oil release. In the same laboratory setting, the injection of surfactant-producing microorganisms and the application of biosurfactants were compared (one example is shown in Fig. (**2**).

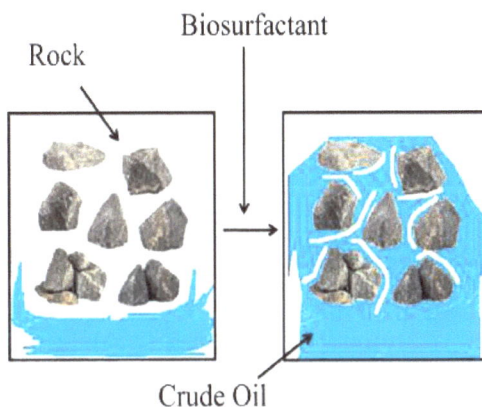

Fig. (2). Biosurfactants' increased oil recovery mechanism.

The oil recovery performance of biosurfactants generated by *Bacillus subtilis* PT2 and *Pseudomonas aeruginosa* SP4 was compared to three synthetic surfactants: Tween 80, sodium dodecyl benzene sulfonate (SDBS), and alkyl polypropylene oxide [57,58]. Sodium sulphate (Alfoterra) in a study by Pornsunthorntawee *et al.* [59]. We employed a sand-filled column seeded with a motor oil compound for this. A surfactant solution was placed into the packed column to test if it improved oil recovery.

The biosurfactants generated by *Bacillus subtilis* PT2 and *Pseudomonas aeruginosa* SP4 demonstrated remarkable oil recovery efficiency, removing around 62% and 57% of the tested oil, respectively. *Pseudomonas aeruginosa* SP4 biosurfactant recovered oil more successfully than *Bacillus subtilis* PT2 biosurfactant. The synthetic surfactants tested had an oil recovery rate of 53-55%. Biosurfactants can also be utilised to recover hydrocarbon molecules from oil shale as a petroleum energy alternative fuel. *Rhodococcus erythropolis* and *Rhodococcus ruber* biosurfactants were effectively employed by Hadadin and colleagues for hydrocarbon desorption from El Lajjun oil shale [60].

Soil Cleaning Technology

The physio-chemical features of biosurfactants are distinctive of soil cleaning procedures, not their impact on bacterial metabolic activity or surface qualities [61]. However, these approaches can enhance the bioavailability of bioremediation. Aqueous biosurfactant solutions can also be used to leach low-solubility chemicals from soil and other media.

According to Urum *et al.* [62], synthetic surfactants such as sodium dodecyl sulphate (SDS) and rhamnolipid biosurfactants remove more crude oil than natural surfactants such as saponins (27%) [63]. Kang *et al.* investigated the application of soil detergents such as Sophorolipid, Tween (80/60/20), and Span (20/80/85) to liberate 2-methylnaphthalene from experimentally polluted soil. In removing stains, sophorolipid beats all other nonionic surfactants except Tween 80. This might be owing to its high hydrophilic-lipophilic balance (HLB). Surfactants with higher HLB appear to have increased 2-methylnaphthalene solubility. Lai *et al.* [64] used rhamnolipid, surfactin, tween80 and tritonX-100 to investigate natural and synthetic biosurfactant's ability to remove total petroleum hydrocarbons (TPH) from the soil.

TPH effectiveness was examined by washing (bio)surfactant solutions through low (LTC) and high (HTC) TPH contaminated soils. It was observed that 0.2 mass percent additions of rhamnolipids, surfactin, Triton X 100, and Tween 80 to LTC and HTC soils resulted in removal efficiencies of 23%, 14%,6%, 4%, and 63%, 62%, 40%, and 35% respectively. This indicates that of the four (biological) surfactants tested, rhamnolipids and sulactin were the most effective in their TPH removal and can be used as biostimulators for bioremediation against contaminated oil and gas surfaces.

METAL REMOVAL

The presence of heavy metals in the soil environment is highly hazardous to people and other species in the ecosystem. Heavy metals are so toxic that even low concentrations of heavy metals in soil can have serious consequences. Today, there are many ways to remediate soil contaminated with heavy metals. Non-biological remediation methods include excavation and disposal of contaminated soil at landfills and biological methods [65]. Biological processes remove metals from soil using plants (phytoremediation) or microorganisms (bioremediation). Metal pollution has traditionally been reduced using microorganisms. Heavy metals are not biodegradable; they can only change chemical states, which results in altered mobility and toxicity. Microbes may influence metals in several ways. Certain metals can be altered by redox or alkylation reactions. Microorganisms can collect metals through metabolism-independent (passive) or metabolism-dependent (active) absorption. Microorganisms can indirectly impact metal mobility by changing pH or generating or releasing chemicals that modify metal mobility [66, 67].

Studies have been undertaken to investigate the possibility of metal removal using biosurfactant anionic characteristics. Jwarkar and colleagues used the biosurfactant *Pseudomonas aeruginosa* BS2 to remove cadmium and lead [68].

Rhamnolipid scavenging of Cd and Pb was investigated using column experiments. Cadmium removal is more costly than lead removal. Within 36 hours, rhamnolipid (0.1%) removed over 92% Cd and 88% Pb. Rhamnolipids reduced toxicity and generated microbial activity (Azotobacter and Rhizobium) without affecting soil quality. Their profitability should also be evaluated. Assi *et al.* [69] compared the cadmium uptake capacity of two soil components, sepiolite and feldspar. Sepiolite has been determined to accumulate cadmium better than feldspar. Feldspar desorption (96%) was significantly higher than sepiolite desorption (10%).

Sorption can impair biosurfactants' capacity to remove pollutants from diverse soil components [70]. As a result, a rhamnolipids sorption assay was performed. The efficacy of rhamnolipid removal was reduced by sorption. Metal concentration was required for the absorption of the mono-rhamnolipid (R1). Sorption in organic decomposition followed the order hematite- kaolinite-MnO2 iris Camontmorillonite- gibbsite- humic acid-coated silica at low R1 concentrations. It was found that R1 is more adsorptive than R2 but removes metals more effectively than R2. This information helps in predicting the feasibility of rhamnolipid therapy and the needed dose of rhamnolipid formation. Utilizing rhamnolipids as the R2/R2 combination enhances the concentration of R1 in the repair solution. Kim and Vipulanandan [71] investigated the lead removal from water and soil (kaolinite). A linear isotherm represented desorption of lead from kaolinite. Vegetable oil was used for the production of biosurfactants. Using 10x CMC can remove approximately 75% of the lead in 100 mg/contaminated water. For best lead removal, the ratio of biosurfactant to lead was 100:1. FTIR spectroscopy demonstrated that the biosurfactant's carboxyl group was involved in the removal process. Micelle splitting may be described using the Langmuir and Freundlich models. Biosurfactant micelles distributed better than synthetic surfactants such as sodium dodecyl sulphate and Triton X-100.

The performance of Rhamnolipid was investigated by Dahrazma and Mulligan [72] in a continuous flow configuration (CFC) for heavy metal removal (copper, zinc, and nickel) from sediment samples of the Canadian Lachine Canal. Flow was simulated inside a column through the remediation technique. A continuous flow of rhamnolipid solution was pushed through a sediment sample. The rhamnolipid and additive contents, as well as the time and flow rate, were investigated. When applied, Rhamnolipid removed up to 37% of Cu, 13% of Zn, and 27% of Ni from sediments. When 1% NaOH was added to 0.5% rhamnolipid, copper removal increased up to fourfold compared to 0.5% rhamnolipid alone.

Metal Removal by Biosurfactants - Process Mechanism

Biosurfactants have clear benefits because of strains of bacteria that can produce surface-active chemicals that do not need to be able to live in soil contaminated with heavy metals. On the other hand, using just biosurfactants, regular fresh portions of these compounds are continuously added. The capacity of biosurfactants to form compounds with metals is helpful for bioremediation of heavy metal-contaminated soils. Ionic bonding allows anionic biosurfactants to create non-ionic compounds with more vital metals than metal-soil bonds.

Metal-biosurfactant complexes get de-absorbed with the decrease in interfacial tension. Cationic biosurfactants compete with surfaces charged negatively to displace same-charged metal ions. Biosurfactant micelles remove metal ions from the soil surface by binding metals with the polar head groups of micelles, causing metal mobilization in water refer to Fig. (3).

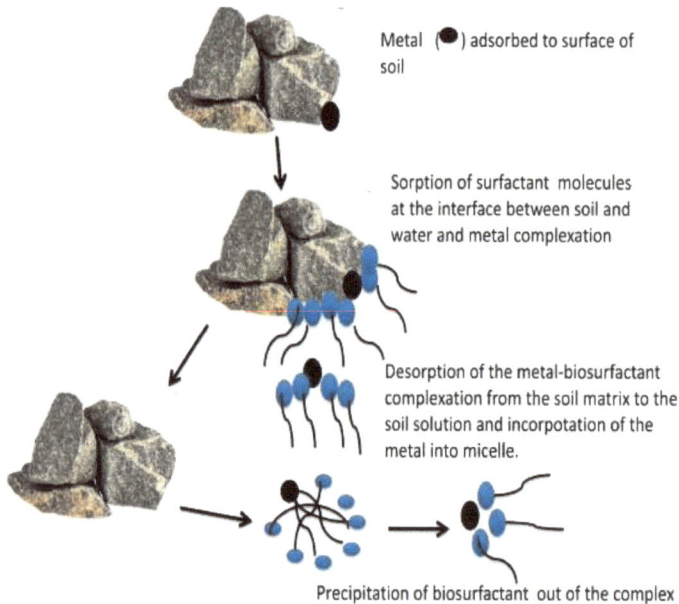

Metal ( ) adsorbed to surface of soil

Sorption of surfactant molecules at the interface between soil and water and metal complexation

Desorption of the metal-biosurfactant complexation from the soil matrix to the soil solution and incorpotation of the metal into micelle.

Precipitation of biosurfactant out of the complex

Fig. (3). Biosurfactants' activity mechanisms in metal-contaminated soil.

Biosurfactants are used to trap trivalent chromium within micelles, enabling bacteria to tolerate and resist high concentrations of Cr (III). Nyanamani and colleagues [73] investigated biosurfactants of a marine strain, Bacillus sp. MTCC 5514 for chromium (VI) bioremediation. Sanitation of this strain was achieved by two processes: extracellular chromium reductase conversion of Cr(VI) to Cr(III) and biosurfactant removal of Cr(III). The first process transforms chromium from a hazardous to a less toxic state, whereas the second shields bacterial cells against chromium exposure (III). Both methods preserve bacterial cell function while conferring tolerance and resistance to high hexavalent and trivalent chromium levels.

Biosurfactants in Phyto-remediation

Plants inoculated with bacteria tolerant to heavy metals and producing biosurfactants can boost the efficacy of soil phytoremediation. Bacillus species strain J119 was tested for its potential to improve plant growth and cadmium uptake in soils loaded with varying amounts of Cd [74]. The tested strain colonized the rhizospheres of all studied plants, although the treatment only enhanced biomass and Cd absorption in tomato plant tissues. This shows that plant type influences the root colonization activity of the imported strains. Further research into the interaction of plants with the biosurfactant-producing bacterial strain J119, on the other hand, might lead to the development of new microbial-assisted phytoremediation approaches for metal-contaminated soils. Further study is needed on the use of biosurfactants and biosurfactant-producing bacteria in phytoremediation, particularly in areas polluted with organic and metallic pollutants.

Biosurfactants in Agriculture

The different roles of biosurfactants make them attractive candidates for future crop protection applications due to their different properties and involvement in biological regulation, antifungal drugs, and inducing systemic tolerance. Biosurfactants generally allow biological control by creating channels in the cell wall and disrupting the pathogen's cell surface [75]. Plants are primarily protected against phytopathogenic fungi by glycolipids such as cellobiose lipids, rhamnolipids, and cyclic lipopeptides, like surfactin, iturin, fengycin, and a variety of other biosurfactants [10].

Teichmann *et al.* [76] demonstrated that co-inoculation with wild-type Ustilago mydis sporidien prevented Botrytis cinerea infection of tomato leaves. Ustilagic acid, the first cellobiose lipid identified, is produced by this plant pathogen. Cellobiose lipids are natural surfactants with membrane-damaging characteristics that can induce cell death in yeast and mycelium at low doses [77].

Stanghellini and Miller [78] described how rhamnolipids function in various oomycete plant diseases by rupturing zoospore membranes and inducing zoospore lysis. Meanwhile, various research works on the relevance of rhamnolipids have been published in combating various phytopathogenic fungi. Devode *et al.* [79] discovered that the *Pseudomonas* species helps inhibit the viability of *Verticillium microsclerotia*. Piljac *et al.* [80] expected the collapse of Pythium hyphae incubation in a liquid medium containing both phenazine and *P. aeruginosa* PNA1-produced rhamnolipids. Both metabolites were shown to be crucial in the inhibition of Pythium species. The causative soil-borne diseases act synergistically.

A biosurfactant produced from rhamnolipids was discovered to have a direct antimicrobial effect, which suppressed *Bacillus subtilis* spore germination and mycelial growth. In addition, Ca^{2+} influx, the activation of mitogen-activated protein kinases, and the generation of reactive oxygen species were identified as early events in the grapevine response to the disease. The induction of plant defenses, accompanied by the production of various defense genes and hypersensitivity responses, has explained some of the mechanisms behind plant tolerance. Furthermore, rhamnolipid enhanced the defensive responses elicited by chitosan triggers and her B culture filtrate. Cinerea shows that a combination of rhamnolipids and additional effectors helps grapes resist botrytis [81].

Cyclic lipopeptides, like rhamnolipids, are another type of biosurfactant with antifungal activity against plant pathogens. Non-ribosomal peptide synthetases are guided by cyclic lipopeptide production, which is encoded by a vast gene cluster and consists of one module for each amino acid required for the oligopeptide. Each module is composed of several conserved domains, which involve the recognition, activation, transport, and binding of amino acids to peptide chains. The thioesterase domain governs peptide cyclization and release in the final module. This novel biosynthetic approach enables the incorporation of rare amino acids [82].

The structure, genes, biosynthesis, and control of known cyclic lipopeptides generated by numerous bacterial taxa have been disclosed by genome sequencing. *Bacillus amyloliquefaciens* GA1 genome has been sequenced, and four gene clusters promoting the synthesis of the cyclic lipopeptides surfactin, iturin A, and fengycin have been found (Arguelles-Arias *et al.* 2009). Current research has shown the presence of biosynthetic gene clusters that create secondary metabolites, including lipopeptides (surfactin, iturin, fengycin), which are involved in Bacillus' bioregulatory functions [83, 84].

Pseudomonas species are another major category of plant-associated bacteria with biological control capabilities *via* cyclic lipopeptide synthesis. Numerous research works focus on glycolipid-type biosurfactants and fungus interaction, and the zoospore-killing capacity of cyclic lipopeptides has also been found [82, 85], indicating their potential application as biocontrol agents.

Nanotechnology

and biosurfactants are advancing. Using a water-in-oil microemulsion, we created NiO nanorods [86]. A biosurfactant was added to heptane. The nickel chloride solution was then mixed into the biosurfactant/heptane mixture to create two different microemulsions, and adding ammonium hydroxide to the same hydrocarbon mixture created other microemulsions. To remove biosurfactants and heptanes, the microemulsions were pooled, and centrifuged, and the precipitate was washed with ethanol. The nanorods were 22 nm in diameter and 150-250 nm in length (pH 9.6). The form of the particles was affected by pH. This might be related to pH's influence on biosurfactants' morphology. Biosurfactants are a more ecologically friendly option.

In a study by Reddy *et al.*, surfactin was shown to stabilize the formation of silver nanoparticles [87]. Sodium borohydrate was combined with silver nitrate. Surfactin was added to $HAuCl^4$, followed by a dropwise addition of sodium borohydrate mixture. These nanoparticles exhibit various physical, chemical, magnetic, and structural characteristics. Various pH and temperature settings were tried. Surfactin was utilized to stabilize the nanoparticles for two months. Surfactin is a low-toxicity, renewable, and biodegradable stabilizer, and an ecologically beneficial ingredient.

Rhamnolipids have been examined for their influence on zirconia nanoparticles' electrokinetic and rheological behavior [88]. As indicated by zeta potential studies, the biosurfactant adsorbed zirconia particles with increasing concentration and could scatter the zirconia particles above pH 7.

Biosurfactants have a variety of roles, including hydrophobic water insoluble substrate bioavailability, increased surface area, metal-binding property, quorum sensing, bacterial pathogenicity, biofilm destruction, and so on. This diverse behavior of biosurfactants, diverse chemical groups, and diverse structures have opened the gate to many fields. In the medical field, they are popular due to their broad range of structures, and antimicrobial properties, antibacterial properties. Some act as drug delivery agents, and interact with pathogens. In comparison, biocompatibility and digestibility allow them to play a crucial role in the cosmetics and food industries [89]. Surfactants are synthetic/ chemically available, but nowadays, biosurfactants are replacing them. Chemically synthesized surfactants are non-biodegradable and toxic to human health as well as to the environment; therefore, this chapter discusses the surfactants of microbial origin and their applications in healthcare.

Many yeasts and bacteria grow on n-alkanes, contributing to fatty acids, phospholipids, and neural lipids. Lipopeptides and lipoproteins involve gramicidin as lipopeptides and polymyxins as lipopeptide antibiotics. Polymeric biosurfactants, such as emulsan, liposan, mannoprotein, and polymeric protein complex, are also available. Microemulsions are formed by particulate biosurfactants, which are crucial for alkane uptake by microbial cells [90, 91].

Biosurfactants show different biological activities, such as the inhibition of bacterial growth, toxic effects on harmful cells, tumor growth inhibitors, antibiotics, cell lysis, fungicidal properties, food digestion, anticancer agents, antibiofilm agents, antiadhesive agents, *etc.* Some of these activities, along with microorganisms, are mentioned in Table **2**.

Table 2. Biological activity of some common Biosurfactants.

S. No.	Biosurfactants	Biological Activities/ Applications	References
1.	Glycolipids	Antimicrobial activity	[92]
2.	Cyclic depsipeotides	Antimicrobial activity	[93]
3.	Rhamnolipids	Antimicrobial activity and antiadhesive activity.	[94]
4.	Surfactin	Antimicrobial, antifungal, antitumor and antiviral, anti-inflammatory, apoptosis.	[95, 96]
5.	Iturin	Antimicrobial and antifungal	[97, 98]
6.	Mannosylerythritol lipid	Antimicrobial, apoptosis, growth inhibition, and immunological properties	[99, 100]
7.	Sphingolipid	Antimicrobial activity	[101]
8.	Trehalose lipid	Antiviral activity	[102]
9.	Viscosinamide	Antifungal activity	[103]
10.	Pumilacidine	Antiviral activity	[104]

Biosurfactants as Antimicrobial Agents

Literature reports much experimental evidence, *i.e.*, research and reviews of the antimicrobial activities of various biosurfactants. It is ubiquitous to use antimicrobial agents to cure bacterial infections, but the constant use of these different kinds of antibiotics has created resistance in the microbial community. Resistance organisms generate antimicrobial-resistant strains or bacterial resistance strains by natural processes such as conjugation, transformation, or transduction. These medicines work by inhibiting the cell wall, and nucleic acid or protein, interfering with metabolic pathways, or damaging the bacterial membrane. While resistant bacteria defend themselves against antimicrobial compounds by diverse mechanisms such as enzymatic degradation, target modification, inactivation, or change of antimicrobial substances [105, 106]. Biosurfactants, such as *Lactococcus lactis 53* and *Streptococcus thermophilus* A, have recently been widely employed as antimicrobial agents, even in low concentrations against the yeast *Candida tropicalis* GB9/9, which is responsible for prosthesis failure. When these were tested in high concentrations, both were highly active against different bacterial and yeast cells [107].

In another study, it was found that biosurfactants-producing lactobacilli which are probiotically active help in maintaining a healthy gut or intestinal tract as well as protect against pathogens [108]. *In vivo* investigations have demonstrated that *lactobacillus plantarum* 299v and *lactobacillus rhamnosus* GG (both probiotic strains) may prevent *E. coli* adherence to intestinal epithelial cells by expressing mucin, implying that they do so through creating biosurfactant [13]. These results suggest that biosurfactants may include signalling molecules that hinder bacteria adherence [109, 110]. By microdilution, *L. paracasei* produces biosurfactants against *Staphylococcus aureus*, pathogenic *C. Albicans, Staphylococcus epidermidis, Staphylococcus agalactiae,* and *E.coli*, which have been proposed as an alternative antibiotic against clinical infections [111, 112].

Citobacter and *Enterobacter* genera produce antimicrobial biosurfactants like surfactins, kurstakins, iturins, and fengycins which are used in pharma industries and cosmetic industries [113]. Saravankumari and Mani studied glycolipids from *L.lactis* which can be safely administered orally or dermally, because they show a wide range of antimicrobial activities [114]. *Lactobacillus rhamnosus* and *Lactobacillus jensei* which have *in vitro* antimicrobial properties, *i.e.*, cell-bound biosurfactants play an important role in MDR (multi-drugresistance). These damage the cell wall of *S. aureus* and the membrane in *A.baumanii* [115].

Biosurfactants have a wide variety of actions such as antibacterial, antiviral, antimycobacterial, and antimycoplasmal properties. Surfactins, iturin, surfactin, and lichenysin are active against human diseases such as Candida species. Lipopeptides generated by Marine *B. circulans* have antibacterial action against both gram-positive and gram-negative pathogens. Lipopeptides also inhibit the growth of pathogenic microorganism in the gastrointestinal tract produced from *Bacillus subtilis*. Rhamnolipid biosurfactants can act against several bacteria and fungi. Therefore, biosurfactants are very useful as antimicrobial agents [107, 111].

Biosurfactants as Antibiofilm Agents

Biofilms are described as colonies of microbial communities. Different colonies reside together on different biotic and abiotic surfaces. These colonies are known to have severe effects on human health. Antimicrobial/antibiofilm agents are known to reduce biofilm by inhibiting their growth. Looking towards the current scenario of resistance increasing in bacterial communities, antibiofilm agents can be a way to recover from these damages [116].

Tooth decay or tooth surface infections are one of the examples. *Streptococcus mutans* one of the causative agents of dental plague, invades the teeth surface and forms dental caries. In this case, using antibiotics eliminates the bad microbes as well as the good microbes causing harm to the teeth surface. Hence, *Lactobacillus acidophilus*-derived biosurfactant is seen to downregulate the gene responsible for adherence of *S. mutans* to the tooth surface. Coryxin a *Corynebacterium xerosis* strain N55 derived biosurfactant, can easily inhibit and destroy the biofilm generated by *staphylococcus aureus, S. mutans* about 82% and 80% respectively, and 66% of *E.coli* and *P.aeruginosa* about 30% [117, 118].

Two *candida albican* strains of biofilms are approximately 80-82% destroyed by the biosurfactants produced from *Lactobacillus sp* CV8LAC [119]. From these studies, we can state that biosurfactants can destroy many such pathogenic biofilms.

Biosurfactants as Antitumor or Anticancer Agents

The emergence of various types of cancer is constantly mounting every day worldwide. Synthetic drugs are very useful for cancer treatment but many of them have toxicity toward the normal cells. Tumors are formed by the constant/regular growth of cells, hence disrupting the natural apoptotic process. Biosurfactant holds the properties of inhibiting their growth. Mannosylerythritol lipid is a type of glycolipid that can initiate the growth arrest of mouse melanoma cells [99, 120]. In a study, it is observed that *L. paracasei* derived glycoproteins act as an anticancer agent for breast cancer cell lines. By using the glycoprotein derived from *L paracasei*, a reduction in cell viability was seen in 48hrs [121]. Some act as novel agents for human promyelocytic leukemia cell lines and kill the cell by the cell differentiation process [122]. Biosurfactants are also spotted in initiating apoptosis in mouse malignant melanoma B16 cells, hence, cell death was observed [123]. It is studied that by increasing the concentration of MEL (mannosyl erythritol lipid) the B16 cells start accumulating in the G0/G1 phase causing their death. MEL can induce apoptosis too by activating PKC *i.e.* protein kinase C and cell differentiation process and cause HL60(human promyelocytic leukemia) cells to be engulfed by granulocytes. MEL has both the properties of initiating apoptosis as well as cell-differentiating mechanism [123].

Wicker hamielladomercqiae produces sophorolipids also have both cell differentiation and apoptosis initiation quality [124]. They induce apoptosis in H7402 human liver cancer cells. Their mechanism halts the cell cycle in the G1 phase, proceeding with activating caspease 3 and increasing calcium concentration in the cytoplasm. Similarly, surfactin stops cell proliferation by arresting cell growth in the G2/M phase and initiating preapoptotic activity.

Surfactins are known as the best anticancer agents for MCF-7 cells *i.e.* human breast cancer genes by JNK-mediated caspease pathway [125]. The also induce a protein kinase pathway for HepG2 cells [126]. There is a difference between the normal cell membrane lipid composition and cancerous cell membrane composition. Biosurfactants are shown to interact with the lipids which are surface-active and modify them leading to cell death. Phosphatidylcholine and spingomyelin help in stabilizing the lipid bilayer. Biosurfactants tend to change the phosphatidylcholine and sphingomyelin concentration in cancerous cell lipid bilayer. We can say that this strategy can be very helpful in eradicating many different kinds of cancers. This cell bilayer weakening strategy will help disrupt the cell and then the cell can be engulfed by macrophages through the apoptosis process [127].

BIOSURFACTANTS IN OTHER THERAPEUTICS/HEALTHCARE SECTORS

There are many other ways by which biosurfactants can be used as therapeutics apart from antimicrobial, antibiofilm, and anticancer agents such as:

Adjuvants

Emulsan is a biosurfactant used as an adjuvant in the human body. It has been approved to have immunomodulatory potential. Adjuvants are mixed with antigens for use.

Antiviral Activity

One example is sophorolipids diacetate ethyl ester, which is a potent anti-HIV, virucidal agent and has cytotoxic activities.

Antifungal Activity

Flocculosin, Rhamnolipids, surfactin, cyclic lipopeptides, fengycin and iturin show anti-fungal activities.

Gene Delivery/Gene Therapy and Vaccines

There are methods of gene delivery such as transfection or lipofection, *etc.* Biosurfactant-mediated liposome gene delivery has good efficiency. *In vivo* or *in vitro*, different nano vectors are also produced containing biosurfactants, which can also increase the efficacy of transfection. Surfactin-mediated gold nanoparticles are used for gene delivery, hence opening a new door for healthcare sectors.

FUTURE OUTLOOK

Despite their considerable potential, biosurfactants are rarely used in bioremediation. Current and future research on the processes of interactions between hydrocarbons, surfactants, and cells may be more beneficial to this area than case-specific studies using well-known biosurfactants compounds [128]. The cost penalty may not be as severe in the food, biomedical, and cosmetic industries where expensive items are manufactured. Different combinations of diverse components generated by organisms complicate applications and necessitate additional research to solve unique issues. This effort would benefit from the present emphasis on isolating and characterising biosurfactants generated by extremophilic bacteria such as thermophilic and halophilic bacteria [129 - 132], which is targeted at identifying novel compounds with distinctive characteristics.

Several biosurfactants, such as rhamnolipids and saponins, have demonstrated the ability to remediate soils and water. Other biosurfactants should be investigated. Desorption and biodegradation processes can be used to remediate both organic and inorganic pollutants. Biosurfactants appear to improve contaminant solubilisation and emulsification. The low biodegradability and toxicity of biosurfactants like rhamnolipids make them attractive to incorporate into repair processes. *In situ* synthesis has a substantial advantage over manufactured surfactants.

This requires additional studies. Further research is needed to anticipate the fate of pollutants and their behaviour during transportation. To allow model predictions of transport and repair, the process of solubilization of hydrocarbons and heavy metals by biosurfactants has to be explored further. Biosurfactants are finding new uses in nanoparticles' applications. Future studies should focus on the stability of biosurfactant-stabilized nanoparticles prior to their addition during repair.

Biosurfactants demonstrated antibacterial, antiadhesive, and immunomodulatory properties, and are successfully used in the safety of gene therapy, immunotherapy, and medical insertion, thereby making study in this field worthwhile. This indicates that biomedical advances may lead the way as the potential economic benefits are greater. Furthermore, due to their self-assembly capabilities, biosurfactants are expected to find new and exciting applications in nanotechnology [133, 134]. Detailed investigation of cell-to-cell communication, pathogenicity, motility, microbial competitive interactions, and their natural roles in biofilm development and maintenance may lead to new and exciting applications in the future.

Microbial surfactants' commercial viability is now limited by high manufacturing costs. Improved growth/production conditions combining low-cost renewable substrates and a novel efficient multi-step downstream processing technique might make biosurfactant manufacturing more advantageous and cost-effective. Additionally, recombinant and mutant high-producing microbial strains that may be cultivated on a variety of low-cost substrates may create biosurfactants in high yields and provide the breakthrough required for economical production.

REFERENCES

[1]   Rosenberg, E.; Ron, E.Z. Bioemulsans: Microbial polymeric emulsifiers. *Curr. Opin. Biotechnol.,* **1997**, *8*(3), 313-316.
        [http://dx.doi.org/10.1016/S0958-1669(97)80009-2] [PMID: 9206012]

[2]   Neu, T.R. Significance of bacterial surface active compounds in interaction of bacteria with interfaces. *Microbiol. Rev.,* **1996**, *60*(1), 151-166.
        [http://dx.doi.org/10.1128/mr.60.1.151-166.1996] [PMID: 8852899]

[3]   Smyth, T.J.P.; Perfumo, A.; Marchant, R.; Banat, I.M. Isolation and analysis of low molecular weight microbial glycolipids. In: *Handbook of hydrocarbon and lipid microbiology*; Timmis, K.N., Ed.; Springer, **2010**; pp. 3705-3723.
        [http://dx.doi.org/10.1007/978-3-540-77587-4_291]

[4]   Smyth, T.J.P.; Perfumo, A.; McClean, S.; Marchant, R.; Banat, I.M. Isolation and analysis of lipopeptides and high molecular weight biosurfactants. In: *Handbook of hydrocarbon and lipid microbiology*; Timmis, K.N., Ed.; Springer, **2010**; pp. 3689-3704.
        [http://dx.doi.org/10.1007/978-3-540-77587-4_290]

[5]   Mulligan, C.N. Environmental applications for biosurfactants. *Environ. Pollut.,* **2005**, *133*(2), 183-198.
        [http://dx.doi.org/10.1016/j.envpol.2004.06.009] [PMID: 15519450]

[6]   Lang, S.; Wagner, F. Structure and properties of biosurfactants. In: *Biosurfactants and Biotechnology*; Kosaric, N.; Cairns, W.L., Eds.; Marcel Dekker, **1987**; pp. 21-45.

[7]   Cooper, D.G. Biosurfactants. *Microbiol. Sci.,* **1986**, *3*(5), 145-149.
        [PMID: 3153155]

[8]   Desai, J.D.; Banat, I.M. Microbial production of surfactants and their commercial potential. *Microbiol. Mol. Biol. Rev.,* **1997**, *61*(1), 47-64.
        [PMID: 9106364]

[9]   Franzetti, A.; Caredda, P.; Ruggeri, C.; Colla, P.L.; Tamburini, E.; Papacchini, M.; Bestetti, G. Potential applications of surface active compounds by Gordonia sp. strain BS29 in soil remediation technologies. *Chemosphere,* **2009**, *75*(6), 801-807.
        [http://dx.doi.org/10.1016/j.chemosphere.2008.12.052] [PMID: 19181361]

[10]   Banat, I.M.; Franzetti, A.; Gandolfi, I.; Bestetti, G.; Martinotti, M.G.; Fracchia, L.; Smyth, T.J.; Marchant, R. Microbial biosurfactants production, applications and future potential. *Appl. Microbiol. Biotechnol.,* **2010**, *87*(2), 427-444.
        [http://dx.doi.org/10.1007/s00253-010-2589-0] [PMID: 20424836]

[11]   Makkar, R.S.; Cameotra, S.S.; Banat, I.M. Advances in utilization of renewable substrates for biosurfactant production. *AMB Express,* **2011**, *1*(1), 5.
        [http://dx.doi.org/10.1186/2191-0855-1-5] [PMID: 21906330]

[12]   Perfumo, A.; Rancich, I.; Banat, I.M. Possibilities and challenges for biosurfactants uses in petroleum industry. In: *Biosurfactants*; Sen, R., Ed.; Advances in Experimental Medicine and Biology. Landes Bioscience, **2010**; pp. 135-145.
        [http://dx.doi.org/10.1007/978-1-4419-5979-9_10]

[13]   Banat, I.M.; Makkar, R.S.; Cameotra, S.S. Potential commercial applications of microbial surfactants. *Appl. Microbiol. Biotechnol.,* **2000**, *53*(5), 495-508.
[http://dx.doi.org/10.1007/s002530051648] [PMID: 10855707]

[14]   Desai, J.D.; Banat, I.M. Microbial production of surfactants and their commercial potential. *Microbiol. Mol. Biol. Rev.,* **1997**, *61*(1), 47-64.
[PMID: 9106364]

[15]   Christofi, N.; Ivshina, I.B. Microbial surfactants and their use in field studies of soil remediation. *J. Appl. Microbiol.,* **2002**, *93*(6), 915-929.
[http://dx.doi.org/10.1046/j.1365-2672.2002.01774.x] [PMID: 12452947]

[16]   Al-Tahhan, R.A.; Sandrin, T.R.; Bodour, A.A.; Maier, R.M. Rhamnolipid-induced removal of lipopolysaccharide from *Pseudomonas aeruginosa*: Effect on cell surface properties and interaction with hydrophobic substrates. *Appl. Environ. Microbiol.,* **2000**, *66*(8), 3262-3268.
[http://dx.doi.org/10.1128/AEM.66.8.3262-3268.2000] [PMID: 10919779]

[17]   Maier, R.M.; Soberón-Chávez, G. *Pseudomonas aeruginosa* rhamnolipids: Biosynthesis and potential applications. *Appl. Microbiol. Biotechnol.,* **2000**, *54*(5), 625-633.
[http://dx.doi.org/10.1007/s002530000443] [PMID: 11131386]

[18]   Franzetti, A.; Gandolfi, I.; Bestetti, G.; Smyth, T.J.P.; Banat, I.M. Production and applications of trehalose lipid biosurfactants. *Eur. J. Lipid Sci. Technol.,* **2010**, *112*(6), 617-627.
[http://dx.doi.org/10.1002/ejlt.200900162]

[19]   Whang, L.M.; Liu, P.W.G.; Ma, C.C.; Cheng, S.S. Application of biosurfactants, rhamnolipid, and surfactin, for enhanced biodegradation of diesel-contaminated water and soil. *J. Hazard. Mater.,* **2008**, *151*(1), 155-163.
[http://dx.doi.org/10.1016/j.jhazmat.2007.05.063] [PMID: 17614195]

[20]   Gerson, O.F.; Zajic, J.E. Surfactant production from hydrocarbons by Corynebacterium lepus, sp. nov. and *Pseudomonas asphaltenicus*, sp. nov. *Dev. Ind. Microbiol.,* **1978**, *19*, 577-599.

[21]   Ishigami, Y.; Yamazaki, S.; Gama, Y. Surface active properties of biosoap from spiculisporic acid. *J. Colloid Interface Sci.,* **1983**, *94*(1), 131-139.
[http://dx.doi.org/10.1016/0021-9797(83)90242-4]

[22]   Jennema, G.E.; McInerney, M.J.; Knapp, R.M.; Clark, J.B.; Feero, J.M.; Revus, D.E.; Menzie, D.E. A halotolerant, biosurfactants-producing Bacillus species potentially useful for enhanced oil recovery. *Dev. Ind. Microbiol.,* **1983**, *24*, 485-492.

[23]   Thomas, CP; Duvall, ML; Robertson, EP; Barrett, KB; Bala, GA Surfactant-based EOR mediated by naturally occurring microorganisms *Spe Reserv. Eng.,* **1993**, *8*, 285-291.
[http://dx.doi.org/10.2118/22844-PA]

[24]   Zosim, Z.; Gutnick, D.; Rosenberg, E. Properties of hydrocarbon-in-water emulsions stabilized by *Acinetobacter* RAG-1 emulsan. *Biotechnol. Bioeng.,* **1982**, *24*(2), 281-292.
[http://dx.doi.org/10.1002/bit.260240203] [PMID: 18546302]

[25]   Rosenberg, E.; Rubinovitz, C.; Legmann, R.; Ron, E.Z. Purification and chemical properties of *Acinetobacter calcoaceticus* A2 Biodispersan. *Appl. Environ. Microbiol.,* **1988**, *54*(2), 323-326.
[http://dx.doi.org/10.1128/aem.54.2.323-326.1988] [PMID: 16347545]

[26]   Cirigliano, M.C.; Carman, G.M. Purification and characterization of liposan, a bioemulsifier from Candida lipolytica. *Appl. Environ. Microbiol.,* **1985**, *50*(4), 846-850.
[http://dx.doi.org/10.1128/aem.50.4.846-850.1985] [PMID: 16346917]

[27]   Henkel, M.; Müller, M.M.; Kügler, J.H.; Lovaglio, R.B.; Contiero, J.; Syldatk, C.; Hausmann, R. Rhamnolipids as biosurfactants from renewable resources: Concepts for next-generation rhamnolipid production. *Process Biochem.,* **2012**, *47*(8), 1207-1219.
[http://dx.doi.org/10.1016/j.procbio.2012.04.018]

[28]   Müller, M.M.; Kügler, J.H.; Henkel, M.; Gerlitzki, M.; Hörmann, B.; Pöhnlein, M.; Syldatk, C.; Hausmann, R. Rhamnolipids Next generation surfactants? *J. Biotechnol.,* **2012**, *162*(4), 366-380.
[http://dx.doi.org/10.1016/j.jbiotec.2012.05.022] [PMID: 22728388]

[29]   Marchant, R.; Banat, I.M. Microbial biosurfactants: challenges and opportunities for future exploitation. *Trends Biotechnol.,* **2012**, *30*(11), 558-565.
[http://dx.doi.org/10.1016/j.tibtech.2012.07.003] [PMID: 22901730]

[30]   Marchant, R.; Banat, I.M. Biosurfactants: A sustainable replacement for chemical surfactants? *Biotechnol. Lett.,* **2012**, *34*(9), 1597-1605.
[http://dx.doi.org/10.1007/s10529-012-0956-x] [PMID: 22618240]

[31]   Perfumo, A.; Smyth, T.J.P.; Marchant, R.; Banat, I.M. Production and roles of biosurfactants and bioemulsifiers in accessing hydrophobic substrates. In: *Handbook of hydrocarbon and lipid microbiology*; Timmis, K.N., Ed.; Springer, **2010**; pp. 1501-1512.
[http://dx.doi.org/10.1007/978-3-540-77587-4_103]

[32]   Miller, R.M.; Zhang, Y. Measurement of biosurfactant-enhanced solubilization and biodegradation of hydrocarbons. In: *Bioremediation Protocols*; Humana Press, **1997**; pp. 59-66.
[http://dx.doi.org/10.1385/0-89603-437-2:59]

[33]   Volkering, F.; Breure, A.M.; Rulkens, W.H. Microbiological aspects of surfactant use for biological soil remediation. *Biodegradation,* **1997-1998**, *8*(6), 401-417.
[http://dx.doi.org/10.1023/A:1008291130109] [PMID: 15765586]

[34]   Muller-Hurtig, R.; Wagner, F.; Blaszczyk, R.; Kosaric, N. Biosurfactants for environmental control. Biosurfactants. Eds by Kosaric N. Marcel Dekker 1993; pp. 447-469.

[35]   Jack, T.R. Microbial enhancement of oil recovery. *Curr. Opin. Biotechnol.,* **1991**, *2*(3), 444-449.
[http://dx.doi.org/10.1016/S0958-1669(05)80154-5]

[36]   Finnerty, W.R. Fossil resource biotechnology: Challenges and prospects. *Curr. Opin. Biotechnol.,* **1992**, *3*(3), 277-282.
[http://dx.doi.org/10.1016/0958-1669(92)90104-Q]

[37]   Bragg, J.R.; Prince, R.C.; Wilkenson, J.B.; Atlas, R.M. Bioremediation for Shoreline Cleanup Following the 1989 Alaskan Oil Spill; Exxon Company USA: Houston, Texas, **1992**.

[38]   Chakrabarty, A.M. Genetically-manipulated microorganisms and their products in the oil service industries. *Trends Biotechnol.,* **1985**, *3*(2), 32-39.
[http://dx.doi.org/10.1016/0167-7799(85)90056-3]

[39]   Mulligan, C.N.; Gibbs, B.F. Types, production and applications of biosurfactants. *Proc Indian Nat Sci Acad,* **2004**, *1*, 31-55.

[40]   Franzetti, A.; Gandolfi, I.; Bestetti, G.; Smyth, T.J.P.; Banat, I.M. Production and applications of trehalose lipid biosurfactants. *Eur. J. Lipid Sci. Technol.,* **2010**, *112*(6), 617-627.
[http://dx.doi.org/10.1002/ejlt.200900162]

[41]   Franzetti, A.; Caredda, P.; Ruggeri, C.; Colla, P.L.; Tamburini, E.; Papacchini, M.; Bestetti, G. Potential applications of surface active compounds by Gordonia sp. strain BS29 in soil remediation technologies. *Chemosphere,* **2009**, *75*(6), 801-807.
[http://dx.doi.org/10.1016/j.chemosphere.2008.12.052] [PMID: 19181361]

[42]   Déziel, E.; Paquette, G.; Villemur, R.; Lépine, F.; Bisaillon, J. Biosurfactant production by a soil *pseudomonas* strain growing on polycyclic aromatic hydrocarbons. *Appl. Environ. Microbiol.,* **1996**, *62*(6), 1908-1912.
[http://dx.doi.org/10.1128/aem.62.6.1908-1912.1996] [PMID: 16535330]

[43]   Rahman, K.S.M.; Rahman, T.J.; Lakshmanaperumalsamy, P.; Marchant, R.; Banat, I.M. The potential of bacterial isolates for emulsification with a range of hydrocarbons. *Acta Biotechnol.,* **2003**, *23*(4), 335-345.

[http://dx.doi.org/10.1002/abio.200390043]

[44]   Inakollu, S.; Hung, H.C.; Shreve, G.S. Biosurfactant enhancement of microbial degradation of various structural classes of hydrocarbons in mixed waste systems. *Environ. Eng. Sci.,* **2004**, *21*(4), 463-469.
       [http://dx.doi.org/10.1089/1092875041358467]

[45]   Obayori, O.S.; Ilori, M.O.; Adebusoye, S.A.; Oyetibo, G.O.; Omotayo, A.E.; Amund, O.O. Degradation of hydrocarbons and biosurfactant production by *Pseudomonas* sp. strain LP1. *World J. Microbiol. Biotechnol.,* **2009**, *25*(9), 1615-1623.
       [http://dx.doi.org/10.1007/s11274-009-0053-z]

[46]   Reddy, M.S.; Naresh, B.; Leela, T.; Prashanthi, M.; Madhusudhan, N.C.; Dhanasri, G.; Devi, P. Biodegradation of phenanthrene with biosurfactant production by a new strain of *Brevibacillus* sp. *Bioresour. Technol.,* **2010**, *101*(20), 7980-7983.
       [http://dx.doi.org/10.1016/j.biortech.2010.04.054] [PMID: 20627713]

[47]   Kang, S.W.; Kim, Y.B.; Shin, J.D.; Kim, E.K. Enhanced biodegradation of hydrocarbons in soil by microbial biosurfactant, sophorolipid. *Appl. Biochem. Biotechnol.,* **2010**, *160*(3), 780-790.
       [http://dx.doi.org/10.1007/s12010-009-8580-5] [PMID: 19253005]

[48]   Das, K.; Mukherjee, A.K. Crude petroleum oil biodegradation efficiency of *Bacillus subtilis* and *Pseudomonas aeruginosa* strains isolated from a petroleum oil contaminated soil from North-East India. *Bioresour. Technol.,* **2007**, *98*(7), 1339-1345.
       [http://dx.doi.org/10.1016/j.biortech.2006.05.032] [PMID: 16828284]

[49]   Joseph, P.J.; Joseph, A. Microbial enhanced separation of oil from a petroleum refinery sludge. *J. Hazard. Mater.,* **2009**, *161*(1), 522-525.
       [http://dx.doi.org/10.1016/j.jhazmat.2008.03.131] [PMID: 18468790]

[50]   Barkay, T.; Navon-Venezia, S.; Ron, E.Z.; Rosenberg, E. Enhancement of solubilization and biodegradation of polyaromatic hydrocarbons by the bioemulsifier alasan. *Appl. Environ. Microbiol.,* **1999**, *65*(6), 2697-2702.
       [http://dx.doi.org/10.1128/AEM.65.6.2697-2702.1999] [PMID: 10347063]

[51]   Hua, X.; Wu, Z.; Zhang, H.; Lu, D.; Wang, M.; Liu, Y.; Liu, Z. Degradation of hexadecane by Enterobacter cloacae strain TU that secretes an exopolysaccharide as a bioemulsifier. *Chemosphere,* **2010**, *80*(8), 951-956.
       [http://dx.doi.org/10.1016/j.chemosphere.2010.05.002] [PMID: 20537678]

[52]   Sen, R. Biotechnology in petroleum recovery: The microbial EOR. *Pror. Energy Combust. Sci.,* **2008**, *34*(6), 714-724.
       [http://dx.doi.org/10.1016/j.pecs.2008.05.001]

[53]   Ghojavand, H.; Vahabzadeh, F.; Shahraki, A.K. Enhanced oil recovery from low permeability dolomite cores using biosurfactant produced by a *Bacillus mojavensis* (PTCC 1696) isolated from Masjed-I Soleyman field. *J. Petrol. Sci. Eng.,* **2012**, *81*, 24-30.
       [http://dx.doi.org/10.1016/j.petrol.2011.12.002]

[54]   She, Y.H.; Zhang, F.; Xia, J.J.; Kong, S.Q.; Wang, Z.L.; Shu, F.C.; Hu, J.M. Investigation of biosurfactant-producing indigenous microorganisms that enhance residue oil recovery in an oil reservoir after polymer flooding. *Appl. Biochem. Biotechnol.,* **2011**, *163*(2), 223-234.
       [http://dx.doi.org/10.1007/s12010-010-9032-y] [PMID: 20652442]

[55]   Xia, W.J.; Luo, Z.; Dong, H.P.; Yu, L.; Cui, Q.F.; Bi, Y.Q. Synthesis, characterization, and oil recovery application of biosurfactant produced by indigenous *pseudomonas aeruginosa* WJ-1 using waste vegetable oils. *Appl. Biochem. Biotechnol.,* **2012**, *166*(5), 1148-1166.
       [http://dx.doi.org/10.1007/s12010-011-9501-y] [PMID: 22198867]

[56]   Castorena-Cortés, G.; Zapata-Peñasco, I.; Roldán-Carrillo, T.; Reyes-Avila, J.; Mayol-Castillo, M.; Román-Vargas, S.; Olguín-Lora, P. Evaluation of indigenous anaerobic microorganisms from Mexican carbonate reservoirs with potential MEOR application. *J. Petrol. Sci. Eng.,* **2012**, *81*, 86-93.
       [http://dx.doi.org/10.1016/j.petrol.2011.12.010]

[57]   Zhang, F.; She, Y.H.; Li, H.M.; Zhang, X.T.; Shu, F.C.; Wang, Z.L.; Yu, L.J.; Hou, D.J. Impact of an indigenous microbial enhanced oil recovery field trial on microbial community structure in a high pour-point oil reservoir. *Appl. Microbiol. Biotechnol.,* **2012**, *95*(3), 811-821.
[http://dx.doi.org/10.1007/s00253-011-3717-1] [PMID: 22159733]

[58]   Darvishi, P.; Ayatollahi, S.; Mowla, D.; Niazi, A. Biosurfactant production under extreme environmental conditions by an efficient microbial consortium, ERCPPI-2. *Colloids Surf. B Biointerfaces,* **2011**, *84*(2), 292-300.
[http://dx.doi.org/10.1016/j.colsurfb.2011.01.011] [PMID: 21345657]

[59]   Pornsunthorntawee, O.; Arttaweeporn, N.; Paisanjit, S.; Somboonthanate, P.; Abe, M.; Rujiravanit, R.; Chavadej, S. Isolation and comparison of biosurfactants produced by *Bacillus subtilis* PT2 and *Pseudomonas aeruginosa* SP4 for microbial surfactant-enhanced oil recovery. *Biochem. Eng. J.,* **2008**, *42*(2), 172-179.
[http://dx.doi.org/10.1016/j.bej.2008.06.016]

[60]   Haddadin, M.S.Y.; Abou Arqoub, A.A.; Abu Reesh, I.; Haddadin, J. Kinetics of hydrocarbon extraction from oil shale using biosurfactant producing bacteria. *Energy Convers. Manage.,* **2009**, *50*(4), 983-990.
[http://dx.doi.org/10.1016/j.enconman.2008.12.015]

[61]   Banat, I.M.; Franzetti, A.; Gandolfi, I.; Bestetti, G.; Martinotti, M.G.; Fracchia, L.; Smyth, T.J.; Marchant, R. Microbial biosurfactants production, applications and future potential. *Appl. Microbiol. Biotechnol.,* **2010**, *87*(2), 427-444.
[http://dx.doi.org/10.1007/s00253-010-2589-0] [PMID: 20424836]

[62]   Urum, K.; Grigson, S.; Pekdemir, T.; McMenamy, S. A comparison of the efficiency of different surfactants for removal of crude oil from contaminated soils. *Chemosphere,* **2006**, *62*(9), 1403-1410.
[http://dx.doi.org/10.1016/j.chemosphere.2005.05.016] [PMID: 16005939]

[63]   Kang, S.W.; Kim, Y.B.; Shin, J.D.; Kim, E.K. Enhanced biodegradation of hydrocarbons in soil by microbial biosurfactant, sophorolipid. *Appl. Biochem. Biotechnol.,* **2010**, *160*(3), 780-790.
[http://dx.doi.org/10.1007/s12010-009-8580-5] [PMID: 19253005]

[64]   Lai, C.C.; Huang, Y.C.; Wei, Y.H.; Chang, J.S. Biosurfactant-enhanced removal of total petroleum hydrocarbons from contaminated soil. *J. Hazard. Mater.,* **2009**, *167*(1-3), 609-614.
[http://dx.doi.org/10.1016/j.jhazmat.2009.01.017] [PMID: 19217712]

[65]   Aşçı, Y.; Nurbaş, M.; Sağ Açıkel, Y. Investigation of sorption/desorption equilibria of heavy metal ions on/from quartz using rhamnolipid biosurfactant. *J. Environ. Manage.,* **2010**, *91*(3), 724-731.
[http://dx.doi.org/10.1016/j.jenvman.2009.09.036] [PMID: 19850403]

[66]   Bruins, M.R.; Kapil, S.; Oehme, F.W. Microbial resistance to metals in the environment. *Ecotoxicol. Environ. Saf.,* **2000**, *45*(3), 198-207.
[http://dx.doi.org/10.1006/eesa.1999.1860] [PMID: 10702338]

[67]   Ledin, M. Accumulation of metals by microorganisms — processes and importance for soil systems. *Earth Sci. Rev.,* **2000**, *51*(1-4), 1-31.
[http://dx.doi.org/10.1016/S0012-8252(00)00008-8]

[68]   Juwarkar, A.A.; Nair, A.; Dubey, K.V.; Singh, S.K.; Devotta, S. Biosurfactant technology for remediation of cadmium and lead contaminated soils. *Chemosphere,* **2007**, *68*(10), 1996-2002.
[http://dx.doi.org/10.1016/j.chemosphere.2007.02.027] [PMID: 17399765]

[69]   Aşçı, Y.; Nurbaş, M.; Sağ Açıkel, Y. A comparative study for the sorption of Cd(II) by K-feldspar and sepiolite as soil components, and the recovery of Cd(II) using rhamnolipid biosurfactant. *J. Environ. Manage.,* **2008**, *88*(3), 383-392.
[http://dx.doi.org/10.1016/j.jenvman.2007.03.006] [PMID: 17462813]

[70]   Ochoa-Loza, F.J.; Noordman, W.H.; Jannsen, D.B.; Brusseau, M.L.; Maier, R.M. Effect of clays, metal oxides, and organic matter on rhamnolipid biosurfactant sorption by soil. *Chemosphere,* **2007**,

66(9), 1634-1642.
[http://dx.doi.org/10.1016/j.chemosphere.2006.07.068] [PMID: 16965801]

[71]   Kim, J.; Vipulanandan, C. Removal of lead from contaminated water and clay soil using a biosurfactant. *J. Environ. Eng.,* **2006**, *132*(7), 777-786.
[http://dx.doi.org/10.1061/(ASCE)0733-9372(2006)132:7(777)]

[72]   Dahrazma, B.; Mulligan, C.N. Investigation of the removal of heavy metals from sediments using rhamnolipid in a continuous flow configuration. *Chemosphere,* **2007**, *69*(5), 705-711.
[http://dx.doi.org/10.1016/j.chemosphere.2007.05.037] [PMID: 17604818]

[73]   Gnanamani, A.; Kavitha, V.; Radhakrishnan, N.; Suseela Rajakumar, G.; Sekaran, G.; Mandal, A.B. Microbial products (biosurfactant and extracellular chromate reductase) of marine microorganism are the potential agents reduce the oxidative stress induced by toxic heavy metals. *Colloids Surf. B Biointerfaces,* **2010**, *79*(2), 334-339.
[http://dx.doi.org/10.1016/j.colsurfb.2010.04.007] [PMID: 20483569]

[74]   Sheng, X.; He, L.; Wang, Q.; Ye, H.; Jiang, C. Effects of inoculation of biosurfactant-producing *Bacillus* sp. J119 on plant growth and cadmium uptake in a cadmium-amended soil. *J. Hazard. Mater.,* **2008**, *155*(1-2), 17-22.
[http://dx.doi.org/10.1016/j.jhazmat.2007.10.107] [PMID: 18082946]

[75]   Raaijmakers, J.M.; de Bruijn, I.; de Kock, M.J.D. Cyclic lipopeptide production by plant-associated *Pseudomonas* spp.: Diversity, activity, biosynthesis, and regulation. *Mol. Plant Microbe Interact.,* **2006**, *19*(7), 699-710.
[http://dx.doi.org/10.1094/MPMI-19-0699] [PMID: 16838783]

[76]   Teichmann, B.; Linne, U.; Hewald, S.; Marahiel, M.A.; Bölker, M. A biosynthetic gene cluster for a secreted cellobiose lipid with antifungal activity from *Ustilago maydis. Mol. Microbiol.,* **2007**, *66*(2), 525-533.
[http://dx.doi.org/10.1111/j.1365-2958.2007.05941.x] [PMID: 17850255]

[77]   Kulakovskaya, T.V.; Golubev, W.I.; Tomashevskaya, M.A.; Kulakovskaya, E.V.; Shashkov, A.S.; Grachev, A.A.; Chizhov, A.S.; Nifantiev, N.E. Production of antifungal cellobiose lipids by Trichosporon porosum. *Mycopathologia,* **2010**, *169*(2), 117-123.
[http://dx.doi.org/10.1007/s11046-009-9236-2] [PMID: 19757153]

[78]   Stanghellini, M.E.; Miller, R.M. Their identity and potential efficacy in the biological control of zoosporic plant pathogens. *Plant Dis.,* **1997**, *81*(1), 4-12.
[http://dx.doi.org/10.1094/PDIS.1997.81.1.4] [PMID: 30870944]

[79]   Debode, J.; Maeyer, K.D.; Perneel, M.; Pannecoucque, J.; Backer, G.D.; Höfte, M. Biosurfactants are involved in the biological control of *Verticillium* microsclerotia by *Pseudomonas* spp. *J. Appl. Microbiol.,* **2007**, *103*(4), 1184-1196.
[http://dx.doi.org/10.1111/j.1365-2672.2007.03348.x] [PMID: 17897223]

[80]   Piljac, A.; Stipčević, T.; Piljac-Žegarac, J.; Piljac, G. Successful treatment of chronic decubitus ulcer with 0.1% dirhamnolipid ointment. *J. Cutan. Med. Surg.,* **2008**, *12*(3), 142-146.
[http://dx.doi.org/10.2310/7750.2008.07052] [PMID: 18544299]

[81]   Varnier, A.L.; Sanchez, L.; Vatsa, P.; Boudesocque, L.; Garcia-Brugger, A.; Rabenoelina, F.; Sorokin, A.; Renault, J.H.; Kauffmann, S.; Pugin, A.; Clement, C.; Baillieul, F.; Dorey, S. Bacterial rhamnolipids are novel MAMPs conferring resistance to *Botrytis cinerea* in grapevine. *Plant Cell Environ.,* **2009**, *32*(2), 178-193.
[http://dx.doi.org/10.1111/j.1365-3040.2008.01911.x] [PMID: 19021887]

[82]   D'aes, J.; De Maeyer, K.; Pauwelyn, E.; Höfte, M. Biosurfactants in plant– *Pseudomonas* interactions and their importance to biocontrol. *Environ. Microbiol. Rep.,* **2010**, *2*(3), 359-372.
[http://dx.doi.org/10.1111/j.1758-2229.2009.00104.x] [PMID: 23766108]

[83]   Blom, J.; Rueckert, C.; Niu, B.; Wang, Q.; Borriss, R. The complete genome of *Bacillus amyloliquefaciens* subsp. plantarum CAU B946 contains a gene cluster for nonribosomal synthesis of

iturin A. *J. Bacteriol.,* **2012**, *194*(7), 1845-1846.
[http://dx.doi.org/10.1128/JB.06762-11] [PMID: 22408246]

[84] Dunlap, C.A.; Bowman, M.J.; Schisler, D.A. Genomic analysis and secondary metabolite production in *Bacillus* amyloliquefaciens AS 43.3: A biocontrol antagonist of Fusarium head blight. *Biol. Control,* **2013**, *64*(2), 166-175.
[http://dx.doi.org/10.1016/j.biocontrol.2012.11.002]

[85] Höfte, M.; Altier, N. Fluorescent pseudomonads as biocontrol agents for sustainable agricultural systems. *Res. Microbiol.,* **2010**, *161*(6), 464-471.
[http://dx.doi.org/10.1016/j.resmic.2010.04.007] [PMID: 20457252]

[86] Palanisamy, P. Biosurfactant mediated synthesis of NiO nanorods. *Mater. Lett.,* **2008**, *62*(4-5), 743-746.
[http://dx.doi.org/10.1016/j.matlet.2007.06.053]

[87] Reddy, A.S.; Chen, C.Y.; Baker, S.C.; Chen, C.C.; Jean, J.S.; Fan, C.W.; Chen, H.R.; Wang, J.C. Synthesis of silver nanoparticles using surfactin: A biosurfactant as stabilizing agent. *Mater. Lett.,* **2009**, *63*(15), 1227-1230.
[http://dx.doi.org/10.1016/j.matlet.2009.02.028]

[88] Biswas, M.; Raichur, A.M. Electrokinetic and rheological properties of nano zirconia in the presence of rhamnolipid biosurfactant. *J. Am. Ceram. Soc.,* **2008**, *91*(10), 3197-3201.
[http://dx.doi.org/10.1111/j.1551-2916.2008.02617.x]

[89] Monteiro, S.A.; Sassaki, G.L.; de Souza, L.M.; Meira, J.A.; de Araújo, J.M.; Mitchell, D.A.; Ramos, L.P.; Krieger, N. Molecular and structural characterization of the biosurfactant produced by *Pseudomonas aeruginosa* DAUPE 614. *Chem. Phys. Lipids,* **2007**, *147*(1), 1-13.
[http://dx.doi.org/10.1016/j.chemphyslip.2007.02.001] [PMID: 17382918]

[90] Mukherjee, S.; Das, P.; Sen, R. Towards commercial production of microbial surfactants. *Trends Biotechnol.,* **2006**, *24*(11), 509-515.
[http://dx.doi.org/10.1016/j.tibtech.2006.09.005] [PMID: 16997405]

[91] Hossain, H.; Wellensiek, H.J.; Geyer, R.; Lochnit, G. Structural analysis of glycolipids from Borrelia burgdorferi. *Biochimie,* **2001**, *83*(7), 683-692.
[http://dx.doi.org/10.1016/S0300-9084(01)01296-2] [PMID: 11522398]

[92] Gerard, J.; Lloyd, R.; Barsby, T.; Haden, P.; Kelly, M.T.; Andersen, R.J. Massetolides A-H, antimycobacterial cyclic depsipeptides produced by two pseudomonads isolated from marine habitats. *J. Nat. Prod.,* **1997**, *60*(3), 223-229.
[http://dx.doi.org/10.1021/np9606456] [PMID: 9157190]

[93] Rodrigues, L.R.; Banat, I.M.; Mei, H.C.; Teixeira, J.A.; Oliveira, R. Interference in adhesion of bacteria and yeasts isolated from explanted voice prostheses to silicone rubber by rhamnolipid biosurfactants. *J. Appl. Microbiol.,* **2006**, *100*(3), 470-480.
[http://dx.doi.org/10.1111/j.1365-2672.2005.02826.x] [PMID: 16478486]

[94] Kim, K.; Jung, S.Y.; Lee, D.K.; Jung, J.K.; Park, J.K.; Kim, D.K.; Lee, C.H. Suppression of inflammatory responses by surfactin, a selective inhibitor of platelet cytosolic phospholipase A2. *Biochem. Pharmacol.,* **1998**, *55*(7), 975-985.
[http://dx.doi.org/10.1016/S0006-2952(97)00613-8] [PMID: 9605421]

[95] Vollenbroich, D.; Pauli, G.; Ozel, M.; Vater, J. Antimycoplasma properties and application in cell culture of surfactin, a lipopeptide antibiotic from *Bacillus subtilis. Appl. Environ. Microbiol.,* **1997**, *63*(1), 44-49.
[http://dx.doi.org/10.1128/aem.63.1.44-49.1997] [PMID: 8979337]

[96] Tanaka, Y.; Tojo, T.; Uchida, K. 5470827 Method of producing iturin a and antifungal agent for profound mycosis. *Biotechnol. Adv.,* **1997**, *15*(1), 234-235.
[http://dx.doi.org/10.1016/S0734-9750(97)88427-3]

[97]   Besson, F.; Peypoux, F.; Michel, G.; Delcambe, L. Characterization of iturin a in antibiotics from various strains of *Bacillus subtilis. J. Antibiot.,* **1976**, *29*(10), 1043-1049.
[http://dx.doi.org/10.7164/antibiotics.29.1043] [PMID: 825665]

[98]   Zhao, X.; Geltinger, C.; Kishikawa, S.; Ohshima, K.; Murata, T.; Nomura, N.; Nakahara, T.; Yokoyama, K.K. Treatment of mouse melanoma cells with phorbol 12-myristate 13-acetate counteracts mannosylerythritol lipid-induced growth arrest and apoptosis. *Cytotechnology,* **2000**, *33*(1/3), 123-130.
[http://dx.doi.org/10.1023/A:1008129616127] [PMID: 19002819]

[99]   Kitamoto, D.; Yanagishita, H.; Shinbo, T.; Nakane, T.; Kamisawa, C.; Nakahara, T. Surface active properties and antimicrobial activities of mannosylerythritol lipids as biosurfactants produced by Candida antarctica. *J. Biotechnol.,* **1993**, *29*(1-2), 91-96.
[http://dx.doi.org/10.1016/0168-1656(93)90042-L]

[100]   Bibel, D.J.; Aly, R.; Shinefield, H.R. Inhibition of microbial adherence by sphinganine. *Can. J. Microbiol.,* **1992**, *38*(9), 983-985.
[http://dx.doi.org/10.1139/m92-158] [PMID: 1464071]

[101]   Uchida, Y.; Tsuchiya, R.; Chino, M. Extracellular accumulation of mono and di succinyl trehalose lipids by a strain of *Rodococcus erythropolis* grown on n-alkanes. *Agric. Biol. Chem.,* **1989**, *53*, 757-766.

[102]   Nielsen, T.H.; Christophersen, C.; Anthoni, U.; Sørensen, J. Viscosinamide, a new cyclic depsipeptide with surfactant and antifungal properties produced by *Pseudomonas fluorescens* DR54. *J. Appl. Microbiol.,* **1999**, *87*(1), 80-90.
[http://dx.doi.org/10.1046/j.1365-2672.1999.00798.x] [PMID: 10432590]

[103]   Niels, G.; Mehmet, N.N.; Jean, M.C.; Jos, M. Membrane interactions of natural cyclic lipodepsipeptides of the viscosin. *Biochimica Biophysica Acta,* **2017**, *1853*(3), 331-339.
[http://dx.doi.org/10.1016/j.bbamem.2016.12.013]

[104]   Naruse, N.; Tenmyo, O.; Kobaru, S.; Kamei, H.; Miyaki, T.; Konishi, M.; Oki, T. Pumilacidin, a complex of new antiviral antibiotics. Production, isolation, chemical properties, structure and biological activity. *J. Antibiot.,* **1990**, *43*(3), 267-280.
[http://dx.doi.org/10.7164/antibiotics.43.267] [PMID: 2157695]

[105]   Giedraitienė, A.; Vitkauskienė, A.; Naginienė, R.; Pavilonis, A. Antibiotic resistance mechanisms of clinically important bacteria. *Medicina,* **2011**, *47*(3), 19.
[http://dx.doi.org/10.3390/medicina47030019] [PMID: 21822035]

[106]   Tenover, F.C. Mechanisms of antimicrobial resistance in bacteria. *Am. J. Med.,* **2006**, *119*(6) 1, S3-S10.
[http://dx.doi.org/10.1016/j.amjmed.2006.03.011] [PMID: 16735149]

[107]   Rodrigues, L.; van der Mei, H.C.; Teixeira, J.; Oliveira, R. Influence of biosurfactants from probiotic bacteria on formation of biofilms on voice prostheses. *Appl. Environ. Microbiol.,* **2004**, *70*(7), 4408-4410.
[http://dx.doi.org/10.1128/AEM.70.7.4408-4410.2004] [PMID: 15240331]

[108]   Reid, G.; Bruce, A.W.; Fraser, N.; Heinemann, C.; Owen, J.; Henning, B. Oral probiotics can resolve urogenital infections. *FEMS Immunol. Med. Microbiol.,* **2001**, *30*(1), 49-52.
[http://dx.doi.org/10.1111/j.1574-695X.2001.tb01549.x] [PMID: 11172991]

[109]   Rodrigues, L.; Banat, I.M.; Teixeira, J.; Oliveira, R. Biosurfactants: Potential applications in medicine. *J. Antimicrob. Chemother.,* **2006**, *57*(4), 609-618.
[http://dx.doi.org/10.1093/jac/dkl024] [PMID: 16469849]

[110]   Mack, D.R.; Michail, S.; Wei, S.; McDougall, L.; Hollingsworth, M.A. Probiotics inhibit enteropathogenic *E. coli* adherence *in vitro* by inducing intestinal mucin gene expression. *Am. J. Physiol.,* **1999**, *276*(4), G941-G950.

[PMID: 10198338]

[111] Fariq, A.; Saeed, A. Production and biomedical applications of probiotic biosurfactants. *Curr. Microbiol.,* **2016**, *72*(4), 489-495.
[http://dx.doi.org/10.1007/s00284-015-0978-4] [PMID: 26742771]

[112] Gudiña, E.J.; Teixeira, J.A.; Rodrigues, L.R. Isolation and functional characterization of a biosurfactant produced by *Lactobacillus paracasei. Colloids Surf. B Biointerfaces,* **2010**, *76*(1), 298-304.
[http://dx.doi.org/10.1016/j.colsurfb.2009.11.008] [PMID: 20004557]

[113] Mandal, S.M.; Sharma, S.; Pinnaka, A.K.; Kumari, A.; Korpole, S. Isolation and characterization of diverse antimicrobial lipopeptides produced by Citrobacter and Enterobacter. *BMC Microbiol.,* **2013**, *13*(1), 152.
[http://dx.doi.org/10.1186/1471-2180-13-152] [PMID: 23834699]

[114] Saravanakumari, P.; Mani, K. Structural characterization of a novel xylolipid biosurfactant from *Lactococcus lactis* and analysis of antibacterial activity against multi-drug resistant pathogens. *Bioresour. Technol.,* **2010**, *101*(22), 8851-8854.
[http://dx.doi.org/10.1016/j.biortech.2010.06.104] [PMID: 20637606]

[115] Sambanthamoorthy, K.; Feng, X.; Patel, R.; Patel, S.; Paranavitana, C. Antimicrobial and antibiofilm potential of biosurfactants isolated from lactobacilli against multi-drug-resistant pathogens. *BMC Microbiol.,* **2014**, *14*(1), 197.
[http://dx.doi.org/10.1186/1471-2180-14-197] [PMID: 25124936]

[116] Dusane, D.H.; Damare, S.R.; Nancharaiah, Y.V.; Ramaiah, N.; Venugopalan, V.P.; Kumar, A.R.; Zinjarde, S.S. Disruption of microbial biofilms by an extracellular protein isolated from epibiotic tropical marine strain of *Bacillus licheniformis. PLoS One,* **2013**, *8*(5), e64501.
[http://dx.doi.org/10.1371/journal.pone.0064501] [PMID: 23691235]

[117] Salehi, R.; Savabi, O.; Kazemi, M.; kamali, S.; Salehi, A.; Eslami, G.; Tahmourespour, A. Effects of *Lactobacillus reuteri*derived biosurfactant on the gene expression profile of essential adhesion genes (gtfB, gtfC and ftf) of Streptococcus mutans. *Adv. Biomed. Res.,* **2014**, *3*(1), 169.
[http://dx.doi.org/10.4103/2277-9175.139134] [PMID: 25221772]

[118] Dalili, D.; Amini, M.; Faramarzi, M.A.; Fazeli, M.R.; Khoshayand, M.R.; Samadi, N. Isolation and structural characterization of Coryxin, a novel cyclic lipopeptide from *Corynebacterium xerosis* NS5 having emulsifying and anti-biofilm activity. *Colloids Surf. B Biointerfaces,* **2015**, *135*, 425-432.
[http://dx.doi.org/10.1016/j.colsurfb.2015.07.005] [PMID: 26280817]

[119] Fracchia, L.; Cavallo, M.; Allegrone, G.; Martinotti, M.G. A Lactobacillus-derived biosurfactant inhibits biofilm formation of human pathogenic *Candida albicans* biofilm producers. *Appl. Microbiol. Biotechnol.,* **2010**, *2*, 827-837.

[120] Cao, X.; Wang, A.; Wang, C.; Mao, D.; Lu, M.; Cui, Y.; Jiao, R. Surfactin induces apoptosis in human breast cancer MCF-7 cells through a ROS/JNK-mediated mitochondrial/caspase pathway. *Chem. Biol. Interact.,* **2010**, *183*(3), 357-362.
[http://dx.doi.org/10.1016/j.cbi.2009.11.027] [PMID: 19954742]

[121] Zhao, X.; Geltinger, C.; Kishikawa, S.; Ohshima, K.; Murata, T.; Nomura, N.; Nakahara, T.; Yokoyama, K.K. Treatment of mouse melanoma cells with phorbol 12-myristate 13-acetate counteracts mannosylerythritol lipid-induced growth arrest and apoptosis. *Cytotechnology,* **2000**, *33*(1/3), 123-130.
[http://dx.doi.org/10.1023/A:1008129616127] [PMID: 19002819]

[122] Duarte, C.; Gudiña, E.J.; Lima, C.F.; Rodrigues, L.R. Effects of biosurfactants on the viability and proliferation of human breast cancer cells. *AMB Express,* **2014**, *4*(1), 40.
[http://dx.doi.org/10.1186/s13568-014-0040-0] [PMID: 24949273]

[123] Gudiña, E.J.; Rangarajan, V.; Sen, R.; Rodrigues, L.R. Potential therapeutic applications of biosurfactants. *Trends Pharmacol. Sci.,* **2013**, *34*(12), 667-675.

[http://dx.doi.org/10.1016/j.tips.2013.10.002] [PMID: 24182625]

[124] Krishnaswamy, M.; Subbuchettiar, G.; Ravi, T.K.; Panchaksharam, S. Biosurfactants properties, commercial production and application. *Curr. Sci.,* **2008**, *94*, 736-747.

[125] Lin, C-T. Characterization of biosurfactant from a diesel-oil degradation bacterium and application potential in beauty care products. *2011 3rd International Conference on Chemical, Biological, and Environmental Engineering.,* IPCBEE vol. 20 IACSIT Press, Singapore, 2011.

[126] Cao, X.; Wang, A.; Wang, C.; Mao, D.; Lu, M.; Cui, Y.; Jiao, R. Surfactin induces apoptosis in human breast cancer MCF-7 cells through a ROS/JNK-mediated mitochondrial/caspase pathway. *Chem. Biol. Interact.,* **2010**, *183*(3), 357-362.
[http://dx.doi.org/10.1016/j.cbi.2009.11.027] [PMID: 19954742]

[127] Wang, C.; Liu, C.; Niu, L.; Wang, L.; Hou, L.; Cao, X. Surfactin-induced apoptosis through ROS-ERS-Ca2+-ERK pathways in HepG2 cells. *Cell Biochem. Biophys.,* **2013**, *67*(3), 1433-1439.
[http://dx.doi.org/10.1007/s12013-013-9676-7] [PMID: 23733672]

[128] Naughton, P.J.; Marchant, R.; Naughton, V.; Banat, I.M. Microbial biosurfactants: Current trends and applications in agricultural and biomedical industries. *J. Appl. Microbiol.,* **2019**, *127*(1), 12-28.
[http://dx.doi.org/10.1111/jam.14243] [PMID: 30828919]

[129] Banerjee, R.; Preetha, A.; Huilgol, N. Surface activity, lipid profiles and their implications in cervical cancer. *J. Cancer Res. Ther.,* **2005**, *1*(3), 180-186.
[http://dx.doi.org/10.4103/0973-1482.19600] [PMID: 17998650]

[130] Mnif, S.; Chamkha, M.; Sayadi, S. Isolation and characterization of *Halomonas* sp. strain C2SS100, a hydrocarbon-degrading bacterium under hypersaline conditions. *J. Appl. Microbiol.,* **2009**, *107*(3), 785-794.
[http://dx.doi.org/10.1111/j.1365-2672.2009.04251.x] [PMID: 19320948]

[131] Joshi, S.; Bharucha, C.; Jha, S.; Yadav, S.; Nerurkar, A.; Desai, A.J. Biosurfactant production using molasses and whey under thermophilic conditions. *Bioresour. Technol.,* **2008**, *99*(1), 195-199.
[http://dx.doi.org/10.1016/j.biortech.2006.12.010] [PMID: 17321739]

[132] Kumar, M.; León, V.; De Sisto Materano, A.; Ilzins, O.A.; Luis, L. Biosurfactant production and hydrocarbon degradation by halotolerant and thermotolerant *Pseudomonas* sp. *World J. Microbiol. Biotechnol.,* **2008**, *24*(7), 1047-1057.
[http://dx.doi.org/10.1007/s11274-007-9574-5]

[133] Reddy, A.S.; Chen, C.Y.; Chen, C.C.; Jean, J.S.; Fan, C.W.; Chen, H.R.; Wang, J.C.; Nimje, V.R. Synthesis of gold nanoparticles *via* an environmentally benign route using a biosurfactant. *J. Nanosci. Nanotechnol.,* **2009**, *9*(11), 6693-6699.
[http://dx.doi.org/10.1166/jnn.2009.1347] [PMID: 19908586]

[134] Kitamoto, D.; Morita, T.; Fukuoka, T.; Konishi, M.; Imura, T. Self assembling properties of glycolipid biosurfactants and their potential applications. *Curr. Opin. Colloid Interface Sci.,* **2009**, *14*(5), 315-328.
[http://dx.doi.org/10.1016/j.cocis.2009.05.009]

CHAPTER 2

Biosurfactants: Screening, Production and their Applications

Truptirekha Das[1], **Sameer Ranjan Sahoo**[2,*] and **Arun Kumar Pradhan**[2]

[1] *Rautara Govt. High School, Binjharpur, Jajapur, Odisha-755004, India*

[2] *Centre for Biotechnology, Siksha 'O' Anusandhan (Deemed to be University), Bhubaneswar, Odisha-751003, India*

Abstract: Biosurfactants are commonly recognised as biologically derived surface active agents. The most significant microbes have been studied for the production of glycolipid biosurfactants, including *Pseudomonas aeruginosa, Bacillus subtilis,* and Candida spp. Microbial derived biosurfactants are found commercially superior to chemical derivatives due to their biodegradability, renewability, and good performance under harsh working conditions. It has been found that (isolated) hydrocarbon-degrading bacteria produce far more biosurfactants than that predicted from oil spills. This is because all genomes are regulated in lipid metabolism synthesis. The oil and petroleum sector use biosurfactants as an emulsifier for both recovery and removal from contaminated sites. They also play a role in the removal of heavy metals in metallurgy. In this work, we have provided an overview of the screening of microorganisms that produce biosurfactants, production techniques, and variables that affect the production of biosurfactants. Several analytical approaches for crude metabolite processing are also given. Hence, the importance of biosurfactants in environmental cleaning is simply understood from this review.

Keywords: Antiviral, Biodegradability, Biosurfactants, Metallurgical, Petrochemical industry.

INTRODUCTION

Biosurfactants are microorganism-produced surface active chemicals with critical environmental applications. Biosurfactants are classified into three groups: glycolipids, phospolipids, and lipopeptides. It was observed that glycolipid biosurfactants comprising sugar molecules and hydroxyl fatty acids had hydrophilic and hydrophobic characteristics. The latter has been demonstrated to have the properties of a surfactant, an emulsifier, and a bioactive material. Biological surfactants are often biodegradable, secure, and renewable. They have

---

* **Corresponding author Sameer Ranjan Sahoo:** Centre for Biotechnology, Siksha 'O' Anusandhan (Deemed to be University), Bhubaneswar, Odisha-751003, India; E-mail: sahu.samir070@gmail.com

Arun Kumar Pradhan and Manoranjan Arakha (Eds.)

the potential to function better than artificial surfactants because of their strong interfacial tension, high surface tension, and critical micelle concentration. They are readily and quickly manufactured.

Biological surfactants function well as detergents, foaming agents, wetting agents, and microemulsifiers [1]. They can function in high pH, salinity, and temperature environments [2].

The critical micelle concentration of surfactants ranges from 5-380 mg/l, and they can generally lower the surface tension of water to 25-40 mN/m. Additionally, it lowers the interfacial tension between oil and water to 1 dyn/cm. Due to its capacity to degrade a variety of substrates, *Pseudomonas aeruginosa* is a perfect microbe for biosurfactant production. For production, economically feasible raw materials like oil waste, soap residues, waste from food firms and vegetable oil refineries are being used. Of all carbon sources, vegetable-based oil has the highest biosurfactant yield [3]. Also, the characteristics of biosurfactants are comparable. Glycolipid biosurfactants, on the other hand, have been identified as potential biosurfactants with a variety of benefits. Environmental cleanup, non-toxicity, and biodegradability are only a few of the benefits. Biosurfactants have a wide range of uses, including medicines, therapies, cosmetics, detergents, agriculture, heavy metal removal, and recovery of oil [4].

Biosurfactant Classification

Synthetically produced biosurfactants are categorised according to their polar groups, which are determined by distinct molecules constituting the hydrophobic (efficient emulsion stabilisers) and hydrophilic (reduce surface and interfacial tension) moieties along with their microbial origin. Amino acids, peptides, mono, and di-polysaccharides are the primary constituents for hydrophilic moiety and saturated or unsaturated fatty acids for hydrophobic moiety [2, 5]. The nutritional environment of the developing bacterium influences the type of surfactant synthesis. Important biosurfactants, as per their types, are listed in Table **1**.

Mechanism of Hydrocarbon Utilization

For early enzymatic catabolism within cells, a wide range of adaptive mechanisms by microbes are followed for collecting and delivering hydrocarbons into the cell [6]. Bacteria are speculated to use a passive process to transfer and digest soluble alkanes dissolved in water. Initially, it was assumed they could only use solubilized hydrocarbons [7]. Until observed, the rate of alkanes destroyed surpassed hydrocarbon solubility in the aqueous phase, indicating the use of an alternative hydrocarbon absorption mechanism by microbes [8]. Several pathways for aliphatic hydrocarbon absorption have been postulated but were ruled out due to the limited solubility of long-chain alkanes in water [9]. The following stages have been discovered to occur during hydrocarbon absorption. a) Little hydrocarbon droplets (micelles) are confined within the cells. b) Cells can encapsulate hydrocarbon within their cells due to direct interaction with the larger hydrocarbon phase. A study revealed that microbes degrading hydrocarbon are designed to flourish in oily environments, that can have a significant effect on biological remediation [10]. Bacteria that break down hydrocarbons produce biosurfactants with a variety of chemical characteristics and molecular sizes. One of the process limitations is the bioavailability of different oil fractions at low temperatures.

Table 1. Biosurfactants type and micro-organism involved.

Microbial Surfactants	Organisms
Glycolipids	*Serratia marcescens, Alcanivorax borkumensis*, Arthrobacter sp., corynebacterium sp., Pseudomonas sp., *Pseudomonas aeruginosa, Serratia rubidea, Torulopsis apicola, T. bombicola, T. petrophilium, Candida apicola, Candida bombicola, Candida bogoriensis, Candida lipolytica, Rhodococcuserythropolis, Nocardia erythropolis*, Mycobacterium sp., *Arthrobacter paraffineus*, Corynebacterium sp.
Fatty Acids (Spiculisporic Acids, Corynomycolic Acids, *etc.*,)	*Candida lepus*, Capnocytophaga sp., *Corynebacterium lepus, Penicillium, spiculisporum, Nocardia erythropolis*
Carbohydrate-lipid-protein	*Pseudomonas fluorescens*
Mannan-lipid-protein	*Candida tropicalis*
Particulate Surfactants	*Pseudomonas marginalis*

Screening of Microorganism

The first step in the selection stage is the isolation of the strains from their natural environments, which are then subjected to screening for the desired product manufacturing in microbial bio-processing. Primary screening is the collection of specified selective methods that permit the identification of isolate-generating targeted metabolites. Ideally, it should be fast, economical, and specific, which is effective for a wide variety of substances with large-scale usability. Screening to find a few suitable ones from a vast number of microbes is time-consuming and labor-intensive. The numerous screening procedures used for bioprocessing biosurfactant-generating microorganisms are explained briefly below.

One effective approach is the hydrocarbon overlay agar test, in which colonies produced on oil are emulsified with halo zones, suggesting potential biosurfactant producers [11]. CTAB agar plate is an appropriate screening procedure for rhamnolipids forming a dark blue halo zone around the colony. This is due to the anionic biosurfactant released into the media forming an insoluble ion pair with the cationic CTAB [12]. Another way is hemolytic activity, which detects red blood cells rupturing in the presence of the biosurfactant. Unfortunately, it is an unreliable criteria for detection [13]. Drop collapse method is a well-known common screening method. The presence of biosurfactants [14] causes the collapse of Pennzoil (hydrocarbon source). This approach yields negative findings when the biosurfactant concentration is very low [15]. Emulsification activity is a critical metric for assessing biosurfactant generating bacteria. Rosenberg *et al.* [16] established the first technique for measuring emulsification activity using optical density, which was later refined by Neu and Poralla [17]. The degree of emulsification may be determined by comparing the optical densities of the culture medium with and without hydrocarbons. The emulsification index is another way to emulsification activity. The emulsion layer that forms between the aqueous and kerosene layers is determined and used to compute the emulsification index. The stability of the emulsification index provides information on the strength of the biosurfactant [18, 19].

Production of Biosurfactant

Numerous studies used culture medium to produce biosurfactants from various types of bacteria. The majority of microorganisms utilised are obtained from industrial waste lands contaminated with petroleum hydrocarbon byproducts [20].

Fermentation Strategies for Production

For the manufacturing of biosurfactants, fermentation techniques like: flask shake, continuous fed-batch, integrated microbial/enzymatic, genetic engineering and culture immobilization have been employed. For the synthesis of biosurfactants such as Rhamnolipid, a carbon source is included, keeping nitrogen and phosphorous at balance. It is a secondary metabolite released after restricting microbial growth. Nitrate salts, including sodium nitrate, ammonium nitrate, and potassium nitrate [21], have been proven to promote biosurfactant synthesis. Glucose, n-alkanes, ethanol, and glycerolipids are the primary carbon sources. In continuous batch culture, glucose and hydrocarbon are employed as substrates. Glycerol/Plant oils are employed in batch cultures, which act as a growth-limiting substrate in biosurfactant production. Solid state culture by Camilos-Neto *et al.* was reported in the continuous fermentation process [22].

Cooper *et al.* [23] used foam fractionation to separate the product from the reactor after employing glucose for surfactin production. In order to produce surfactin from *Bacillus subtilis*, Noah *et al.* [24] investigated the use of chemostat-stirred tank reactor. *Bacillus subtilis* produced surfactin in an airlift fermenter in a continuous foam collection utilising potato process effluent as a carbon source.

After that, a new model of the reactor was developed by Yeh *et al*. [25] to prevent foam spilling during synthesis. *Bacillus subtilis* [26] in a bubbleless reactor was reported for the production of surfactin and fengycin where an air-liquid contactor is attached made of a hollow fiber membrane.

Factors Affecting the Biosurfactant Production

Table **2** shows how several parameters influence the yield in the manufacturing of biosurfactants. Some of the most critical criteria are described further below:

Table 2. Various factors that influence biosurfactant production.

S. No.	Microorganism	Biosurfactant	pH	Temp.	Carbon Source	Yield	References
1	*Bacillus brevis*	Lipopeptide	8	33°C	8.5g/l of glucose	-	[44]
2	*Pleurotus djamor*		5.5	29°C	5g/l of sunflower seed shell	8.9±0.5 g/l	[45]
3	*Pseudomonas aeruginosa* KVD-HR42	Rhamnolipids	7.8	37°C	23.85g/l karanja oil	5.90±2.1 g/l	[46]
4	*Bacillus* ICA 56		8	33°C	Glycerol and sunflower oil	1290 mg/l	[47]
5	*Pseudomonas aeruginosa* KVD-HR42 F23		8	30°C	1% coconut oil	2.8 g/l	[48]

Effect of Carbon Sources

A range of carbon sources are needed by microbes as an energy source to proliferate surfactant. Rhamnolipid producing *Pseudomonas aeruginosa* can use carbon sources like glycerol, mannitol, glucose, and ethanol, which are water-soluble only [27]. Glycerol is an exception than others as with the concentration exceeding 2%, a sharp decline in the rhamnolipid is found. Safi *et al*. observed [28], 2g/l of rhamnolipids was produced on the use of 3% glycerol / 6% grape seed oil / sunflower oil in fermentation. Rhamnolipid yield with a 6% glucose was determined to be 1400-1500 mg/l. A 6% and 5% concentrations of diesel and kerosene oil, respectively yielded, 1.3 and 2.1 g/L of the compound [29]. Soybean lecithin and crude oil has been estimated as a good carbon source. Changjun Zoua [29] concluded in his study that soybean lecithin is more efficient than crude oil for synthesis. Huy *et al*. [30] stated with their experiment that crude oil can also be proven an excellent carbon source for Acenitobacter-related bacteria. Only water-soluble carbon sources other than paraffin and n-hexadecane, according to Jorge F.B. Pereira, are practical for the synthesis of surfactants [31]. A research by Onwosi and Odibo [31] showed that 2% of glucose concentration acts as an excellent carbon source for the synthesis of rhamnolipids, giving 5.28 g/l.

Effect of Nitrogen Source

Nitrogen is essential for biomass development, which results in the development of a biosurfactant. *Pseudomonas aeruginosa* is an excellent biosurfactant synthesis strain. However, a depletion of nitrogen supply results in the entry into the stationary phase, causing a decline in biosurfactant production [33]. Increased nitrogen supply hindered the biosurfactant-generating microbe, resulting in reduced biosurfactants' production [34]. 4.38 g/l yield was observed on introducing sodium nitrate [32]. For synthesis, ammonium nitrate is stated as the best nitrogen source by Joshi and Shekhawat [35]. Johnson *et al.* [36] found potassium nitrate to be a superior nitrogen for *Rhodotorula glutinis IIP*-30 for the synthesis of biosurfactants over ammonium sulphate and urea. Nitrogen sources like beef extract and yeast extract, analysed by Jorge F.B. Pereira, were found to have a successful impact on biosurfactant production [31].

Effect of Temperature

It is another significant component in the formation of biosurfactants. From 25 to 30 °C, rhamnolipid synthesis increased, stayed stable between 30 and 37 °C, and abruptly decreased at 42 °C. In a brief research, in *Pseudomonas aeruginosa*, Vollbrechtet *et al.* investigated the effect of temperature on the growth and synthesis of rhamnolipids. Higher temperatures, such as 47°C, were shown to be unfavourable for culture development, resulting in lower rhamnolipid production. Similarly results were seen on raising temperature, Tsukamurella sp. cell aggregation occurs leading to the reduction of glycolipid synthesis. Changjun Zoua's work revealed microbes, like *Acinetobacter baylyi* ZJ2, which could resist temperature within the range of 40-45°C. The ideally proposed temperature is 30°C that stimulates biomass development and increases glycolipid synthesis. Joice and Parthasarathi demonstrated *Pseudomonas aeruginosa* PBSCI at 30°C producing the most biosurfactants [31, 37, 38].

Effect of pH

pH is a key element influencing biosurfactant synthesis [39]. pH 6.0-6.5 was considered to be an optimal range for synthesis. At pH level above 6.5, the production has declined. At pH 4 - 4.5, the bacteria became incapable of lowering growth media surface tension, resulting in a drop in yield. A study suggested that increasing pH from 6.5 to 7.0 did not inhibit organism growth for synthesis. On the other hand, lowering pH had an adverse effect [40]. Biosurfactant development was slowed in an alkaline environment, as Changjun Zoua discovered when examining biosurfactant synthesis with *Acinetobacter baylyi* ZJ2 [29]. The pH has been discovered to influence microbial metabolism [40]. According to research by Joice and Parthasarathi [37], emulsification was 75.12% at neutral pH, and production varied with pH (5.0 to 8.5). At pH 6.5, surface tension was lowered by around 29.19 mN/m.

Effect of Aeration and Agitation

Agitation influences oxygen mass transfer and medium component mass transfer [41]. As a result, aeration and agitation must be regarded as a significant aspect of cell development and biosurfactant formation, particularly in aerobic species. This improved the air flow rate for biosurfactant generation utilising the response surface approach to 0.75 vvm. Agitation impact was investigated by an increase in the rate (50 to 200 ppm) resulting in enhanced growth rate, and an increase in the possibility of maximum production up to 80% [42]. This happens due to a rise in dissolved oxygen level from 0.1 to 0.55 mg/l. As a result, increased dissolved oxygen levels affected cell development and hence enhanced biosurfactant production [43].

Purification Methods for Biosurfactants

Traditional methods were used to extract crude biosurfactants from microbial biomass using strong hydrochloric acid. The creation of several methods, including membrane-based ones and others that use foam fractionation, extraction, and adsorption, has led to the isolation and purification of crude biosurfactants [49]. Sen and Swaminathan [49] pioneered membrane separation for surfactant recovery. For the effective synthesis of biosurfactants, bubbleless membrane bioreactors have recently been developed [50]. Microfiltration and ultrafiltration are used to increase separation efficiency in a bubbleless membrane bioreactor. Biosurfactant is precipitated using acidified hydrochloric acid in the biosurfactant separation process known as foam fractionation. A solvent can be used to extract the precipitates [51]. Foam fractionation, according to Davis *et al.* [52], is an integrated technique for surfactin separation.

Due to its ease of operation, extraction is becoming increasingly popular among academics. For the extraction of biosurfactants, a variety of solvents are employed, such as acetic acid, butanol, butane, pentane, hexane, diethyl ether, and methanol. Through the solvent extraction process, it was discovered that hydrophobic moieties might dissolve in some solvents, aiding in the extraction of the raw material [2]. For the adsorption and desorption operations to purify biosurfactants, amberlite XAD2 or polystyrene resins are used. The rate of agitation, the size of the activated carbon particles, the pH, the temperature, the initial adsorbent concentration, the volume of adsorbent, and the ionic strength of the adsorbent are some of the factors that influence the biosurfactant recovery in this process. In newly developed technologies, biosurfactant is absorbed using polymer resins, whilst desorption is accomplished using organic solvents. For surfactin [53] recovery, activated carbon is used as an adsorbent. Regenerated activated carbon can also be used to recover biosurfactants [54].

Analytical Methods

Several researchers have used and reported on a variety of analytical approaches to characterise biosurfactants. Table **3** lists the biosurfactant type, micro-organisms, solvents, and analytical technique type.

Table 3. Type of biosurfactants, bacteria, solvent and analytical methods involved.

Biosurfactants Producing Bacteria	Biosurfactants	Analytical Methods	Chemicals/Solvents Required	References
Pseudomonas aeruginosa	**Rhamnolipids**	HPLC	CH_3CN-H_2O	[55]
		TLC	$CHCl_3$/ CH_3OH/ CH_3COOH	[56]
		TLC	CH_3OH/H_2O	[57]
Pseudomonas fluorescens		TLC	CH_3CN/H_2O	[58]
P. aeruginosa MTCC 2297		HPLC	CH_3CN (Contain 2-bromoacetophenone and triethylamine)	[59]
Acinetobacter baylyi ZJ2	**Lipopeptide**	FTIR	$CHCl_3$/ CH_3OH/ CH_3COOH	[60]
Candida bombicola	**Sophorolipid**	HPLC with ELSD	CH_3CN/H_2O	[61]
Acinetobacter sp.	**Phospholipid**	GC-MS	$CHCl_3$/ CH_3OH (Extraction Method)	[62]
Rhodococcus sp. P32C1	**Trehalose lipid**	HPLC	CH_3CN	[63]
Bacillus Subtilis ATCC 21332	**Surfactin**	HPLC	CH_3CN/TFA	[64]

APPLICATION OF BIOSURFACTANTS

Biosurfactants in Metallurgical Industry

As a result of widespread industrialisation, many contaminants are being emitted into the environment. One form of contamination is heavy metals produced by the metallurgical sector. Heavy metal contaminates land and water and also get accumulated in the food chain as hazardous pollutants. The persistent presence of heavy metals in nature causes serious environmental issues. Excavation and dumping of contaminated soil to land areas have been described as methods for cleaning up heavy metal-affected soil [65]. In the process of bioreduction, microorganisms can act as a whole-cell biocatalyst to change the state of heavy metals [66]. For treating heavy metal-contaminated soil with biosurfactants, soil washing and flushing are well-known bioremediation procedures. *In-situ* remediation [67] involves injecting a biosurfactant into the soil using trenches and drain pipes. *Ex-situ*, on the other hand, involves transporting the dirt to a wash column from off-site, where it receives a biosurfactant solution cleaning. At high concentrations and crucial micelle concentrations, biosurfactants might significantly increase heavy metal solubility. Strong ionic interactions between negatively charged surfactants and positively charged metals lead to the creation of a surfactant-metal complex. The metal-biosurfactant mixture is desorbed from the soil by reducing surface tension. The term "bioleaching" refers to a method wherein naturally occurring bacteria or by-products of those microbes that employ biosurfactants dissolve metals from mineral sources.

A biosurfactant is a substance that turns solid metals into soluble metals. Heavy metal elimination may involve processes such as binding, complexation, desorption, and precipitation. Heavy metal precipitation in water has been employed as a key treatment strategy in industrial wastewater for many years. Heavy metal removal has been successfully achieved by combining biosurfactant precipitation with chemical treatment methods like ion-exchange. Metals from the soil that have been contaminated with several metals have been immobilised using *Pseudomonas aeruginosa* di-rhamnolipids [68]. They are also used for removing heavy metals like Cu, Pb, Cd, and Cr from the soil. Marine biosurfactants are a form of biosurfactant derived from sea bacteria that is commonly utilised in polyaromatic hydrocarbon remediation [69]. Nevertheless, there is no study report on heavy metal cleanup. Hazardous heavy metals can be chelated using biosurfactants made from marine creatures. As a result, it is employed in the treatment of wastewater containing heavy metals. The use of alkali improves heavy metal removal [70]. Another kind of biosurfactant-based bioremediation advancement is foam technology.

Wang and Mulligan looked at the efficiency of rhamnolipids in removing Cd and Ni from sand. Usually, the generated foam becomes uniform and then disperses into a porous media, enabling metals to interact with one another successfully. The elimination of Cd and Ni using only the bare rhamnolipid solution is 61.7% and 51%, respectively. Rhamnolipid and foam, however, were demonstrated to increase the effectiveness of Cd and Ni removal by 73.2% and 68.1%, respectively [71]. Massara and associates [72] analysed the removal of Cr (III) from chromium-contaminated kaolinite. Metal removal may benefit from a high pH and the usage of NaOH. The pH enhanced the biosurfactants' chelating activity, resulting in greater metal removal. The use of NaOH improves the solubility of biosurfactants, which enhances metal removal [47, 60]. As stated by many writers, heavy metal removal is shown in Table **4** [64].

Table 4. Removal of heavy metals by biosurfactant producing organism.

S. No.	Metals	Microorganism	Removal (%)	References
1.	Cr	*Pseudomonas aeruginosa*	46	[73]
		Aspergillus niger	21-36	[74]
2.	Cd	*Bacillus strain* H9	36	[75]
		Aspergillusterreus	70	[76]
		Pseudomonas aeruginosa	73.2	[71]
3.	Cu	*Thiobacillus ferrooxidans*	25	[77]
		Schizosaccharomyces pombe	11-25	[78]
4.	Pb	*Pseudomonas aeruginosa* PU21	80	[79]
		Aspergillus niger	13-88	[74]
5.	Ni	*Pseudomonas spp.*	98	[80]
		Candida spp	29-57	[81]
		Pseudomonas aeruginosa	68.1	[71]

Biosurfactants in the Petroleum Industry

In oil-producing wells, oil recovery is aided by the employment of organisms that produce biosurfactants (either indigenous or injected). The microbial enhanced oil recovery technique makes use of biosurfactants, injections of a particular microbe, and direct nutrition injections with microorganisms that can produce desirable products for oil mobilisation. Interfacial tension/oil viscosity reduction and reservoir repressurizations come after this procedure. Oil recovery was enhanced by 30-200% with the addition of biosurfactants, fertilisers, and microorganisms such as *Pseudomona aeruginosa* and *Bacillus licheniformis* [82]. The most effective approach for recovering oil from sources with poor permeability or high viscosity crude oil is this oil recovery procedure.

In the petroleum sector, oil field emulsions are a severe problem. It occurs at numerous stages of crude oil processing. One of the most effective methods for extracting oil from oil field emulsions is de-emulsification. A conventional de-emulsification technique makes use of centrifugation, heat treatment, and chemicals. Biosurfactants can take the place of chemical de-emulsifiers while also offering an ecologically friendly substitute. *Acinetobacter* and *Pseudomonas* are a couple of the most common bacterial species that de-emulsify in mixed cultures [83].

By taking advantage of the hydrophobic cell surface or the amphiphilic characteristics of biosurfactants, microorganisms destabilise emulsions. Glycolipids, glycoproteins, phospholipids, and polysaccharides are examples of microbiological tools that are used to remove emulsifiers from the oil-water interface [1]. Oil may be extracted from petroleum tank bottom sludges using biosurfactants, which also allow for the pipeline transit of heavy crude. Rhamnolipids [43] can be used to recover absorbed oil from the used oil sorbents. Parameters such as sorbent pore size and washing duration affect oil removal. 95% of the oil was removed with commercial rhamnolipids. Fermentation broth can remove crude oil and motor oil from polluted regions by 85% and 90%, respectively, in addition to employing crude biosurfactant. Table **5** displays the rate of oil recovery as stated by various publications.

Table 5. Recovery of oil by biosurfactant producing organism.

S. No.	Biosurfactants Producing Organism	Biosurfactants	Biosurfactant Yield	Recovery of Oil from Contaminated Soil (%)	References
1.	*Bacillus subtilis* CN2	Lipopeptide	7150mg/l	84.6 ± 7.1	[84]
2.	*Bacillus subtilis* BS-37	Surfactin isoform	585mg/l	96	[85]
3.	*Bacillus* strain	-	Crude BS 0.081- 1 g/l CMC Value19.439mg/l	30.22 – 34.19	[86]
4.	*Bacillus subtilis* B 30	Surfactin	Crude BS 0.3 – 0.5 g/l CMC Value 1:8	17-26	[87]
5.	*Candida sphaerica*	Anionic biosurfactants	4.5g/l	75 (Clay soil) 92 (Silty Soil)	[88]
6.	*Candida tropicalis*	-	3.61 ± 2.1	78 - 97	[89]
7.	*Candida glabrata* UCP 1002	-	7.52g/l	92.6	[90]
8.	*Candida sphaerica* UCP 0995	Biosurfactant Lunasan	9g/l	95	[91]

Application of Biosurfactants in Agriculture

Surfactants can be used to increase the solubility of biohazardous compounds like PAH by acting as mobilizers. Consequently, hydrophobic organic pollutants' (HOC) apparent solubility rises. By shortening the length of the diffusion channel between the absorption site and the region of microbial biouptake, surfactants may also facilitate microbe adsorption to contaminated soil particles. In agriculture, surfactants are used to hydrophilize thick soils and ensure excellent water solubility and fertiliser dispersion. They also keep some fertilisers from caking during storage and improve insecticide dispersion and penetration [4]. The species *Pseudomonas* is known to produce the rhamnolipid biosurfactant, which is believed to have potent antibacterial properties. Rhamnolipid biosurfactant exposure over time should not harm either individuals or the environment. Due to their antifungal properties, fengycins may also be used in the biocontrol of plant diseases [92].

Biosurfactants' Applications in Commercial Laundry Detergents

Commercial washing detergents of today include a lot of surfactants, which are almost always created chemically and are bad for freshwater life. The search for natural alternatives to synthetic surfactants in laundry detergents has been fuelled by the public's growing knowledge of the risks and environmental hazards associated with them. High temperatures did not result in the biosurfactants' wide pH range (7.0-12.0) causing Cyclic Lipopeptide (CLP), which possesses surface-active characteristics, to degrade [93]. They showed good compatibility and stability with widely used laundry detergents as well as significant emulsion-forming abilities with vegetable oils, which encourages their usage in the production of laundry detergents [94].

Biosurfactants as Biopesticides

The standard arthropod control method employs broadspectrum insecticides and bicides, which usually have unfavourable side effects. Moreover, the growth of pesticide-resistant insect populations, as well as the growing cost of new chemical pesticides, has fueled the hunt for more ecologically friendly vector control technologies. Several bacteria produce lipopeptide biosurfactants that kill the fruitfly *Drosophila melanogaster*. As a result, randhence appears to be promising as a biopesticide [95].

Application of Biosurfactants in Medicine

Mukherjee *et al.* [1] further elaborated on the several uses of biosurfactants in medicine, which include:

Antibacterial Activity

Biosurfactants' diverse structures allow them to function in a number of ways, such as having detergent-like impact on cell membrane permeability because of their structure. Certain biosurfactants show notable antibacterial, antifungal, and antiviral activities, according to Gharaei-Fathabad [96]. These surfactants work to keep microorganisms from sticking to them, making them useful as probiotics and therapeutic agents for treating a variety of illnesses. There is evidence that the biosurfactant created by marine *B.circulans* has antibacterial efficacy against pathogenic microbial strains, including MDR strains.

Anti-cancer Activity

In the human promyelocytic leukaemia cell line, a number of microbial extracellular glycolipids promote cell differentiation as opposed to cell proliferation. Furthermore, MEL increased facetylcholine esterase activity, which disturbs the G1 phase of the cell cycle, resulting in neurite proliferation and incomplete cellular differentiation. This suggests that MEL promotes the formation of neurons in PC12 cells, opening the door for its use.

Biosurfactants have been shown to prevent dangerous organisms' adherence to infected locations or solid surfaces, according to Rodrigues and team [97]. They have shown the value of pre-coating prior to media inoculation, where vinyl urethral catheters were soaked in surfactin solution, which minimised the formation of biofilm. by *Proteus mirabilis, Salmonella typhimurium, Salmonella enterica,* and *E. coli.* According to Muthusamy [98], pre-treating silicone rubber with a surfactant from the *S. thermophilus* bacteria reduced *C. albicans* adhesion by 85%. *Lactobacillus fermentum* and *Lactobacillus acidophilus* derived surfactants reduce the quantity of adhering uropathogenic cells of *E. faecalis* by 77%.

Immunological Adjuvants

Bacterial lipopeptides are potent, non-toxic, and non-pyrogenic immunological adjuvants when combined with conventional antigens. The humoral human response enhanced when the low molecular mass antigens Iturin A and Landherbicolin A were employed [96].

Antiviral Activity

There have been reports of antimicrobial effects and a decrease in the development of the human immunodeficiency virus in leucocytes [2, 98]. Moreover, Muthusamy *et al.* [98] emphasised that the development of a female-controlled, effective, and secure vaginal, topical microbicide was necessary given the increased incidence of HIV in women. The most effective spermicidal and virucidal drugs have been found to be sophorolipid from *C. bombicola,* and its structural analogues, such as sophorolipid diacetate ethylester, have been an effective compound with virucidal properties action against human semen similar to nonoxynol - 9.

Gene Delivery

According to Gharaei-Fathabad [96], for fundamental science and therapeutic purposes, it is critical to provide a reliable and safe method for introducing foreign nucleotides into mammalian cells.

CONCLUSION

In order to harness environmentally beneficial processes and accelerate production rates, biosurfactants are required for use in environmental applications, which are explored in this research. The study went into great detail on the development of screening several strains for the large-scale production of biosurfactants. There is a brief discussion of various operating factors, analytical methods, product purification, and the influence on the metal and oil-related sectors.

ACKNOWLEDGMENTS

The authors wish to express sincere thanks to their teacher and parents, for her invaluable patience and feedback.

REFERENCES

[1]   Mukherjee, S.; Das, P.; Sen, R. Towards commercial production of microbial surfactants. *Trends Biotechnol.,* **2006**, *24*(11), 509-515.
        [http://dx.doi.org/10.1016/j.tibtech.2006.09.005] [PMID: 16997405]

[2]   Desai, J.D.; Banat, I.M. Microbial production of surfactants and their commercial potential. *Microbiol. Mol. Biol. Rev.,* **1997**, *61*(1), 47-64.
        [PMID: 9106364]

[3]   Jarvis, F.G.; Johnson, M.J. A glyco-lipid produced by *Pseudomonas aeruginosa. J. Am. Chem. Soc.,* **1949**, *71*(12), 4124-4126.
        [http://dx.doi.org/10.1021/ja01180a073]

[4]   Mulligan, C N; Gibbs, B F Types, production and applications of biosurfactants. *Proc Indian natn Sci Acad,* **2004**, *70*, 31-55.

[5]   Rosenberg, E.; Ron, E.Z. High- and low-molecular mass microbial surfactants. *Appl. Microbiol. Biotechnol.,* **1999**, *52*(2), 154-162.
        [http://dx.doi.org/10.1007/s002530051502] [PMID: 10499255]

[6]   Hommel, R.; Ratledge, C. Evidence for two fatty alcohol oxidases in the biosurfactant-producing yeast Candida (Torulopsis) bombicola. *FEMS Microbiol. Lett.,* **1990**, *70*(2), 183-186.

[http://dx.doi.org/10.1111/j.1574-6968.1990.tb13975.x] [PMID: 2227354]

[7]   Britton, L.N. *Microbial degradation of aliphatic hydrocarbons, in microbial degradation of organic compound*; Gibson, D.T., Ed.; Marcel Dekker: New York, **1984**, pp. 89-131.

[8]   Leahy, J.G.; Colwell, R.R. Microbial degradation of hydrocarbons in the environment. *Microbiol. Rev.,* **1990**, *54*(3), 305-315.
       [http://dx.doi.org/10.1128/mr.54.3.305-315.1990] [PMID: 2215423]

[9]   Singer, M.E.; Finnerty, W.R. Microbial metabolism of straight-chain and branched alkanes. In: *Petroleum microbiology*; Atlas, R.M., Ed.; Macmillan Publishing Company: New York, **1984**; pp. 1-59.

[10]   Ron, E.Z.; Rosenberg, E. Biosurfactants and oil bioremediation. *Curr. Opin. Biotechnol.,* **2002**, *13*(3), 249-252.
       [http://dx.doi.org/10.1016/S0958-1669(02)00316-6] [PMID: 12180101]

[11]   Morikawa, M.; Ito, M.; Imanaka, T. Isolation of a new surfactin producer *Bacillus pumilus* A-1, and cloning and nucleotide sequence of the regulator gene, psf-1. *J. Ferment. Bioeng.,* **1992**, *74*(5), 255-261.
       [http://dx.doi.org/10.1016/0922-338X(92)90055-Y]

[12]   Siegmund, I.; Wagner, F. New method for detecting rhamnolipids excreted by Pseudomonas species during growth on mineral agar. *Biotechnol. Tech.,* **1991**, *5*(4), 265-268.
       [http://dx.doi.org/10.1007/BF02438660]

[13]   Banat, I.M. The isolation of a thermophilic biosurfactant producing *Bacillus SP. Biotechnol. Lett.,* **1993**, *15*(6), 591-594.
       [http://dx.doi.org/10.1007/BF00138546]

[14]   Bodour, A.A.; Miller-Maier, R.M. Application of a modified drop-collapse technique for surfactant quantitation and screening of biosurfactant-producing microorganisms. *J. Microbiol. Methods,* **1998**, *32*(3), 273-280.
       [http://dx.doi.org/10.1016/S0167-7012(98)00031-1]

[15]   Satpute, S.K.; Bhawsar, B.D.; Dhakephalkar, P.K.; Chopade, B.A. Assessment of different screening methods for selecting biosurfactant producing marine bacteria. *Indian J. Geo-Mar. Sci.,* **2008**, *37*, 243-250.

[16]   Rosenberg, E.; Zuckerberg, A.; Rubinovitz, C.; Gutnick, D.L. Emulsifier of *Arthrobacter* RAG-1: isolation and emulsifying properties. *Appl. Environ. Microbiol.,* **1979**, *37*(3), 402-408.
       [http://dx.doi.org/10.1128/aem.37.3.402-408.1979] [PMID: 36840]

[17]   Neu, T.; Poralla, K. Emulsifying agents from bacteria isolated during screening for cells with hydrophobic surfaces. *Appl. Microbiol. Biotechnol.,* **1990**, *32*(5), 521-525.
       [http://dx.doi.org/10.1007/BF00173721]

[18]   Cooper, D.G.; Goldenberg, B.G. Surface-active agents from two *bacillus* species. *Appl. Environ. Microbiol.,* **1987**, *53*(2), 224-229.
       [http://dx.doi.org/10.1128/aem.53.2.224-229.1987] [PMID: 16347271]

[19]   Ellaiah, P.; Prabhakar, T.; Sreekanth, M.; Taleb, A.T.; Raju, P.B.; Saisha, V. Production of glycolipids containing biosurfactant by *Pseudomonas species. Indian J. Exp. Biol.,* **2002**, *40*(9), 1083-1086.
       [PMID: 12587744]

[20]   Benincasa, M. Rhamnolipid produced from agroindustrial wastes enhances hydrocarbon biodegradation in contaminated soil. *Curr. Microbiol.,* **2007**, *54*(6), 445-449.
       [http://dx.doi.org/10.1007/s00284-006-0610-8] [PMID: 17457644]

[21]   Lee, K.M.; Hwang, S.H.; Ha, S.D.; Jang, J.H.; Lim, D.J.; Kong, J-Y. Rhamnolipid production in batch and fed-batch fermentation using *Pseudomonas aeruginosa* BYK-2 KCTC 18012P. *Biotechnol. Bioprocess Eng.; BBE,* **2004**, *9*(4), 267-273.
       [http://dx.doi.org/10.1007/BF02942342]

[22]   Camilios-Neto, D.; Bugay, C.; de Santana-Filho, A.P.; Joslin, T.; de Souza, L.M.; Sassaki, G.L.; Mitchell, D.A.; Krieger, N. Production of rhamnolipids in solid-state cultivation using a mixture of sugarcane bagasse and corn bran supplemented with glycerol and soybean oil. *Appl. Microbiol. Biotechnol.,* **2011**, *89*(5), 1395-1403.
[http://dx.doi.org/10.1007/s00253-010-2987-3] [PMID: 21080163]

[23]   Cooper, D.G.; Macdonald, C.R.; Duff, S.J.B.; Kosaric, N. Enhanced production of surfactin from *Bacillus subtilis* by continuous product removal and metal cation additions. *Appl. Environ. Microbiol.,* **1981**, *42*(3), 408-412.
[http://dx.doi.org/10.1128/aem.42.3.408-412.1981] [PMID: 16345840]

[24]   Noah, K.S.; Bruhn, D.F.; Bala, G.A. Surfactin production from potato process effluent by *Bacillus subtilis* in a chemostat. In: *Proc*; Twenty-Sixth SympBiotechnol Fuels Chemicals: Chattanooga, TN, **2005**; pp. 465-473.
[http://dx.doi.org/10.1007/978-1-59259-991-2_41]

[25]   Yeh, M.S.; Wei, Y.H.; Chang, J.S. Bioreactor design for enhanced carrier-assisted surfactin production with *Bacillus subtilis*. *Process Biochem.,* **2006**, *41*(8), 1799-1805.
[http://dx.doi.org/10.1016/j.procbio.2006.03.027]

[26]   Coutte, F.; Lecouturier, D.; Ait Yahia, S.; Leclère, V.; Béchet, M.; Jacques, P.; Dhulster, P. Production of surfactin and fengycin by *Bacillus subtilis* in a bubbleless membrane bioreactor. *Appl. Microbiol. Biotechnol.,* **2010**, *87*(2), 499-507.
[http://dx.doi.org/10.1007/s00253-010-2504-8] [PMID: 20221757]

[27]   Robert, M.; Mercade, M.E.; Bosch, M.P.; Parra, J.L.; Espuny, M.J. Effect of the carbon source on the biosurfactant production by *P. aeruginosa* 44T. *Biotechnol. Lett.,* **1989**, *11*, 871-874.
[http://dx.doi.org/10.1007/BF01026843]

[28]   Monteiro, S.A.; Sassaki, G.L.; de Souza, L.M.; Meira, J.A.; de Araújo, J.M.; Mitchell, D.A.; Ramos, L.P.; Krieger, N. Molecular and structural characterization of the biosurfactant produced by *Pseudomonas aeruginosa* DAUPE 614. *Chem. Phys. Lipids,* **2007**, *147*(1), 1-13.
[http://dx.doi.org/10.1016/j.chemphyslip.2007.02.001] [PMID: 17382918]

[29]   Zou, C.; Wang, M.; Xing, Y.; Lan, G.; Ge, T.; Yan, X.; Gu, T. Characterization and optimization of biosurfactants produced by *Acinetobacter baylyi* ZJ2 isolated from crude oil-contaminated soil sample toward microbial enhanced oil recovery applications. *Biochem. Eng. J.,* **2014**, *90*, 49-58.
[http://dx.doi.org/10.1016/j.bej.2014.05.007]

[30]   Huy, N.Q.; Jin, S.; Amada, K.; Haruki, M.; Huu, N.B.; Hang, D.T.; Ha, D.T.C.; Imanaka, T.; Morikawa, M.; Kanaya, S. Characterization of petroleum-degrading bacteria from oil-contaminated sites in Vietnam. *J. Biosci. Bioeng.,* **1999**, *88*(1), 100-102.
[http://dx.doi.org/10.1016/S1389-1723(99)80184-4] [PMID: 16232582]

[31]   Jorge, F.B Optimization and characterization of biosurfactant production by *Bacillus subtilis* isolates towards microbial enhanced oil recovery applications. *Fuel,* **2013**, *113*, 259-268.

[32]   Onwosi, C.O.; Odibo, F.J.C. Effects of carbon and nitrogen sources on rhamnolipid biosurfactant production by *Pseudomonas nitroreducens* isolated from soil. *World J. Microbiol. Biotechnol.,* **2012**, *28*(3), 937-942.
[http://dx.doi.org/10.1007/s11274-011-0891-3] [PMID: 22805814]

[33]   Ramana, K.V.; Karanth, N.G. Factors affecting biosurfactant production using *Pseudomonas aeruginosa* CFTR-6 under submerged conditions. *J. Chem. Technol. Biotechnol.,* **1989**, *45*(4), 249-257.
[http://dx.doi.org/10.1002/jctb.280450402]

[34]   Syldatk, C.; Lang, S.; Wagner, F.; Wray, V.; Witte, L. Chemical and physical characterization of four interfacial-active rhamnolipids from *Pseudomonas spec.* DSM 2874 grown on n-alkanes. *Z. Naturforsch. C J. Biosci.,* **1985**, *40*(1-2), 51-60.
[http://dx.doi.org/10.1515/znc-1985-1-212] [PMID: 3993180]

[35]   Joshi, P.A.; Shekhawat, D.B. Effect of carbon and nitrogen source on biosurfactant production by biosurfactant producing bacteria isolated from petroleum contaminated site. *Adv. Appl. Sci. Res.,* **2014**, *5*, 159-164.

[36]   Johnson, V.; Singh, M.; Saini, V.S.; Adhikari, D.K.; Sista, V.; Yadav, N.K. Bioemulsifier production by an oleaginous yeast *Rhodotorula glutinis* IIP-30. *Biotechnol. Lett.,* **1992**, *14*(6), 487-490.
[http://dx.doi.org/10.1007/BF01023172]

[37]   Joice, P.A.; Parthasarathi, R. Optimisation of biosurfactant production from *Pseudomonas aeruginosa* PBSC1. *Int. J. Curr. Microbiol. Appl. Sci.,* **2014**, *3*, 140-151.

[38]   Vollbrecht, E.; Heckmann, R.; Wray, V.; Nimtz, M.; Lang, S. Production and structure elucidation of di- and oligosaccharide lipids (biosurfactants) from *Tsukamurella sp.* nov. *Appl. Microbiol. Biotechnol.,* **1998**, *50*(5), 530-537.
[http://dx.doi.org/10.1007/s002530051330] [PMID: 9866171]

[39]   Gobbert, U.; Lang, S.; Wagner, F. Sophorose lipid formation by resting cells of *Torulopsisbombicola*. *Biotechnol. Lett.,* **1984**, *6*, 225-230.
[http://dx.doi.org/10.1007/BF00140041]

[40]   Guerra-Santos, L.H.; Kappeli, O.; Fletcher, A. Dependence of *Pseudomonas aeruginosa* continuous culture biosurfactant production on nutritional and environmental factors. *Appl. Microbiol. Biotechnol.,* **1986**, *24*, 225-230.
[http://dx.doi.org/10.1007/BF00250320]

[41]   Shaligram, N.S.; Singhal, R.S. Surfactin-A review on biosynthesis, fermentation, purification and applications. *Food Technol. Biotechnol.,* **2010**, *48*, 119-134.

[42]   Sen, R. Response surface optimization of the critical media components for the production of surfactin. *J. Chem. Technol. Biotechnol.,* **1997**, *68*(3), 263-270.
[http://dx.doi.org/10.1002/(SICI)1097-4660(199703)68:3<263::AID-JCTB631>3.0.CO;2-8]

[43]   Wei, Y.H.; Chou, C-L.; Chang, J.S. Rhamnolipid production by indigenous *Pseudomonas aeruginosa* J4 originating from petrochemical wastewater. *Biochem. Eng. J.,* **2005**, *27*(2), 146-154.
[http://dx.doi.org/10.1016/j.bej.2005.08.028]

[44]   Mouafi, F.E.; Abo Elsoud, M.M.; Moharam, M.E. Optimization of biosurfactant production by *Bacillus brevis* using response surface methodology. *Biotechnol. Rep. (Amst.),* **2016**, *9*, 31-37.
[http://dx.doi.org/10.1016/j.btre.2015.12.003] [PMID: 28352589]

[45]   Velioglu, Z.; Ozturk Urek, R. Optimization of cultural conditions for biosurfactant production by *Pleurotus djamor* in solid state fermentation. *J. Biosci. Bioeng.,* **2015**, *120*(5), 526-531.
[http://dx.doi.org/10.1016/j.jbiosc.2015.03.007] [PMID: 25865657]

[46]   Deepika, K.V.; Kalam, S.; Ramu Sridhar, P.; Podile, A.R.; Bramhachari, P.V. Optimization of rhamnolipid biosurfactant production by mangrove sediment bacterium *Pseudomonas aeruginosa* KVD-HR42 using response surface methodology. *Biocatal. Agric. Biotechnol.,* **2016**, *5*, 38-47.
[http://dx.doi.org/10.1016/j.bcab.2015.11.006]

[47]   de França, Í.W.L.; Lima, A.P.; Lemos, J.A.M.; Lemos, C.G.F.; Melo, V.M.M.; de Sant'ana, H.B.; Gonçalves, L.R.B. Production of a biosurfactant by *Bacillus subtilis* ICA56 aiming bioremediation of impacted soils. *Catal. Today,* **2015**, *255*, 10-15.
[http://dx.doi.org/10.1016/j.cattod.2015.01.046]

[48]   Patil, S.; Pendse, A.; Aruna, K. Studies on optimization of biosurfactant production by *Pseudomonas aeruginosa* F23 isolated from oil contaminated soil sample. *Int J CurrBiotechnol,* **2014**, *2*, 20-30.

[49]   Sen, R.; Swaminathan, T. Characterization of concentration and purification parameters and operating conditions for the small-scale recovery of surfactin. *Process Biochem.,* **2005**, *40*(9), 2953-2958.
[http://dx.doi.org/10.1016/j.procbio.2005.01.014]

[50]   Coutte, F.; Lecouturier, D.; Leclère, V.; Béchet, M.; Jacques, P.; Dhulster, P. New integrated

bioprocess for the continuous production, extraction and purification of lipopeptides produced by *Bacillus subtilis* in membrane bioreactor. *Process Biochem.,* **2013**, *48*(1), 25-32.
[http://dx.doi.org/10.1016/j.procbio.2012.10.005]

[51]   Huei LC, Ying SC, Ruey SJ. Recovery of surfactin from fermentation broths by a hybrid salting-out and membrane filtration process. *Separation and Purification Technology*, **2008**; *59*(3): 244–252
[http://dx.doi.org/10.1016/j.seppur.2007.06.010]

[52]   Davis, D.A.; Lynch, H.C.; Varley, J. The application of foaming for the recovery of Surfactin from *B. subtilis* ATCC 21332 cultures. *Enzyme Microb. Technol.,* **2001**, *28*(4-5), 346-354.
[http://dx.doi.org/10.1016/S0141-0229(00)00327-6] [PMID: 11240190]

[53]   Liu, T.; Montastruc, L.; Gancel, F.; Zhao, L.; Nikov, I. Integrated process for production of surfactin. *Biochem. Eng. J.,* **2007**, *35*(3), 333-340.
[http://dx.doi.org/10.1016/j.bej.2007.01.025]

[54]   Dubey, K.V.; Juwarkar, A.A.; Singh, S.K. Adsorption-desorption process using wood-based activated carbon for recovery of biosurfactant from fermented distillery wastewater. *Biotechnol. Prog.,* **2005**, *21*(3), 860-867.
[http://dx.doi.org/10.1021/bp040012e] [PMID: 15932266]

[55]   Schenk, T.; Schuphan, I.; Schmidt, B. High-performance liquid chromatographic determination of the rhamnolipids produced by *Pseudomonas aeruginosa. J. Chromatogr. A,* **1995**, *693*(1), 7-13.
[http://dx.doi.org/10.1016/0021-9673(94)01127-Z] [PMID: 7697163]

[56]   Arino, S.; Marchal, R.; Vandecasteele, J.P. Identification and production of a rhamnolipidic biosurfactant by a *Pseudomonas* species. *Appl. Microbiol. Biotechnol.,* **1996**, *45*(1-2), 162-168.
[http://dx.doi.org/10.1007/s002530050665]

[57]   Rahman, K.S.M.; Vasudevan, N.; Lakshmanaperumalsamy, P. Enhancement of biosurfactant production to emulsify different hydrocarbon. *J Environ Poll,* **1999**, *6*, 87-93.

[58]   Caldini, G.; Cenci, G.; Manenti, R.; Morozzi, G. The ability of an environmental isolate of *Pseudomonas fluorescens* to utilize chrysene and other four-ring polynuclear aromatic hydrocarbons. *Appl. Microbiol. Biotechnol.,* **1995**, *44*(1-2), 225-229.
[http://dx.doi.org/10.1007/BF00164506]

[59]   Venkatesh, N.M.; Vedaraman, N. Remediation of soil contaminated with copper using Rhamnolipids produced from *Pseudomonas aeruginosa* MTCC 2297 using waste frying rice bran oil. *Ann. Microbiol.,* **2012**, *62*(1), 85-91.
[http://dx.doi.org/10.1007/s13213-011-0230-9]

[60]   Wong D, Nielsen TB, Bonomo RA, Pantapalangkoor P, Luna B, Spellberg B. Clinical and Pathophysiological Overview of *Acinetobacter* Infections: a Century of Challenges. *Clin Microbiol Rev,* **2017**, *30*(1): 409-447.
[http://dx.doi.org/10.1128/CMR.00058-16] [PMID: 27974412] [PMCID: PMC5217799]

[61]   Davila, A.M.; Marchal, R.; Vandecasteele, J.P. Sophorose lipid fermentation with differentiated substrate supply for growth and production phases. *Appl. Microbiol. Biotechnol.,* **1997**, *47*(5), 496-501.
[http://dx.doi.org/10.1007/s002530050962]

[62]   Koma, D.; Hasumi, F.; Yamamoto, E.; Ohta, T.; Chung, S.Y.; Kubo, M. Biodegradation of long-chain n-paraffins from waste oil of car engine by *Acinetobacter sp. J. Biosci. Bioeng.,* **2001**, *91*(1), 94-96.
[http://dx.doi.org/10.1016/S1389-1723(01)80120-1] [PMID: 16232955]

[63]   Maghsoudi, S.; Vossoughi, M.; Kheirolomoom, A.; Tanaka, E.; Katoh, S. Biodesulfurization of hydrocarbons and diesel fuels by *Rhodococcus* sp. strain P32C1. *Biochem. Eng. J.,* **2001**, *8*(2), 151-156.
[http://dx.doi.org/10.1016/S1369-703X(01)00097-3]

[64]   Biniarz P, Łukaszewicz M. Direct quantification of lipopeptide biosurfactants in biological samples

via HPLC and UPLC-MS requires sample modification with an organic solvent. *Appl Microbiol Biotechnol*, **2017**, *101*(11): 4747-4759.
[http://dx.doi.org/10.1007/s00253-017-8272-y] [PMID: 28432441]

[65] Aşçı, Y.; Nurbaş, M.; Sağ Açıkel, Y. Investigation of sorption/desorption equilibria of heavy metal ions on/from quartz using rhamnolipid biosurfactant. *J. Environ. Manage.*, **2010**, *91*(3), 724-731.
[http://dx.doi.org/10.1016/j.jenvman.2009.09.036] [PMID: 19850403]

[66] Bruins, M.R.; Kapil, S.; Oehme, F.W. Microbial resistance to metals in the environment. *Ecotoxicol. Environ. Saf.*, **2000**, *45*(3), 198-207.
[http://dx.doi.org/10.1006/eesa.1999.1860] [PMID: 10702338]

[67] Singh, P.; Cameotra, S.S. Enhancement of metal bioremediation by use of microbial surfactants. *Biochem. Biophys. Res. Commun.*, **2004**, *319*(2), 291-297.
[http://dx.doi.org/10.1016/j.bbrc.2004.04.155] [PMID: 15178405]

[68] Juwarkar, A.A.; Dubey, K.V.; Nair, A.; Singh, S.K. Bioremediation of multi-metal contaminated soil using biosurfactant — a novel approach. *Indian J. Microbiol.*, **2008**, *48*(1), 142-146.
[http://dx.doi.org/10.1007/s12088-008-0014-5] [PMID: 23100708]

[69] Das, P.; Mukherjee, S.; Sen, R. Biosurfactant of marine origin exhibiting heavy metal remediation properties. *Bioresour. Technol.*, **2009**, *100*(20), 4887-4890.
[http://dx.doi.org/10.1016/j.biortech.2009.05.028] [PMID: 19505818]

[70] Latorrata S, Balzarotti R, Adami MI, Marino B, Mostoni S, Scotti R, Bellotto M, Cristiani C. Wastewater Treatment Using Alkali-Activated-Based Sorbents Produced from Blast Furnace Slag. *Applied Sciences*, **2021**; *11*(7): 2985.
[http://dx.doi.org/10.3390/app11072985]

[71] Wang, S.; Mulligan, C.N. Rhamnolipid foam enhanced remediation of cadmium and nickel contaminated soil. *Water Air Soil Pollut.*, **2004**, *157*(1-4), 315-330.
[http://dx.doi.org/10.1023/B:WATE.0000038904.91977.f0]

[72] Massara, H.; Mulligan, C.N.; Hadjinicolaou, J. Effect of rhamnolipids on chromium contaminated soil. *Soil Sediment Contam.*, **2007**, *16*(1), 1-14.
[http://dx.doi.org/10.1080/15320380601071241]

[73] Hassen, A.; Saidi, N.; Cherif, M.; Boudabous, A. Effects of heavy metals on *Pseudomonas aeruginosa* and *Bacillus thuringiensis*. *Bioresour. Technol.*, **1998**, *65*(1-2), 73-82.
[http://dx.doi.org/10.1016/S0960-8524(98)00011-X]

[74] Dursun, A.Y.; Uslu, G.; Cuci, Y.; Aksu, Z. Bioaccumulation of copper(II), lead(II) and chromium(VI) by growing *Aspergillus niger*. *Process Biochem.*, **2003**, *38*(12), 1647-1651.
[http://dx.doi.org/10.1016/S0032-9592(02)00075-4]

[75] Roane, T.M.; Josephson, K.L.; Pepper, I.L. Dual-bioaugmentation strategy to enhance remediation of cocontaminated soil. *Appl. Environ. Microbiol.*, **2001**, *67*(7), 3208-3215.
[http://dx.doi.org/10.1128/AEM.67.7.3208-3215.2001] [PMID: 11425743]

[76] Massaccesi, G.; Romero, M.C.; Cazau, M.C.; Bucsinszky, A.M. Cadmium removal capacities of filamentous soil fungi isolated from industrially polluted sediments, in La Plata (Argentina). *World J. Microbiol. Biotechnol.*, **2002**, *18*(9), 817-820.
[http://dx.doi.org/10.1023/A:1021282718440]

[77] Boyer, A.; Magnin, J.P.; Ozil, P. Copper ion removal by *Thiobacillus ferrooxidans*biomass. *Biotechnol. Lett.*, **1998**, *20*(2), 187-190.
[http://dx.doi.org/10.1023/A:1005345011862]

[78] Dönmez, G.; Aksu, Z. The effect of copper(II) ions on the growth and bioaccumulation properties of some yeasts. *Process Biochem.*, **1999**, *35*(1-2), 135-142.
[http://dx.doi.org/10.1016/S0032-9592(99)00044-8]

[79] Chang, J.S.; Law, R.; Chang, C.C. Biosorption of lead, copper and cadmium by biomass of

Pseudomonas aeruginosa PU21. *Water Res.,* **1997**, *31*(7), 1651-1658.
[http://dx.doi.org/10.1016/S0043-1354(97)00008-0]

[80] Magyarosy A, Laidlaw RD, Kilaas R, Echer C, Clark DS, Keasling JD. Nickel accumulation and nickel oxalate precipitation by *Aspergillus niger. Appl. Microbiol. Biotechnol.,* **2002**, *59*(2-3), 382-388.
[http://dx.doi.org/10.1007/s00253-002-1020-x] [PMID: 12111174]

[81] Dönmez, G.; Aksu, Z. Bioaccumulation of copper(ii) and nickel(ii) by the non-adapted and adapted growing CANDIDA SP. *Water Res.,* **2001**, *35*(6), 1425-1434.
[http://dx.doi.org/10.1016/S0043-1354(00)00394-8] [PMID: 11317889]

[82] Singh, S.; Kang, S.H.; Mulchandani, A.; Chen, W. Bioremediation: Environmental clean-up through pathway engineering. *Curr. Opin. Biotechnol.,* **2008**, *19*(5), 437-444.
[http://dx.doi.org/10.1016/j.copbio.2008.07.012] [PMID: 18760355]

[83] Nadarajah, N.; Singh, A.; Ward, O.P. De-emulsification of petroleum oil emulsion by a mixed bacterial culture. *Process Biochem.,* **2002**, *37*(10), 1135-1141.
[http://dx.doi.org/10.1016/S0032-9592(01)00325-9]

[84] Bezza, F.A.; Chirwa, E.M.N. Production and applications of lipopeptide biosurfactant for bioremediation and oil recovery by *Bacillus subtilis* CN2. *Biochem. Eng. J.,* **2015**, *101*, 168-178.
[http://dx.doi.org/10.1016/j.bej.2015.05.007]

[85] Liu, Q.; Lin, J.; Wang, W.; Huang, H.; Li, S. Production of surfactin isoforms by *Bacillus subtilis* BS-37 and its applicability to enhanced oil recovery under laboratory conditions. *Biochem. Eng. J.,* **2015**, *93*, 31-37.
[http://dx.doi.org/10.1016/j.bej.2014.08.023]

[86] Joshi, S.J.; Desai, A.J. Bench-scale production of biosurfactants and their potential in *ex-situ* MEOR application. *Soil Sediment Contam.,* **2013**, *22*(6), 701-715.
[http://dx.doi.org/10.1080/15320383.2013.756450]

[87] Al-Wahaibi, Y.; Joshi, S.; Al-Bahry, S.; Elshafie, A.; Al-Bemani, A.; Shibulal, B. Biosurfactant production by *Bacillus subtilis* B30 and its application in enhancing oil recovery. *Colloids Surf. B Biointerfaces,* **2014**, *114*, 324-333.
[http://dx.doi.org/10.1016/j.colsurfb.2013.09.022] [PMID: 24240116]

[88] Sobrinho, H.B.S.; Rufino, R.D.; Luna, J.M.; Salgueiro, A.A.; Campos-Takaki, G.M.; Leite, L.F.C.; Sarubbo, L.A. Utilization of two agroindustrial by-products for the production of a surfactant by *Candida sphaerica* UCP0995. *Process Biochem.,* **2008**, *43*(9), 912-917.
[http://dx.doi.org/10.1016/j.procbio.2008.04.013]

[89] Batista, R.M.; Rufino, R.D.; Luna, J.M.; de Souza, J.E.G.; Sarubbo, L.A. Effect of medium components on the production of a biosurfactant from Candida tropicalis applied to the removal of hydrophobic contaminants in soil. *Water Environ. Res.,* **2010**, *82*(5), 418-425.
[http://dx.doi.org/10.2175/106143009X12487095237279] [PMID: 20480762]

[90] de Gusmão, C.A.B.; Rufino, R.D.; Sarubbo, L.A. Laboratory production and characterization of a new biosurfactant from *Candida glabrata* UCP1002 cultivated in vegetable fat waste applied to the removal of hydrophobic contaminant. *World J. Microbiol. Biotechnol.,* **2010**, *26*(9), 1683-1692.
[http://dx.doi.org/10.1007/s11274-010-0346-2]

[91] Luna, J.M.; Rufino, R.D.; Sarubbo, L.A.; Rodrigues, L.R.M.; Teixeira, J.A.C.; de Campos-Takaki, G.M. Evaluation antimicrobial and antiadhesive properties of the biosurfactant Lunasan produced by *Candida sphaerica* UCP 0995. *Curr. Microbiol.,* **2011**, *62*(5), 1527-1534.
[http://dx.doi.org/10.1007/s00284-011-9889-1] [PMID: 21327556]

[92] Kachholz, T.; Schlingmann, M. Possible food and agricultural applicationof microbial surfactants: An assessment. **1987**.

[93] Mukherjee, A.K. Potential application of cyclic lipopeptide biosurfactants produced by *Bacillus*

subtilis strains in laundry detergent formulations. *Lett. Appl. Microbiol.,* **2007**, *45*(3), 330-335.
[http://dx.doi.org/10.1111/j.1472-765X.2007.02197.x] [PMID: 17718848]

[94]   Das, K.; Mukherjee, A.K. Crude petroleum-oil biodegradation efficiency of *Bacillus subtilis* and *Pseudomonas aeruginosa* strains isolated from a petroleum-oil contaminated soil from North-East India. *Bioresour. Technol.,* **2007**, *98*(7), 1339-1345.
[http://dx.doi.org/10.1016/j.biortech.2006.05.032] [PMID: 16828284]

[95]   Salihu, A.; Abdulkadir, I.; Almustapha, M.N. An investigation for potential development of biosurfactants. *Microbiol. Mol. Biol. Rev.,* **2009**, *3*, 111-117.

[96]   Eshrat Gharaei-Fathabad. Biosurfactants in pharmaceutical industry: A Mini Review. *American Journal of Drug Discovering and Development*, **2011**, *1*, 58-69.

[97]   Rodrigues, L.; Banat, I.M.; Teixeira, J.; Oliveira, R. Biosurfactants: potential applications in medicine. *J. Antimicrob. Chemother.,* **2006**, *57*(4), 609-618.
[http://dx.doi.org/10.1093/jac/dkl024] [PMID: 16469849]

[98]   Krishnaswamy, M.; Subbuchettiar, G.; Ravi, T.K.; Panchaksharam, S. Biosurfactants properties, commercial production and application. *Curr. Sci.,* **2008**, *94*, 736-747.

CHAPTER 3

Applications of Biosurfactants in Various Cancer Therapies

Twinkle Rout[1], Muchalika Satapathy[2], Pratyasha Panda[2], Sibani Sahoo[3] and Arun Kumar Pradhan[4,*]

[1] *Department of Surgical Oncology; Institute of Medical Sciences and Sum Hospital, Siksha 'O' Anusandhan (Deemed to be University), Bhubaneswar, Odisha-751003, India*

[2] *AIPH University, Bhubaneswar, India*

[3] *Gangadhar Meher University, Sambalpur, India*

[4] *Centre for Biotechnology, Siksha 'O' Anusandhan (Deemed to be University), Bhubaneswar, Odisha-751003, India*

Abstract: Biosurfactants are the naturally-occurring surface-active biomolecules produced by microorganisms having a wide range of applications. Because of their unique characteristics like low toxicity, specificity, biodegradability and relative ease of preparation, these surface active molecules have attracted a wide interest recently. The effective and side-effect-free treatment of cancer remains a top priority for researchers despite various advancements in cancer therapy. To go beyond the drawbacks of chemotherapy, it is necessary to investigate anticancer medications derived from natural sources. Since a wide variety of these compounds have revealed the capacity to elicit cytotoxicity against numerous cancer cell lines, hence modulating cancer growth pathways, biosurfactants have recently come to light as prospective agents for cancer therapy. In this context, microbial biosurfactants offer a potential replacement for existing cancer treatments as well as anti-cancer drug delivery methods. The synthesis, structure, and studies of several cancer cell lines, including breast cancer, cervical cancer, lung cancer, pancreatic cancer, and prostate cancer, are all covered in this chapter, which summarizes the state of the art on microbial surfactants with anti-cancer potential.

Keywords: Anti-cancer, Biosurfactants, Cancer therapy, Surfactin.

INTRODUCTION

Recently, biosurfactants have been amazingly utilized for various applications like greater bioremediation, oil recovery, environmental, food processing and pharmaceutical due to their unique properties such as lower toxicity, higher bio-

---

[*] **Corresponding author Arun Kumar Pradhan:** Centre for Biotechnology, Siksha 'O' Anusandhan (Deemed to be University), Bhubaneswar, Odisha-751003, India; E-mail: arunpradhan@soa.ac.in

Arun Kumar Pradhan and Manoranjan Arakha (Eds.)

degradability, specific action, effectiveness at extreme temperature, PH, salinity, widespread applicability and their unique structures. Biosurfactants can also be used as antimicrobial, antiviral, and anti-adhesive against pathogens [1, 2, 3]. Biosurfactants are the surface-active chemicals made by a variety of microorganisms that are very useful in a wide range of biomedical applications. Glycolipids, glycoproteins, and lipopeptides, among other distinct biosurfactants, which comprise a variety of chemical structures, are anticipated to exhibit a variety of characteristics and physiological activities [4]. This molecule serves as an adjuvant for antigens, an inhibitor of fibrin clot formation, a ligand for binding immunoglobulins, a ligand for gene transfection, and an anti-adhesive biological coating for prosthetic materials.

Chemicals like these undergo interactions with diverse organism's cell membranes and/or the environment because of their surface activity, making them possible cancer treatments or components of drug delivery systems [5]. These are a few surfactants derived from microorganisms, such as rhamnolipids from *Pseudomonas aeruginosa*, surfactants from *Bacillus subtilis*, emulsan from *Acinetobacter calcoaceticus*, and sophorolipid from *Candida bombicola* Table **1**.

Table 1. Classification of biosurfactants on the basis of molecular mass.

High Molecular Mass	Polymeric	Carbohydrate Lipid-protein
		Emulsan
		Mannan-lipid protein
		Biodispersion
		Liposan
	Particulate	Vesicles
Low molecular Mass	Glycolipids	Sophorolipids
		Rhamnolipids
		Trehalose lipids
	Phospholipids	Fatty acid
		Phospholipids
		Corinomiocolic acid
	Lipopeptides	Subtilisin
		Lichenysin
		Gramicidin
		Wisconsin
		Peptide lipid
		Surfactin

Both lipopeptide and glycolipid surfactins are well-known biosurfactants that are often utilised in a variety of applications, such as preventing the growth of cancer cells and rupturing cell membranes to trigger apoptosis pathways, which cause the lysis of the cells [5, 6]. During different cancers' developmental stages, biosurfactants have the potential to function as antitumor agents. Biosurfactants significantly reduce the growth of certain tumour types. Particularly, numerous Bacillus genus bacterial species are used to create the biosurfactants surfactin, iturin, and fengycin lipopeptide. Surfactin's amphiphilic nature makes it simple to include into nanoformulations, including liposomes, micelles, micro-emulsions, and polymeric nanoparticles. Using nano-formulations has the benefit of improving surfactin distribution for more effective cancer treatment. A possible platform for reversing multi-drug resistance in cancer chemotherapy may be a surfactin-based nanocarrier (Fig. **1**, Table **2**).

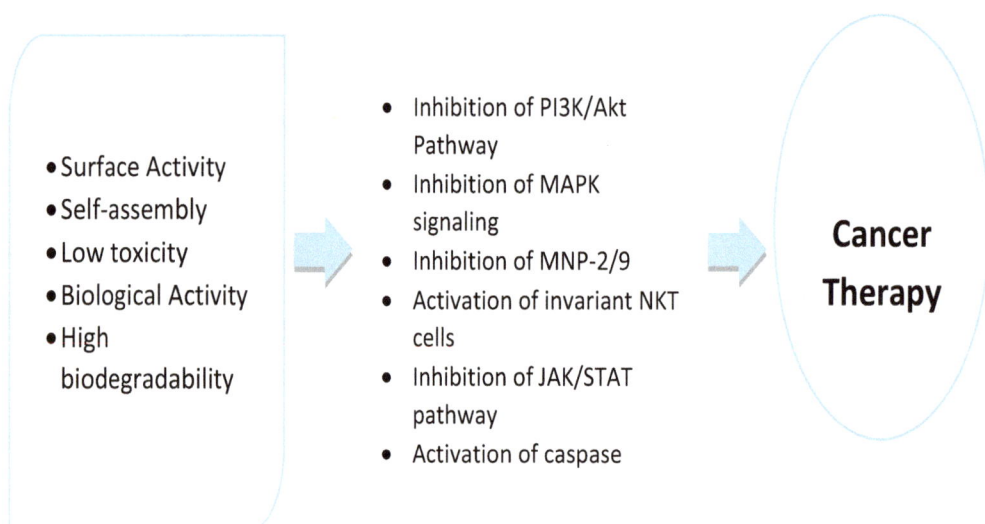

Fig. (1). Function of bio-surfactants targeted with cancer therapy.

Table 2. Activity of biosurfactants in different cancer cell lines.

Biosurfactants	Cell Line	Description	Activity
Mannosylerythritol lipids (MELs)	K562	Leukemia Myelogenous	Growth inhibition & differentiation
Succinoyl trehalose lipids (STLs)	HL60	Promyelocytic leukemia	Growth inhibition & differentiation
	KU812	Basophilic leukemia	Growth inhibition

(Table 2) cont.....

Biosurfactants	Cell Line	Description	Activity
Sophorolipids	HL60	Promyelocytic leukemia	Plasma membrane interaction
	H7402	Liver cancer	Growth inhibition, cell cycle arrest & apoptosis induction
	A549	Lung cancer	Apoptosis induction
	HPAC	Pancreatic cancer	Necrosis
	KYSE109/KYSE450	Esophageal cancer	Growth inhibition
e-poly-L-lysine	HeLaS3	Hepatocellular liver carcinoma	Growth inhibition
	HepG2	Metastatic prostate cancer	Migration inhibition
Glycoprotein from *Lactobacillus paracasei*	T47D/MDA-MB231	Breast cancer	Growth inhibition, cell cycle arrest
Growth inhibition, cell cycle arrest	BEL7402	Hepatocellular carcinoma	Growth inhibition & apoptosis induction
	K562	Myelogenous leukemia	Growth inhibition, cell cycle arrest & apoptosis induction
	LoVo	Colon adenocarcinoma	Growth inhibition, apoptosis induction
	MCF7	Breast cancer	Growth inhibition & apoptosis induction
	T47D/MDA-MB231	Breast cancer	Growth inhibition, cell cycle arrest
	Caco2	Colorectal cancer	Growth inhibition, apoptosis induction
	HCT15/HT29	Colon cancer	Growth inhibition
Viscosin	PC3M	Metastatic prostate cancer	Migration inhibition
Monoolein	HeLa	Cervical cancer	Growth inhibition
	U937	Leukemia cancer	Growth inhibition
Serratamolide	BCLL	B-Chronic lymphocytic leukemia	Apoptosis induction

EFFECT OF BIOSURFACTANTS ON BREAST CANCER

One of the most common malignant diseases in the world is breast cancer. GLOBOCAN 2020 estimates that there are 3,465,951 incidences and 1,121,413 fatalities worldwide. In India in 2020, there were 436,417 fatalities and 1,204,532 new cases [7]. Surfactin generated by *Bacillus subtilis* 573 has anti-tumor properties. Surfactin, an amphipathic molecule, has been shown to affect the stability of cell membranes by destabilising them. Surfactin helps in inducing death in MCF-7 cells through the JNK-mediated caspase pathway [8]. By a mitochondrial caspase pathway driven by ROS/JNK, surfactin triggered apoptosis in the cells [9]. Surfactin causes G1 arrest in cell lines, which leads to a reduction in the number of cells making DNA. Due to the rupture of the plasma membrane, the maximum concentration of surfactin greatly reduces the total number of cells [10]. Cell cycle arrest at the G1 phase can be brought on by lipopeptide. Apoptosis was linked to the lipopeptide's alteration of the cellular fatty acid content of cancer cell lines. The probiotic bacterium *Lactobacillus paracasei sub. paracasei* A20 produces BioEG, a biosurfactant with glycoprotein as its main chemical component [11]. In addition to being a chemotherapeutic drug, bioEG has numerous additional biological uses, including recovery from invasive surgery. Based on the biosurfactant dosage and exposure duration, bioEG reduces the viability of cancer cell lines. A number of viable cells were shown to considerably decrease at four different BioEG doses (0.05, 0.1, 0.15, and 0.2 g l-1) with the longest exposure times (48 and 72 Hrs) [12]. Therefore, longer exposure times and greater BioEG concentrations result in a significant level of necrobiosis. The biosurfactant may be used for tissue regeneration because BioEG has shown to specifically influence the carcinoma cell line, where the results are positive. The glycolipid-based biosurfactant called di-Rhamnolipids from *Pseudomonas aeruginosa* B189 stops the growth of the MCF-7 cancer cell line [13]. A fully original biosurfactant derived from *C. parapsilosis* is frequently used as a suitable carrier for the regulated release of cancer cell regulation using PLA-PEG polymeric nanoparticles [14].

EFFECT OF BIOSURFACTANTS ON CERVICAL CANCER

High molecular weight biosurfactants (polysaccharides, proteins and lipoproteins) and low molecular weight biosurfactants (polypeptides and glycolipids) are two general categories for biosurfactants. Due to their typically simpler structure, biosurfactants (low molecular weight), such as Rhamnolipid (a glycolipid), Sophorolipid, and Surfactin (a lipopeptide), have been thoroughly studied [15]. *Bacillus subtilis* is used in the production of the acidic cyclic non-ribosomallip--heptapeptide known as surfactin. Specifically studied on the HeLa cell line, surfactin has a critical impact on cervical cancer. Cell viability is inhibited by surfactin. It has been demonstrated that Surfactin C15 nanopeptidase causes cytotoxicity in HeLa cell. With respect to nanopeptidase concentration and exposure period, cell viability is decreased. Surfactin had an IC_{50} of 86.9, 73.1, and 50.2 m for inhibiting cellular viability by 50% for 16, 24, and 48 hours, respectively [16]. Certain biosurfactants are essential for the growth and death of cancer cells. *Exophiala dermatitidis* SK80 (fungus) creates a biosurfactant called monoolein that, depending on doses, can suppress the growth of cervical cancer (HeLa cell line) and, even at high doses, did not result in cytotoxicity in normal cells [17]. In order to fluidize the hard cervical cancer tissues and to modify interfacial effects, new therapeutic approaches can also be designed by ingesting biosurfactants that may alter lipid content (especially phosphatidylcholine and sphingomyelin) [17, 18]. Due to their cytotoxic effect, MR01 biosurfactant, a Rhamnolipid type biosurfactant made from PAMR01 (*PA-Pseudomonas aeruginosa*), inhibits the proliferation of the HeLa cell line at a concentration of 5 g/ml [19]. Sophorolipids are biosurfactants that are produced *via* the fermentation of oils utilising a unique species of Candida. It has been demonstrated that sophorolipids have a cytotoxic effect on the human cervical carcinoma (HeLa) cell line [20]. In HeLa cancer cells, SLCA causes cell death through pro-apoptotic chemicals. Apoptosis caused by SLCA in HeLa cells is linked to mitochondrial dysfunction [20]. Amygdalin significantly reduced the viability of the HeLa cell line used to study human cervical cancer by upregulating the protein Bax and down regulating BCL-2. HeLa cell line treated with amygdalin exhibited involvement in the intrinsic route of apoptosis, which also demonstrated the reduction of HeLa cell line growth [21].

EFFECT OF BIOSURFACTANTS ON LUNG CANCER

Biosurfactants are the surface active chemicals made by microorganisms, useful in a variety of biomedical and pharmacological applications. Although there are few results and little knowledge about the processes underlying this interest, the anti-tumor capability of those compounds is being studied. Biosurfactants have many specific roles in the inhibition of cancer cells; such as cytotoxicity against tumor causing factors, growth inhibition of cancerous cells, cell cycle arrest, apoptosis and metastatic arrest [22]. There are many biosurfactants that show many inhibitory roles on A549 cell line of lung cancer. For example, Lactonic sophorolipid at a concentration of 100µg/ml completely killed the A549 cellular line of lung cancer. This process is completely dependent upon incubation time. After 48hr of lactonic sophorolipid treatment, the IC_{50} value was reduced to 40 µg/ml for the A549 cell line. In the instance of the A549 cell line, ROS generation at glycolipid 500 (G500) and lactonic sophorolipid 50 (L-SL50) were confirmed in 16.92% and 22.72% of the cells, respectively [23]. The treatment of A549 cells with G500 and L-SL50 ended in greater necrotic/lifeless cells. L-SL and glucolipid both efficiently inhibited the cell migration in the A549 cell line after 24hr and 48hr of treatments [23]. Biosurfactants of glycolipid nature which are extracted from *B.aryabhattai* strain ZDY2 and lactobacilli have low toxicity against the A549 cell line [24]. *R. erythropolis* SD-74, a species from mycolic acid-containing action bacteria group brought about succinoyl trehalose lipid-1 biosurfactant which has cytotoxic effect against A549 cell line of human lung cancer [25, 26]. For Non-Small Cell Lung Cancer (NSCLC), L-SL, a kind of sophorolipid known to suppress histone deacetylase (HDAC) activity, is employed to assess its anti-cancer properties [27, 28]. After L-SL treatment on the

A549 cell line, more than 80% of the cells' viability was reduced and more than 40% of the cells were rendered dead within 24 hours [29]. The anti-proliferative properties of a biosurfactant isolated from *Aceneto bacterindicus* M6 were validated in the lung cancer cell line A549 [30]. In A549 cells, the G1 phase became stopped, indicating that the method of cell death was an inhibition of DNA synthesis [31]. A biosurfactant produced by Micromonospora exhibits hypoxia-specific toxicity against lung cancer cells. Rakicidins activated caspase-mediated pathways and blocked signalling pathways like MAPK and JNK/P38 to cause apoptosis [32].

EFFECT OF BIOSURFACTANTS ON PANCREATIC CANCER

The high cost of the medications used to treat cancer makes it an expensive process. Most cases of pancreatic cancer are fatal and have a low survival rate. The pancreatic cancer cell lines of various tumours can be significantly inhibited by lipopeptides from *Bacillus subtilis* [33]. Due to the number of deaths from pancreatic adenocarcinoma each year, it continues to be a significant consequence. It is the fourth worst cancer type in the world. C-type lectins or collectins (a soluble collagen) are present in human surfactant protein D [34]. In pancreatic adenocarcinomas, TGF-expression was linked to a highly metastatic phenotype acquired through SMAD signalling. It also controls the expression of the EMT-gene, which enhances cell motility and invasiveness in pancreatic cancer cell lines [35]. For 48 hours, rfhSP-D causes apoptosis in pancreatic cancer cells through a FAS-mediated mechanism that involves the cleavage of caspases 8 and 3. The rfhSP-D reduces TGF- expression in pancreatic cell lines Panc-1, MiaPaCa-2, and Capan-2, lowering their propensity for invasion [36]. By weakening the TGF- signalling pathway, rfhSP-D controls the EMT-related gene signatures Vimentin, Zeb1, and Snail as well as the invasion of pancreatic cancer cell lines [37]. Pancreatic cancer cell lines were grown in DMEM-F12 media supplemented with 2 mM l-glutamine, 10% v/v foetal calf serum (FCS), and penicillin (100 units/ml)/streptomycin (100 g/ml) (Thermo Fisher). To achieve 80-90% confluency, pancreatic cancer cell lines were cultured at 37 °C with 5% v/v CO_2 [38]. To establish the ideal dose of rfhSP-D, pancreatic cell morphological abnormalities, and morphological changes were investigated [39]. Since it is widely known that pancreatic cancer cells over-express TGF-, which plays a significant part in triggering EMT, the expression of TGF- in pancreatic cell lines was examined after being treated with rfhSP-D (20 g/ml) [40]. It has been determined that the MCM7 gene plays a role in DNA replication and cancer cell growth. The MCM7 gene inhibits cell growth in a variety of cancer cells [41]. Cancer cell motility, adhesion, proliferation, and invasion are all significantly influenced by the CXCR4 gene. Levansucrase and levan are produced by *B. subtilis* MT453867. With its powerful antioxidant properties, *B. subtilis*

MT453867 levan significantly reduced the expression of the CXCR4 and MCM7 genes in pancreatic cancer while also successfully fragmenting DNA. Levan, which is produced by *B. subtilis* MT453867, shows anti-tumor and chemo-protective action in pancreatic cancer [42]. At 59.9 ± 1.7 µg/ml in a 24-hour exposure, surfactin molecules reduce the survival of human pancreas cancer cell lines (Sw-1990), resulting in a 50% reduction [43]. In order to develop a treatment approach against pre-malignant stages of solid tumour growth, it is crucial to investigate the EMT suppressor function of rfhSP-D in combination with conventional chemotherapy for pancreatic cancer [44].

EFFECT OF BIOSURFACTANTS ON PROSTATE CANCER

One of the most common cancers affecting the male population worldwide with no single form of acceptable treatment, is prostate cancer. There are various methods for the treatment of prostate cancer, including surgery, internal or external radiation, phototherapy, immunotherapy, and chemotherapy. Prostate cancer treatment heavily relies on the extracellular amphiphilic chemicals that microorganisms create, sometimes known as "biosurfactants". They can activate natural killer cells, stop angiogenesis, and cause cancer cells to undergo apoptosis *via* the death receptor. Surfactant protein D, which induces apoptosis in cancer cells, is composed of an N-terminal collagen-like domain and a calcium-dependent carbohydrate recognition domain region (CRD) from the collectin family [45 - 48]. Using high resolution mass spectrometry, it was discovered that the human SP-D membrane interactome contained 347 proteins in an androgen-free cancer cell [49]. Apoptosis caused by rfhSP-D in prostate cancer cell line PC3 is susceptible to metabolic alterations that replicate those changes. PC3 cells (ATCC, Rockville, MD, USA) were grown in RPMI 1640 supplemented with 10% v/v Foetal Bovine Serum (FBS) and 1% Antibiotics (PenStrep). PC3 cells were grown in 5% v/v CO_2 at 37°C until they were around 90% confluent [50]. The membrane interactome of PC3 cells treated with rfhSP-D included GRP78 (HSPA5), HSPA8, HSP90AB1, HSP90AA1, HSPA1B, HSP90B1, and HSPD1 among others. Pre-incubation of PC3 cells was performed in serum-free media that contained rfhSP-D. The rfhSP-D, which engages in interaction with PC3 membrane proteins, was eliminated by immunodepletion [51]. The LC-MS/MS analysis was performed on the PC3 membrane protein as well as rfhSP-D. Ingenuity and network analysis were used to find the rfhSP-D interactome. The HPSAs/GRP78/rfhSP-D potential binding partner was discovered. *In-silico* validation revealed a relationship between rfhSP-D and GRP78. GRP78 and rfhSP-D interaction was verified. Both GRP78 and HSP90AA1, found in prostate cancer and its metastases, play a major role in cancer cell apoptosis and survival [52]. In contrast to benign prostate samples, the HSPD1/E1 complex is overexpressed in prostate cancer and carcinoma [53]. GRP78 expression is increased in several malignancies and is associated with the aggressiveness, metastasis, and development of chemoresistant mutations in prostate cancer [54-57]. *In silico* investigation has confirmed the link between rFLhSP-D/rfhSP-D and GRP78 *via* the CRD region, confirming GRP78 as a binding partner of SP-D and indicating that it may act as a novel intermediary in the SP-D-mediated innate immune surveillance against prostate cancer (Fig. **2**).

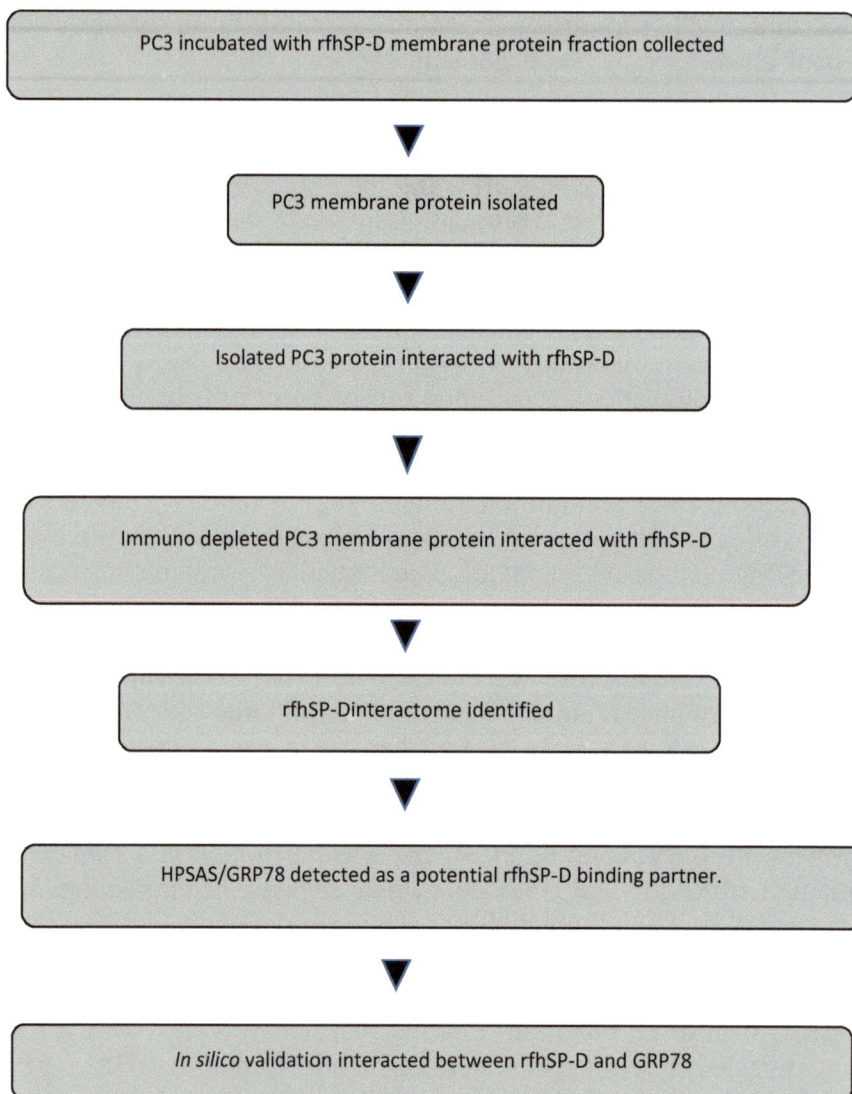

Fig. (**2**). Interaction of rfhSP-D and PC3 with *in-silico*-validation.

CONCLUSION

Surface active chemicals that have been produced chemically are, however widely used in a wide variety of industries and are readily available in the market. However, biosurfactants have advantages over chemical surfactants, including biodegradability and synthesis on substrates made from renewable resources. Biosurfactants can be used in a variety of healthcare settings and have advantages over chemical surfactants. Due to their promising intercellular predicted value, which includes selectively blocking cancer cells, and their ability to proliferate by signal transduction, biosurfactants have been studied as potential cancer-fighting substances. When opposed to chemotherapy, the use of biosurfactants as anti-cancer agents is a better substitute because of their potential to act on cancer cells without harming normal tissue. However, it is still difficult to employ them in this region. So first, the size of microorganisms and their metabolites encourages the continual search for new biosurfactants while unrealizing the potential of those that already exist. Second, there weren't many studies performed specifically to understand how biosurfactants work. Additionally, several research works used semi-purified portions with biosurfactants, which had an impact on how results were interpreted and how the underlying mechanisms were understood. Finally, a large number of ongoing investigations are currently undergoing *in vitro* testing using cell lines. Up until their final clearance by the relevant health regulatory agencies, a large amount of study is still needed before the *in vivo* stage. Overall, there are not many published papers in this field yet, and it is still in its early stages. However, considering the encouraging outcomes that have been described, this area of research will undoubtedly expand in the coming years.

REFERENCES

[1] Gudiña, E.J.; Teixeira, J.A.; Rodrigues, L.R. Isolation and functional characterization of a biosurfactant produced by *Lactobacillus paracasei*. *Colloids Surf. B Biointerfaces,* **2010**, *76*(1), 298-304.
[http://dx.doi.org/10.1016/j.colsurfb.2009.11.008] [PMID: 20004557]

[2] Gudiña, E.J.; Rangarajan, V.; Sen, R.; Rodrigues, L.R. Potential therapeutic applications of biosurfactants. *Trends Pharmacol. Sci.,* **2013**, *34*(12), 667-675.
[http://dx.doi.org/10.1016/j.tips.2013.10.002] [PMID: 24182625]

[3] Rodrigues, L.R. *Inhibition of bacterial adhesion on medical devices. Bacterial Adhesion: Chemistry*; Biology and Physics, **2011**, pp. 351-367.

[4] Banat, I.M.; Franzetti, A.; Gandolfi, I.; Bestetti, G.; Martinotti, M.G.; Fracchia, L.; Smyth, T.J.; Marchant, R. Microbial biosurfactants production, applications and future potential. *Appl. Microbiol. Biotechnol.,* **2010**, *87*(2), 427-444.
[http://dx.doi.org/10.1007/s00253-010-2589-0] [PMID: 20424836]

[5] Rodrigues, L.; Banat, I.M.; Teixeira, J.; Oliveira, R. Biosurfactants: potential applications in medicine. *J. Antimicrob. Chemother.,* **2006**, *57*(4), 609-618.
[http://dx.doi.org/10.1093/jac/dkl024] [PMID: 16469849]

[6] Banat, I.M.; De Rienzo, M.A.D.; Quinn, G.A. Microbial biofilms: biosurfactants as antibiofilm agents.

Appl. Microbiol. Biotechnol., **2014**, *98*(24), 9915-9929.
[http://dx.doi.org/10.1007/s00253-014-6169-6] [PMID: 25359476]

[7] Niamatullah, S.N. Predictors of outcome and survival in prostate cancer – data from tertiary care urology institute in pakistan. *Hematol. Transfus. Cell Ther.,* **2021**, *43*, S1-S2.
[http://dx.doi.org/10.1016/j.htct.2021.10.946]

[8] Sil, J.; Dandapat, P.; Das, S. Health care applications of different biosurfactants. *Int. J. Sci. Res.,* **2017**, *6*(6), 41-50.

[9] Cao, X.; Wang, A.; Wang, C.; Mao, D.; Lu, M.; Cui, Y.; Jiao, R. Surfactin induces apoptosis in human breast cancer MCF-7 cells through a ROS/JNK-mediated mitochondrial/caspase pathway. *Chem. Biol. Interact.,* **2010**, *183*(3), 357-362.
[http://dx.doi.org/10.1016/j.cbi.2009.11.027] [PMID: 19954742]

[10] Liu, X.; Tao, X.; Zou, A.; Yang, S.; Zhang, L.; Mu, B. Effect of themicrobial lipopeptide on tumor cell lines: apoptosis induced by disturbing the fatty acid composition of cell membrane. *Protein Cell,* **2010**, *1*(6), 584-594.
[http://dx.doi.org/10.1007/s13238-010-0072-4] [PMID: 21204010]

[11] Pinto, S.; Alves, P.; Santos, A.C.; Matos, C.M.; Oliveiros, B.; Gonçalves, S.; Gudiña, E.; Rodrigues, L.R.; Teixeira, J.A.; Gil, M.H. Poly(dimethyl siloxane) surface modification with biosurfactants isolated from probiotic strains. *J. Biomed. Mater. Res. A,* **2011**, *98A*(4), 535-543.
[http://dx.doi.org/10.1002/jbm.a.33146] [PMID: 21681946]

[12] Duarte, C.; Gudiña, E.J.; Lima, C.F.; Rodrigues, L.R. Effects of biosurfactants on the viability and proliferation of human breast cancer cells. *AMB Express,* **2014**, *4*(1), 40.
[http://dx.doi.org/10.1186/s13568-014-0040-0] [PMID: 24949273]

[13] Thanomsub, B.; Pumeechockchai, W.; Limtrakul, A.; Arunrattiyakorn, P.; Petchleelaha, W.; Nitoda, T.; Kanzaki, H. Chemical structures and biological activities of rhamnolipids produced by *Pseudomonas aeruginosa* B189 isolated from milk factory waste. *Bioresour. Technol.,* **2006**, *97*(18), 2457-2461.
[http://dx.doi.org/10.1016/j.biortech.2005.10.029] [PMID: 16697639]

[14] Wadhawan, A.; Singh, J.; Sharma, H.; Handa, S.; Singh, G.; Kumar, R.; Barnwal, R.P.; Pal Kaur, I.; Chatterjee, M. Anticancer Biosurfactant-Loaded PLA–PEG Nanoparticles Induce Apoptosis in Human MDA-MB-231 Breast Cancer Cells. *ACS Omega,* **2022**, *7*(6), 5231-5241.
[http://dx.doi.org/10.1021/acsomega.1c06338] [PMID: 35187338]

[15] Ceresa C, Fracchia L, Sansotera AC, De Rienzo MAD, Banat IM. Harnessing the Potential of Biosurfactants for Biomedical and Pharmaceutical Applications. *Pharmaceutics.* **2023** Aug 18; *15*(8): 2156.
[http://dx.doi.org/10.3390/pharmaceutics15082156] [PMID: 37631370] [PMCID: PMC10457971]

[16] Nozhat, Z.; Asadi, A.; Zahri, S. Properties of surfactin C-15 nanopeptide and its cytotoxic effect on human cervix cancer (HeLa) cell line. *J. Nanomater.,* **2012**, 1-5.
[http://dx.doi.org/10.1155/2012/526580]

[17] Chiewpattanakul, P.; Phonnok, S.; Durand, A.; Marie, E.; Thanomsub, B.W. Bioproduction and anticancer activity of biosurfactant produced by the dematiaceous fungus Exophiala dermatitidis SK80. *J. Microbiol. Biotechnol.,* **2010**, *20*(12), 1664-1671.
[PMID: 21193821]

[18] Letizia Fracchia, Jareer J. Banat, Massimo Cavallo, Chiara Ceresa, Ibrahim M. Banat. Potential therapeutic applications of microbial surface-active compounds[J]. *AIMS Bioengineering,* **2015**, *2*(3): 144-162.
[http://dx.doi.org/10.3934/bioeng.2015.3.144]

[19] Lotfabad, T.B.; Abassi, H.; Ahmadkhaniha, R.; Roostaazad, R.; Masoomi, F.; Zahiri, H.S.; Ahmadian, G.; Vali, H.; Noghabi, K.A. Structural characterization of a rhamnolipid-type biosurfactant produced by *Pseudomonas aeruginosa* MR01: Enhancement of di-rhamnolipid proportion using gamma

irradiation. *Colloids Surf. B Biointerfaces,* **2010**, *81*(2), 397-405.
[http://dx.doi.org/10.1016/j.colsurfb.2010.06.026] [PMID: 20732795]

[20] Nawale, L.; Dubey, P.; Chaudhari, B.; Sarkar, D.; Prabhune, A. Anti-proliferative effect of novel primary cetyl alcohol derived sophorolipids against human cervical cancer cells HeLa. *PLoS One,* **2017**, *12*(4), e0174241.
[http://dx.doi.org/10.1371/journal.pone.0174241] [PMID: 28419101]

[21] Chen, Y.; Ma, J.; Wang, F.; Hu, J.; Cui, A.; Wei, C.; Yang, Q.; Li, F. Amygdalin induces apoptosis in human cervical cancer cell line HeLa cells. *Immunopharmacol. Immunotoxicol.,* **2013**, *35*(1), 43-51.
[http://dx.doi.org/10.3109/08923973.2012.738688] [PMID: 23137229]

[22] Rahimi K, Lotfabad TB, Jabeen F, Mohammad Ganji S. Cytotoxic effects of mono- and di-rhamnolipids from *Pseudomonas aeruginosa* MR01 on MCF-7 human breast cancer cells. *Colloids Surf B Biointerfaces.* **2019** Sep 1; *181*: 943-952. Epub 2019 Jun 27.
[http://dx.doi.org/10.1016/j.colsurfb.2019.06.058] [PMID: 31382344]

[23] Haque, F.; Khan, M.S.A.; AlQurashi, N. ROS-mediated necrosis by glycolipid biosurfactants on lung, breast, and skin melanoma cells. *Front. Oncol.,* **2021**, *11*, 622470.
[http://dx.doi.org/10.3389/fonc.2021.622470] [PMID: 33796459]

[24] Yaraguppi, D.A.; Bagewadi, Z.K.; Muddapur, U.M.; Mulla, S.I. Response surface methodology-based optimization of biosurfactant production from isolated *Bacillus aryabhattai* strain ZDY2. *J. Pet. Explor. Prod. Technol.,* **2020**, *10*(6), 2483-2498.
[http://dx.doi.org/10.1007/s13202-020-00866-9]

[25] Kuyukina, M.S.; Ivshina, I.B.; Baeva, T.A.; Kochina, O.A.; Gein, S.V.; Chereshnev, V.A. Trehalolipid biosurfactants from nonpathogenic *Rhodococcus actinobacteria* with diverse immunomodulatory activities. *N. Biotechnol.,* **2015**, *32*(6), 559-568.
[http://dx.doi.org/10.1016/j.nbt.2015.03.006] [PMID: 25796474]

[26] Retamal-Morales, G.; Heine, T.; Tischler, J.S.; Erler, B.; Gröning, J.A.D.; Kaschabek, S.R.; Schlömann, M.; Levicán, G.; Tischler, D. Draft genome sequence of *Rhodococcus erythropolis* B7g, a biosurfactant producing actinobacterium. *J. Biotechnol.,* **2018**, *280*, 38-41.
[http://dx.doi.org/10.1016/j.jbiotec.2018.06.001] [PMID: 29879458]

[27] Li H, Guo W, Ma XJ, Li JS, Song X. *In Vitro* and *in Vivo* Anticancer Activity of Sophorolipids to Human Cervical Cancer. *Appl Biochem Biotechnol.* **2017** Apr; *181*(4): 1372-1387. Epub 2016 Oct 29.
[http://dx.doi.org/10.1007/s12010-016-2290-6] [PMID: 27796874]

[28] Adu, S.A.; Twigg, M.S.; Naughton, P.J.; Marchant, R.; Banat, I.M. Biosurfactants as anticancer agents: glycolipids affect skin cells in a differential manner dependent on chemical structure. *Pharmaceutics,* **2022**, *14*(2), 360.
[http://dx.doi.org/10.3390/pharmaceutics14020360] [PMID: 35214090]

[29] Naz, S.; Banerjee, T.; Totsingan, F.; Woody, K.; Gross, R.A.; Santra, S. Therapeutic efficacy of lactonic sophorolipids: Nanoceria-assisted combination therapy of NSCLC using HDAC and Hsp90 inhibitors. *Nanotheranostics,* **2021**, *5*(4), 391-404.
[http://dx.doi.org/10.7150/ntno.57675] [PMID: 33912379]

[30] Karlapudi, A.P.; Venkateswarulu, T.C.; Srirama, K.; Kota, R.K.; Mikkili, I.; Kodali, V.P. Evaluation of anti-cancer, anti-microbial and anti-biofilm potential of biosurfactant extracted from an *Acinetobacter* M6 strain. *J. King Saud Univ. Sci.,* **2020**, *32*(1), 223-227.
[http://dx.doi.org/10.1016/j.jksus.2018.04.007]

[31] Sawant, S.S.; Patil, S.M.; Gupta, V.; Kunda, N.K. Microbes as medicines: harnessing the power of bacteria in advancing cancer treatment. *Int. J. Mol. Sci.,* **2020**, *21*(20), 7575.
[http://dx.doi.org/10.3390/ijms21207575] [PMID: 33066447]

[32] Gudiña, E.; Teixeira, J.; Rodrigues, L. Biosurfactants produced by marine microorganisms with therapeutic applications. *Mar. Drugs,* **2016**, *14*(2), 38.
[http://dx.doi.org/10.3390/md14020038] [PMID: 26901207]

[33]   Zhao, H.; Shao, D.; Jiang, C.; Shi, J.; Li, Q.; Huang, Q.; Rajoka, M.S.R.; Yang, H.; Jin, M. Biological activity of lipopeptides from Bacillus. *Appl. Microbiol. Biotechnol.,* **2017**, *101*(15), 5951-5960.
[http://dx.doi.org/10.1007/s00253-017-8396-0] [PMID: 28685194]

[34]   Maier, H.J.; Wirth, T.; Beug, H. Epithelial-mesenchymal transition in pancreatic carcinoma. *Cancers (Basel),* **2010**, *2*(4), 2058-2083.
[http://dx.doi.org/10.3390/cancers2042058] [PMID: 24281218]

[35]   Kaur, A.; Riaz, M.S.; Murugaiah, V.; Varghese, P.M.; Singh, S.K.; Kishore, U. A recombinant fragment of human surfactant protein D induces apoptosis in pancreatic cancer cell lines *via* fas-mediated pathway. *Front. Immunol.,* **2018**, *9*, 1126.
[http://dx.doi.org/10.3389/fimmu.2018.01126] [PMID: 29915574]

[36]   Beuran, M.; Negoi, I.; Paun, S.; Ion, A.D.; Bleotu, C.; Negoi, R.I.; Hostiuc, S. The epithelial to mesenchymal transition in pancreatic cancer: A systematic review. *Pancreatology,* **2015**, *15*(3), 217-225.
[http://dx.doi.org/10.1016/j.pan.2015.02.011] [PMID: 25794655]

[37]   Mahajan, L.; Madan, T.; Kamal, N.; Singh, V.K.; Sim, R.B.; Telang, S.D.; Ramchand, C.N.; Waters, P.; Kishore, U.; Sarma, P.U. Recombinant surfactant protein-D selectively increases apoptosis in eosinophils of allergic asthmatics and enhances uptake of apoptotic eosinophils by macrophages. *Int. Immunol.,* **2008**, *20*(8), 993-1007.
[http://dx.doi.org/10.1093/intimm/dxn058] [PMID: 18628238]

[38]   Kishore, U.; Bernal, A.L.; Kamran, M.F.; Saxena, S.; Singh, M.; Sarma, P.U.; Madan, T.; Chakraborty, T. Surfactant proteins SP-A and SP-D in human health and disease. *Arch. Immunol. Ther. Exp. (Warsz.),* **2005**, *53*(5), 399-417.
[PMID: 16314824]

[39]   Teraoka, H.; Sawada, T.; Yamashita, Y.; Nakata, B.; Ohira, M.; Ishikawa, T.; Nishino, H.; Hirakawa, K. TGF-β1 promotes liver metastasis of pancreatic cancer by modulating the capacity of cellular invasion. *Int. J. Oncol.,* **2001**, *19*(4), 709-715.
[http://dx.doi.org/10.3892/ijo.19.4.709] [PMID: 11562745]

[40]   Yu, S.; Wang, G.; Shi, Y.; Xu, H.; Zheng, Y.; Chen, Y. MCMs in cancer: prognostic potential and mechanisms. *Anal. Cell. Pathol.,* **2020**, *2020*, 1-11.
[http://dx.doi.org/10.1155/2020/3750294] [PMID: 32089988]

[41]   Gamal, A.A.; Abbas, H.Y.; Abdelwahed, N.A.M.; Kashef, M.T.; Mahmoud, K.; Esawy, M.A.; Ramadan, M.A. Optimization strategy of *Bacillus subtilis* MT453867 levansucrase and evaluation of levan role in pancreatic cancer treatment. *Int. J. Biol. Macromol.,* **2021**, *182*, 1590-1601.
[http://dx.doi.org/10.1016/j.ijbiomac.2021.05.056] [PMID: 34015407]

[42]   Liu, X.; Tao, X.; Zou, A.; Yang, S.; Zhang, L.; Mu, B. Effect of themicrobial lipopeptide on tumor cell lines: apoptosis induced by disturbing the fatty acid composition of cell membrane. *Protein Cell,* **2010**, *1*(6), 584-594.
[http://dx.doi.org/10.1007/s13238-010-0072-4] [PMID: 21204010]

[43]   Arumugam, T.; Ramachandran, V.; Fournier, K.F.; Wang, H.; Marquis, L.; Abbruzzese, J.L.; Gallick, G.E.; Logsdon, C.D.; McConkey, D.J.; Choi, W. Epithelial to mesenchymal transition contributes to drug resistance in pancreatic cancer. *Cancer Res.,* **2009**, *69*(14), 5820-5828.
[http://dx.doi.org/10.1158/0008-5472.CAN-08-2819] [PMID: 19584296]

[44]   Holmskov, U.; Thiel, S.; Jensenius, J.C. Collections and ficolins: humoral lectins of the innate immune defense. *Annu. Rev. Immunol.,* **2003**, *21*(1), 547-578.
[http://dx.doi.org/10.1146/annurev.immunol.21.120601.140954] [PMID: 12524383]

[45]   Jakel, A.; Qaseem, A.S.; Kishore, U.; Sim, R.B. Ligands and receptors of lung surfactant proteins SP-A and SP-D. *Front. Biosci.,* **2013**, *18*(3), 1129-1140.
[http://dx.doi.org/10.2741/4168] [PMID: 23747872]

[46] Lu, J.; Teh, C.; Kishore, U.; Reid, K.B. Collectins and ficolins: sugar pattern recognition molecules of the mammalian innate immune system. *Biochim. Biophys. Acta, Gen. Subj.,* **2002**, *1572*(2-3), 387-400.
[http://dx.doi.org/10.1016/S0304-4165(02)00320-3] [PMID: 12223281]

[47] Crouch, E.; Persson, A.; Chang, D.; Heuser, J. Molecular structure of pulmonary surfactant protein D (SP-D). *J. Biol. Chem.,* **1994**, *269*(25), 17311-17319.
[http://dx.doi.org/10.1016/S0021-9258(17)32556-5] [PMID: 8006040]

[48] Szklarczyk, D.; Gable, A.L.; Lyon, D.; Junge, A.; Wyder, S.; Huerta-Cepas, J.; Simonovic, M.; Doncheva, N.T.; Morris, J.H.; Bork, P.; Jensen, L.J.; Mering, C. STRING v11: protein–protein association networks with increased coverage, supporting functional discovery in genome-wide experimental datasets. *Nucleic Acids Res.,* **2019**, *47*(D1), D607-D613.
[http://dx.doi.org/10.1093/nar/gky1131] [PMID: 30476243]

[49] Thakur, G.; Prakash, G.; Murthy, V.; Sable, N.; Menon, S.; Alrokayan, S.H.; Khan, H.A.; Murugaiah, V.; Bakshi, G.; Kishore, U.; Madan, T. Human SP-D Acts as an Innate Immune Surveillance Molecule Against Androgen-Responsive and Androgen-Resistant Prostate Cancer Cells. *Front. Oncol.,* **2019**, *9*, 565.
[http://dx.doi.org/10.3389/fonc.2019.00565] [PMID: 31355132]

[50] Santos, T.G.; Martins, V.; Hajj, G. Unconventional secretion of heat shock proteins in cancer. *Int. J. Mol. Sci.,* **2017**, *18*(5), 946.
[http://dx.doi.org/10.3390/ijms18050946] [PMID: 28468249]

[51] Uhlen, M; Fagerberg, L; Hallström, BM; Lindskog, C; Oksvold, P; Mardinoglu, A Tissue-based map of the human proteome. *Science,* **2015**, *347*(6220), 1260419.
[http://dx.doi.org/10.1126/science.1260419]

[52] Ciocca, D.R.; Fanelli, M.A.; Cuello-Carrion, F.D.; Castro, G.N. Heat shock proteins in prostate cancer: from tumorigenesis to the clinic. *Int. J. Hyperthermia,* **2010**, *26*(8), 737-747.
[http://dx.doi.org/10.3109/02656731003776968] [PMID: 20858068]

[53] Pootrakul, L.; Datar, R.H.; Shi, S.R.; Cai, J.; Hawes, D.; Groshen, S.G.; Lee, A.S.; Cote, R.J. Expression of stress response protein Grp78 is associated with the development of castration-resistant prostate cancer. *Clin. Cancer Res.,* **2006**, *12*(20), 5987-5993.
[http://dx.doi.org/10.1158/1078-0432.CCR-06-0133] [PMID: 17062670]

[54] Misra, U.K.; Gonzalez-Gronow, M.; Gawdi, G.; Pizzo, S.V. The role of MTJ-1 in cell surface translocation of GRP78, a receptor for α 2-macroglobulin-dependent signaling. *J. Immunol.,* **2005**, *174*(4), 2092-2097.
[http://dx.doi.org/10.4049/jimmunol.174.4.2092] [PMID: 15699139]

[55] Sokolowska, I.; Woods, A.G.; Gawinowicz, M.A.; Roy, U.; Darie, C.C. Identification of a potential tumor differentiation factor receptor candidate in prostate cancer cells. *FEBS J.,* **2012**, *279*(14), 2579-2594.
[http://dx.doi.org/10.1111/j.1742-4658.2012.08641.x] [PMID: 22613557]

[56] Nakatsuka, A.; Wada, J.; Iseda, I.; Teshigawara, S.; Higashio, K.; Murakami, K.; Kanzaki, M.; Inoue, K.; Terami, T.; Katayama, A.; Hida, K.; Eguchi, J.; Horiguchi, C.S.; Ogawa, D.; Matsuki, Y.; Hiramatsu, R.; Yagita, H.; Kakuta, S.; Iwakura, Y.; Makino, H. Vaspin is an adipokine ameliorating ER stress in obesity as a ligand for cell-surface GRP78/MTJ-1 complex. *Diabetes,* **2012**, *61*(11), 2823-2832.
[http://dx.doi.org/10.2337/db12-0232] [PMID: 22837305]

[57] Kant Misra, U.; Gonzalez-Gronow, M.; Gawdi, G.; Wang, F.; Vincent Pizzo, S. A novel receptor function for the heat shock protein Grp78: silencing of Grp78 gene expression attenuates α2M*-induced signalling. *Cell. Signal.,* **2004**, *16*(8), 929-938.
[http://dx.doi.org/10.1016/j.cellsig.2004.01.003] [PMID: 15157672]

CHAPTER 4

Forecasting the Parallel Interaction between Biosurfactants and Neurons: A Challenge for Clinicians

Soumyashree Rout[1], **Srikanta Kumar Sahoo**[1,*] and **Arun Kumar Pradhan**[2]

[1] Department of Neurology, Siksha 'O' Anusandhan (Deemed to be University), Bhubaneswar, Odisha-751003, India

[2] Center for Biotechnology, Siksha 'O' Anusandhan (Deemed to be University), Bhubaneswar, Odisha-751003, India

Abstract: Bacteria that are associated with human health are receiving a growing amount of attention, particularly those that inhabit the body's niches, such as the neural stem, neurons, gastrointestinal tract, skin, vaginal environment, and lungs. Biosurfactants are molecules that are both hydrophobic and hydrophilic, and receive little attention among the secondary metabolites that are released by microorganisms that are associated with human health. Not only do they serve as biosurfactants, but they also have the potential to control the microbiota through their antimicrobial activity and quorum sensing system in the complex human environment. They and the human body as a whole are shielded from microbial and fungal pathogens by these functions. Because of their diverse structures, biological functions, low toxicity, higher biodegradability, and adaptability, biosurfactants are now emerging as promising bioactive molecules. As a result, biosurfactants with antimicrobial activity, which are produced by bacteria that are associated with the human body and are related to everything that humans come into contact with, such as food, beverages, and comestics, are the subject of this comprehensive review.

Keywords: Blood brain barrier, Cytotoxic, Cancer therapy, Neurons, Rhamnolipid.

INTRODUCTION

Biosurfactants (BS) become an excellent alternative to synthetic surfactants because of their low toxicity, pH stability, biodegradability, thermal resistance, and effective critical micelle concentration. Because of its unique surface activity,

---

** **Corresponding author Srikanta Kumar Sahoo:** Department of Neurology, Siksha 'O' Anusandhan (Deemed to be University), Bhubaneswar, Odisha-751003, India; E-mail: drsrikanta25@gmail.com*

Arun Kumar Pradhan and Manoranjan Arakha (Eds.)

it has become an extraordinarily unique biomaterial applied to different sectors like food, agriculture, waste treatment, environmental issues, and most importantly human health.

The basic principle of producing biosurfactants is either by excretion or adhesion of microorganisms to cells. The major functional characteristics of biosurfactants are to expand microbial cells on the insoluble substrates by decreasing the surface tension between the two states, and increasing hydrophobic substrates to the maximum for uptake and metabolism. Microbial cells can uptake insoluble hydrocarbons through cleaving hydrocarbons at the interface stage of liquid water-hydrocarbon or solid. Microbial cells can exceed the extension of their adhesion by facilitating cell surface regarding its hydrophobicity through customizing cell surface components by hydrocarbon-degrading bacteria [1, 2, 3]. The type and concentration of biosurfactant production processes are variable and depend upon the different sources like carbon and nitrogen sources, and the amount of lipophilic substrate and nutrients. Various physical and chemical parameters of oxygen, temperature, pH, aeration and agitation speed also influence the production of biosurfactants [2, 3].

The biosurfactants introduce remarkable physio-chemical properties. The composition of biosurfactants makes them more efficient and a candidate for the development of economically viable bioprocess. Due to diverse structural and chemical properties of biosurfactants, their action mechanisms are considered for use as an emulsifier, or a de-emulsifier, with saturating, expanding and foaming properties (Table **1**). Their advantage can also be reserved and they can be used as detergents in different refined manufacturation processes such as petroleum and petrochemicals, organic chemicals, beauty products and pharmaceutical, foodstuffs and refreshments, mining and metallurgy and organic manures [4]. In medical science, biosurfactants appear as a drug delivery system for enhancing the presence of bioavailability of drugs. In addition to this, microbial surfactants possess anti-bacterial, antifungal, and antiviral properties, which make them helpful to cure several diseases [5, 6].

Table 1. Bioemulsifier biosurfactants and their applications.

High Molecular Weight Biosurfactants	Types of Biosurfactants	Source	Property	Applications	Components	References
Polymeric biosurfactant	Emulsan	*Acinetobacter calcoaceticus*	Bioemulsifier	Cosmetics, food, pharmaceutical & petroleum industry	Lipopolysacharides	[25]

(Table 1) cont.....

| - | Liposan | *Candidalipolytica* | bioemulsifier | Food & cosmetic industries. | Carbohydrate & protein | [25] |
| - | Alasan | *Acinetobacter Radioresistens KA53.* | Bioemulsifier & solubilization activity. | Petroleum industry, cosmetic & pharmaceutical | Anionic polysaccharide & a protein. | [25] |

The human body is very sensitive to toxic substances which are commonly produced by synthetic surfactants resulting in bad health. As biosurfactant chemical composition is unique and rare, it performs on humans as an anti-microbial, anti-adhesive, anti-immunomodulator, and anti-tumor agent. In the human body, biosurfactant plays an important role in molecular recognition phases like signal transduction, cell differentiation, and immune resistance (Table 2). The morphology of a membrane protein can be modified by using biosurfactants' anti-adhesive and anti-microbial characteristics which cause disruption of the membrane as well as leakage of the metabolites (Fig. 1). As a result, decreasing cell adhesion and biofilm formation occur by influencing energy production and converting the lipopolysaccharide system to another stage. BS also stimulates the body for the generation of cationic proteins, lysozyme, and reactive oxygen species (ROS) in response to inflammatory activities. In addition to this, glycosphingolipid (GSL) biosurfactants are fully expressed in the nervous system, comprising up to 12% of the total lipid content of neuronal membranes while in other tissues, it constitutes only 2%. As per a recent review (2016], glucose-6-phosphate, a glucose sugar, is the key factor for the manufacture of biosurfactants carbohydrates *i.e.* found in the hydrophilic part [2].

Table 2. Low molecular weight biosurfactants and their applications in the health area.

Low Molecular Weight Biosurfactants	Types of Biosurfactant	Source	Property	Applications	References
Glycolipids	Rhamnolipids	*Pseudomonas aeruginosa*	Anti-biofilm & antiadhesive agent	Possess significant anti-microbial activity against mycobacterium tuberculosis organisms. (ii) Rhamnolipid opposes several bacterial & yeast strains isolated from trachea oesophageal puncture.	[25, 26, 27]

(Table 2) cont.....

-	Sophorolipids	*Candida batistae, Candida apicola*	Lowering surface tension, degreasing & bioemulsifier.	Work as a moisturizer for the skin. (ii) Applied as an inhibitor against H7402 Liver cancer.	[28, 29]
-	Trehalose lipids	*Rhodococcus erythropolis*	Antiadhesive against *C. albicans*, & *Escheria. coli*	This type of lipid opposes appreciable levels of anti-viral activity against HSV & influenza virus.	[30, 31]
Fatty acid	Phospholipids	*Klebsiella pneumoniae*	Anti-inflammatory, anti-oxidant, & emulsifier	Manage pulmonary innate and adaptive immunity during respiratory infection.	[32]
Lipopeptides	Surfactin	*Bacillus subtilis*	Anti-bacterial, anti-viral, Anti-mycoplasma	This type of biosurfactant possesses antiviral activity against HIV-1 & HSV-1. Anti-tumor activity against Ehrlich ascites carcinoma cells.	[33]
-	Lichenysin	*Bacillus lichenformis*	Anti-bacterial, Anti-adhesive	A chelating agent that might describe the membrane disruption & permeability as the effect of lipopeptides.	[33, 34]
-	Iturin	*Bacillus subtilis*	Anti-biotic, anti-fungal, anti-biofilm.	Anti-microbial & anti-fungal activity against profound mycosis. Can elevate the specific conductance of biomolecular lipid membranes.	[33]

While Sydatk & Wagner researched biosurfactants and established four different pathways for their production namely, (i) biosynthesis of lipids and carbohydrates; (ii) combined synthesis of half carbohydrates and half-lipids, while lipids are dependent on the diameter of the carbon chain that is present in the medium; (iii) synthesis of lipid half and carbon half while half of the carbon part depends upon the substrate used; (iv) synthesis of the carbon and lipid halves where both carbon and lipid became dependent on the substrate [2]. Hence, during microbial fermentation, the chemical properties of carbon sources can allow for altering the biosynthesis of surfactants. Therefore, insufficient biosurfactants in the human body can cause different neurological, immunological, and dermatological diseases. In this review, we will discuss the different effects of biosurfactants on neurons.

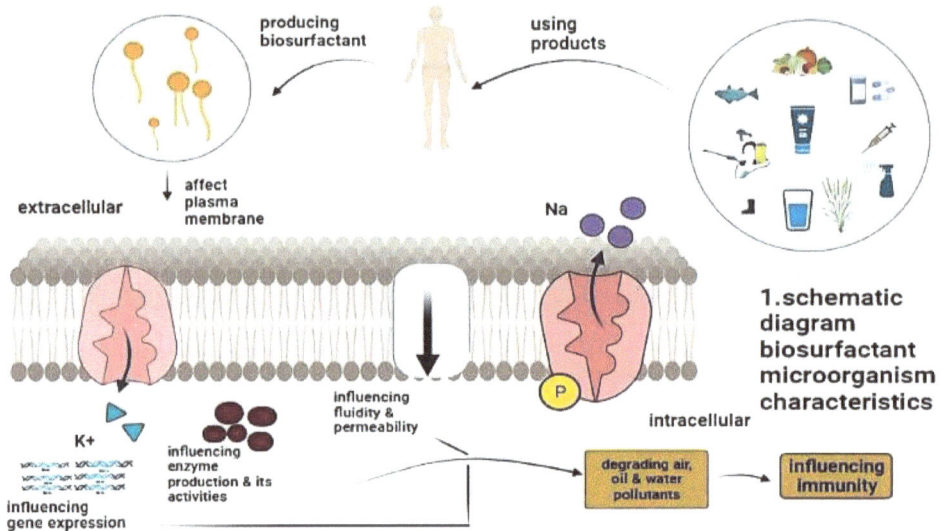

Fig. (1). Function of bio-surfactants targeted with cancer therapy.

POSITIVE EFFECTS OF BIOSURFACTANTS ON NEURONS

In 2013, Dr. Stipcevic researched the effect of di-Rhamnolipid BAC3 on neural stem cells and found that this type of glycolipid not only promotes the proliferation of neural tissue cells but also expands the endogenous pool for which neurodegenerative and neuropsychiatric disorders can be managed through the regeneration of nervous tissues [7].

Staphylococcus aureus, is a bacteria that cause oral infection in humans, resulting in neuropathic pain through activating peripheral sensory neurons. During infections, TRPV1 cation channels are also stimulated by noxious stimuli and cause hyperalgesia [8]. Biosurfactants generated from *Lactobacillus casei* can avoid this severe pain by triggering their microbial activities [9] (Fig. **2**).

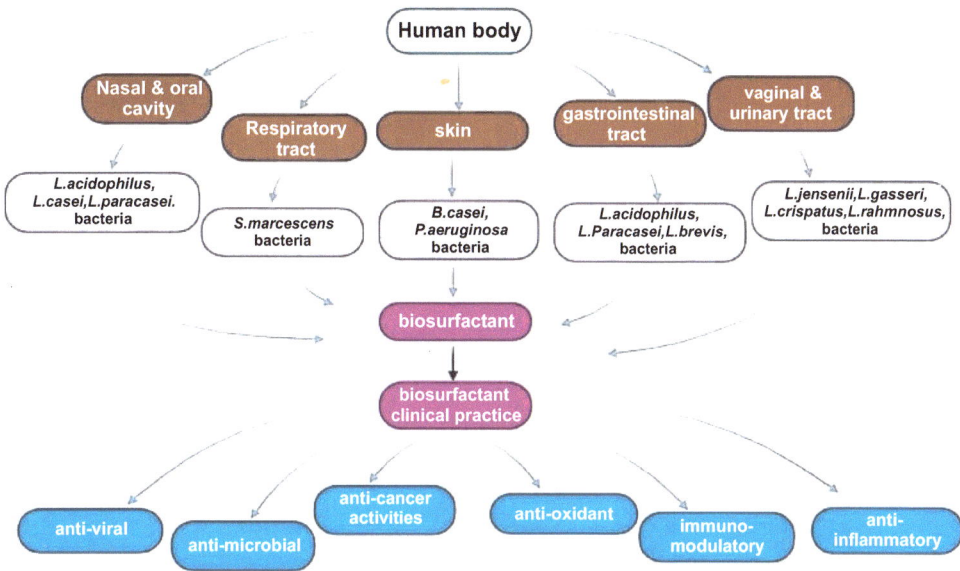

Fig. (2). Diagrammatic presentation of biosurfactants present in the human body and their applications.

Another bacteria named *Cryptococcus* can also cause various disorders like meningitis, encephalitis, skin lesions, rapid visual loss, and prostatic cryptococcosis in humans. According to recent research, the Caenorhabditis elegan model with Parkinsonism after *Cryptococcus* infection successfully showed dopaminergic neuronal degeneration [10]. Cryptococcal infection also affects the brain to cause meningitis by penetrating the BBB (Blood Brain Barrier). BBB prepares a barrier around all the capillaries in the CNS for maintaining a normal neural microenvironment. Pathogens can invade the BBB transcellularly or paracellularly and infect neuronal functions [11,12,13]. Not only *Cryptococcus* species *(Cryptococcus neoformans)*, but other fungal agents like *Histoplasma capsulatum, Coccidioides immitis, Paracoccidioides brasiliensis, Blastomyces dermatitidis, Aspergillus* species and *Zygomycetes* can cause fungal meningitis. This type of fungal pathogen is frequently identified in the respiratory tract of HIV and AIDS patients. Cryptococcal pulmonary infection is established through calcineurin and the protein kinase C1(PKC1]-activated MAP kinase pathway [11, 12]. The calcineurin pathway affects neuronal intracellular calcium concentration, which can also disturb synaptic activities. Based on Dr. Steven's study, surfactant protein A has the ability to resist *Cryptococcus neoformans* by defense mechanisms [14]. Even cellobiose lipids produced from *Sympodiomy Copsispaphiopedili* work effectively against *Cryptococcus terreus* [15].

A study conducted by Lin proposed that infectious keratitis can be caused due to *Pseudomonas aeruginosa*. This bacterial pathogen activates trigeminal ganglia-derived nociceptive sensory neurons, which are dependent on the 3 virulence factors pili, flagella and T3SS (Type III secretory system), which stimulate TRPV1(Transient receptor potential cation channel, subfamily v, member 1) which is decreased, resulting in the development of excitability in the peripheral terminal membrane. When the TRPV1 is activated, then sodium flows through that channel to depolarize nociceptive neurons, and action potential is produced that travels along to the axon for which calcium influxes and CGRP (calcitonin gene-related protein) is released and mediates neurogenic inflammation. As per *P.aeruginosa* infection, the frequency of CGRP increases and affects the concentration of ICAM1 neutrophils in the cornea, leading to visual acuity and even blindness [16]. However, sophorolipids have enough microbial activities to defend against these neuronal problems [15] (Fig. **2**).

Some biosurfactants like MEL-A, MEL-B, polyol lipid, sophorolipids, and succinoyl trehalose lipids STL-1 and STL-3 show anti-cancer activities. According to these biosurfactants' characterizations, all the trial glycolipids induce cell differentiation contrary to cell proliferation in the human promyelocytic leukaemia cell line HL60. Various glycolipids were also tested to know their effects on the initiation of the neurite process in PC12 cells extracted from a rat pheochromocytoma. A notable and remarkable neurite outgrowth was found as a result of inserting MEL-A, MEL-B & SL into PC12 cells. MEL-A biosurfactants not only increase the activities of acetylcholine esterase but also interfere with the continuity of the cell cycle at the G1 stage, which leads to the development of neurite growth and partial cellular differentiation. These research results demonstrated the neuronal differential capability of MEL in PC12 cells as well as introduced it as a reagent for therapeutic activities for cancer treatment [17, 18].

Not only biosurfactants but surfactant proteins SP-A & SP-D are also able to remove various pathogens from the lungs. These proteins can eliminate *Mycobacterium tuberculosis,* which enters the body through inspiratory neurons, and alveolar macrophages, as well as regulate different neuronal and cellular processes for killing this invading pathogen [19]. *Pseudomonas aeruginosa* has the capacity to prohibit surfactant production, lowering host immunity and releasing elastase to decrease surfactant proteins. When lipopolysaccharide (LPS) is released, the SP-A level is increased in human alveolar epithelial cells, while SP-D promotes phagocytosis of *P. aeruginosa*, *Haemophilus influenza*, and *Klebsiella pneumoniae* by the activation of specific ligands [19, 20].

Nanostructured lipid carriers (NLC) that are composed of physiological and biocompatible lipids, surfactants and co-surfactants describe different neuropathways for drug transport directly to the brain *via* the nasal route. For improving the treatment of Alzheimer's disease, two neuro-pathways have been developed: (i) the olfactory nerve pathway and (ii) the trigeminal nerve pathway. The olfactory nerve pathway is separated into two pathways, namely (i) intraneuronal transportation occurs along axons and (ii) extraneuronal transportation occurs through perineural channels. Alzheimer's medicine can be supplied to the brain *via* the trigeminal nerve branches, which innervate the anterior, dorsal and lateral parts of the nasal mucosa. These trigeminal nerve branches go through the pons and get connected with the remaining part of the forebrain and hindbrain. This pathway also supports intracellular (along the axon) as well as extracellular (through the perineural channel) transportation, which is directly connected to the brain tissue and CSF. Another indirect pathway is also present in the respiratory region that supplies the drugs through the bloodstream to the BBB, but crossing the BBB to reach the CNS is very difficult because of the presence of numerous blood vessels in the olfactory and respiratory regions.

Based on Anand's research, sucrose stearate, being a non-ionic surfactant, can be used to develop rivastigmine, which contains nanostructured lipid carriers in *C.elegan* model. Rivastigmine is a cholinesterase inhibitor for better treatment of dementia in people with Alzheimer's disease. It is a class of medication that improves mental functionality. This research suggests that using natural biosurfactants to treat different diseases can avoid neurodegeneration [21].

NEGATIVE EFFECTS OF BIOSURFACTANT ON NEURONS

Very few research works are accessible in regard to biosurfactant toxicity because non-toxic characteristics are found in their products, which can be utilized by pharmaceutical, cosmetic and food applications (Fig. **1**). In comparison with cationic surfactants (Cetyltrimethylammonium bromide, Tetradecylmethy-lammonium bromide, BC) and anionic sodium dodecyl sulfate (SDS), biosurfactants show lower activities of hemolytic to human erythrocyte than synthetic surfactants [22]. Even biosurfactants do not exhibit any harmful and disastrous effects on the heart, lungs, kidneys, or liver and interfere with blood coagulation in normal clotting time.

But still, some biosurfactants possess cytotoxic activities. In 2021, Voulgarido researched the toxicity profile of biosurfactants derived from two novel marine bacterial strains: (a) *MCTG107b* and (b) *MCTG214(3b1),* and found that both biosurfactants possess cytotoxicity at the concentration of up to 0.25 mg/ml on HaCat cells [23]. It has also been proved that in the absence of camptothecin,

T.thermophilus rhamnolipids exert some mutagenic activities on human lymphocytes [23, 24]. As per the recent study, *Pseudomonas aeruginosa B189* Rhamnolipid displayed remarkable cytotoxicity against MCF-7 cells, with a lower inhibitory capacity concentration of 6.25 µg/ml [23].

CONCLUSION AND FUTURE PERSPECTIVES

Biosurfactants are regarded as multifunctional biomolecules because of their widespread applications covering daily life to industrial uses. Many microbial strains have been recognized and analyzed for biosurfactant capacity, and every day, new biosurfactants have been detected. Recognizing the characteristics of certain potential microorganisms that are producing antimicrobial biosurfactants is generally beneficial for human health and is also broadly utilized in the medical field and food industry. Biosurfactants attracted attention as agents for neuronal, metabolic and immune functions. Biosurfactants showed different potentials in the neural network by facilitating antibacterial, antifungal and antiviral activities on various sites of the body. A slight anti-mutagenic effect was demonstrated; however, additional experiments are required to clarify their potential cytoprotective properties.

Clarifying the chemical characteristics of the BSs that have already been extracted but not yet defined would help to increase the opportunities for potential treatment methods. That will further study novel antimicrobial drugs, antioxidant molecules, and antiproliferative agents, a thorough investigation of biosurfactant molecules is also required. This prevents the body from its hazardous side effects and is also economical. However, a substantial study is required for the use of biosurfactants in cancer therapy and drug delivery for the treatment of chronic disorders. Additionally, despite improvements in technology and resource availability, producing biosurfactants at a high cost and producing a small amount of them remain difficult challenges.

REFERENCES

[1]   Whyte, L.G.; Slagman, S.J.; Pietrantonio, F.; Bourbonnière, L.; Koval, S.F.; Lawrence, J.R.; Inniss, W.E.; Greer, C.W. Physiological adaptations involved in alkane assimilation at a low temperature by *Rhodococcus* sp. strain Q15. *Appl. Environ. Microbiol.,* **1999**, *65*(7), 2961-2968.
[http://dx.doi.org/10.1128/AEM.65.7.2961-2968.1999] [PMID: 10388690]

[2]   Santos, D.; Rufino, R.; Luna, J.; Santos, V.; Sarubbo, L. Biosurfactants: Multifunctional Biomolecules of the 21st Century. *Int. J. Mol. Sci.,* **2016**, *17*(3), 401.
[http://dx.doi.org/10.3390/ijms17030401] [PMID: 26999123]

[3]   Sarubbo, L.A.; Silva, M.G.C.; Durval, I.J.B.; Bezerra, K.G.O.; Ribeiro, B.G.; Silva, I.A.; Twigg, M.S.; Banat, I.M. Biosurfactants: Production, properties, applications, trends, and general perspectives. *Biochem. Eng. J.,* **2022**, *181*, 108377.
[http://dx.doi.org/10.1016/j.bej.2022.108377]

[4]   Kourmentza, C.; Freitas, F.; Alves, V.; Reis, M.A.M. Microbial conversion of Waste & surplus

material into high value added products: the case of biosurfactant. *Microbial Applications,* Vol-1, 29-77. **2022**, *1*, 29-77.

[5]   Saha, P.; Nath, D.; Choudhary, M.D.; Talukdar, A.D. Probiotic biosurfactants: A potential therapeutic exercises in biomedical sciences. *Microbial biotechnology,* **2022**, 499-514.

[6]   Smith, M.L.; Gandolfi, S.; Coshall, P.M.; Rahman, P.K.S.M. Biosurfactants: A Covid-19 Perspective. *Front. Microbiol.,* **2020**, *11*, 1341.
       [http://dx.doi.org/10.3389/fmicb.2020.01341]

[7]   Stipcevic, T.; Knight, C.P.; Kippin, T.E. Stimulation of adult neural stem cells with a novel glycolipid biosurfactant. *Acta Neurol. Belg.,* **2013**, *113*(4), 501-506.
       [http://dx.doi.org/10.1007/s13760-013-0232-4] [PMID: 23846482]

[8]   Blake, K.J.; Baral, P.; Voisin, T.; Lubkin, A.; Pinho-Ribeiro, F.A.; Adams, K.L.; Roberson, D.P.; Ma, Y.C.; Otto, M.; Woolf, C.J.; Torres, V.J.; Chiu, I.M. *Staphylococcus aureus* produces pain through pore-forming toxins and neuronal TRPV1 that is silenced by QX-314. *Nat. Commun.,* **2018**, *9*(1), 37.
       [http://dx.doi.org/10.1038/s41467-017-02448-6] [PMID: 29295977]

[9]   Giani.A.D, Zampolli.J, &Gennaro.P.D, Recent trends on Biosurfactants with anti-microbial activity produced by bacteria associated with human health: different perspectives on their properties, challenges & potential applications.. *Front. Microbiol.,* **2021**, *12*, 655150.
       [http://dx.doi.org/10.3389/fmicb.2021.655150] [PMID: 33967992]

[10]   Kitisin, T.; Muangkaew, W.; Sukphopetch, P. Infections of *Cryptococcus* species induce degeneration of dopaminergic neurons and accumulation of α-Synuclein in *Caenorhabditis elegans. Front. Cell. Infect. Microbiol.,* **2022**, *12*, 1039336.
       [http://dx.doi.org/10.3389/fcimb.2022.1039336] [PMID: 36389163]

[11]   Liu, T.B.; Perlin, D.S.; Xue, C. Molecular mechanisms of *cryptococcal meningitis. Virulence,* **2012**, *3*(2), 173-181.
       [http://dx.doi.org/10.4161/viru.18685] [PMID: 22460646]

[12]   Antinori, S. Antinori.S, New insights HIV/AIDS-associated cryptococcosis. *ISRN AIDS,* **2013**, *2013*, 1-22.
       [http://dx.doi.org/10.1155/2013/471363] [PMID: 24052889]

[13]   Casadevall, A.; Coelho, C.; Alanio, A. Mechanisms of *Cryptococcus neoformans* mediated host damage. *Front. Immunol.,* **2018**, *9*, 855.
       [http://dx.doi.org/10.3389/fimmu.2018.00855] [PMID: 29760698]

[14]   Giles, S.S.; Zaas, A.K.; Reidy, M.F.; Perfect, J.R.; Wright, J.R. *Cryptococcus neoformans* is resistant to surfactant protein A mediated host defense mechanisms. *PLoS One,* **2007**, *2*(12), e1370.
       [http://dx.doi.org/10.1371/journal.pone.0001370] [PMID: 18159253]

[15]   Da Silva, A.F.; Banat, I.M.; Giachini, A.J.; Robl, D. Fungal biosurfactants, from nature to biotechnological product: bioprospection, production and potential applications. *Bioprocess Biosyst. Eng.,* **2021**, *44*(10), 2003-2034.
       [http://dx.doi.org/10.1007/s00449-021-02597-5] [PMID: 34131819]

[16]   Lin, T.; Quellier, D.; Lamb, J.; Voisin, T.; Baral, P.; Bock, F.; Schönberg, A.; Mirchev, R.; Pier, G.; Chiu, I.; Gadjeva, M. *Pseudomonas aeruginosa*-induced nociceptor activation increases susceptibility to infection. *PLoS Pathog.,* **2021**, *17*(5), e1009557.
       [http://dx.doi.org/10.1371/journal.ppat.1009557] [PMID: 33956874]

[17]   Fakruddi.J, Biosurfactant: Production & application.. *J. Pet. Environ. Biotechnol.,* **2012**, *3*, 4.

[18]   Rodrigues, L.R.; Teixeira, J.A. Biomedical and therapeutic applications of biosurfactants. *Adv. Exp. Med. Biol.,* **2010**, *672*, 75-87.
       [http://dx.doi.org/10.1007/978-1-4419-5979-9_6] [PMID: 20545275]

[19]   Schob, S.; Schicht, M.; Sel, S.; Stiller, D.; Kekulé, A.; Paulsen, F.; Maronde, E.; Bräuer, L. The detection of surfactant proteins A, B, C and D in the human brain and their regulation in cerebral

infarction, autoimmune conditions and infections of the CNS. *PLoS One,* **2013**, *8*(9), e74412.
[http://dx.doi.org/10.1371/journal.pone.0074412] [PMID: 24098648]

[20]   Han, S.; Mallampalli, R.K. The role of surfactant in lung disease & host defense against pulmonary infections. *Ann. Am. Thorac. Soc.,* **2015**, *12*(5), 765-774.
[http://dx.doi.org/10.1513/AnnalsATS.201411-507FR] [PMID: 25742123]

[21]   Anand, A.; Arya, M.; Kaithwas, G.; Singh, G.; Saraf, S.A. Sucrose stearate as a biosurfactant for development of rivastigmine containing nanostructured lipid carriers and assessment of its activity against dementia in C. elegans model. *J. Drug Deliv. Sci. Technol.,* **2019**, *49*, 219-226.
[http://dx.doi.org/10.1016/j.jddst.2018.11.021]

[22]   Cunha, S.; Forbes, B.; Sousa Lobo, J.M.; Silva, A.C. Improving drug delivery for Alzheimer's disease through the nose to brain delivery using nanoemulsions & nanostructured lipid carriers(NLC) and *in situ*-hydrogels. *Int. J. Nanomedicine,* **2021**, *16*, 4373-4390.
[http://dx.doi.org/10.2147/IJN.S305851] [PMID: 34234432]

[23]   Klosowska-Chomiczewska, I.E.; Medrzycka, K.; Karpenko, E. Biosurfactants-Biodegradability, toxicity, efficiency in comparison with synthetic surfactants **2011**.

[24]   Voulgaridou, G.P.; Mantso, T.; Anestopoulos, I.; Klavaris, A.; Katzastra, C.; Kiousi, D.E.; Mantela, M.; Galanis, A.; Gardikis, K.; Banat, I.M.; Gutierrez, T.; Sałek, K.; Euston, S.; Panayiotidis, M.I.; Pappa, A. Toxicity profiling of biosurfactants produced by novel marine bacterial strains. *Int. J. Mol. Sci.,* **2021**, *22*(5), 2383.
[http://dx.doi.org/10.3390/ijms22052383] [PMID: 33673549]

[25]   Gerard, J.; Lloyd, R.; Barsby, T.; Haden, P.; Kelly, M.T.; Andersen, R.J. Massetolides A-H, antimycobacterial cyclic depsipeptides produced by two pseudomonads isolated from marine habitats. *J. Nat. Prod.,* **1997**, *60*(3), 223-229.
[http://dx.doi.org/10.1021/np9606456] [PMID: 9157190]

[26]   Rodrigues, L.R.; Banat, I.M.; Mei, H.C.; Teixeira, J.A.; Oliveira, R. Interference in adhesion of bacteria and yeasts isolated from explanted voice prostheses to silicone rubber by rhamnolipid biosurfactants. *J. Appl. Microbiol.,* **2006**, *100*(3), 470-480.
[http://dx.doi.org/10.1111/j.1365-2672.2005.02826.x] [PMID: 16478486]

[27]   Maier. R, Soberon-chavez. G, *Pseudomonas aeruginosa* rahmnolipids: biosynthesis & potential application.. *Appl. Microbiol. Biotechnol.,* **2000**, *54*, 625-633.
[http://dx.doi.org/10.1007/s002530000443] [PMID: 11131386]

[28]   Oliveira, M.R.D; Magri, A; Baldo, C; Camilios-Neto, D Review: Sophorolipids A promising biosurfactant & its applications. *Int. J. Adv. Biotechnol. Res.,* **2015**, *6*, 161-174.

[29]   Cho.W.Y, Ng.J.F, Yap.W.H, Goh.B.H, Sophorolipids – biobased antimicrobial formulating agents for applications in food & health.. *Molecules,* **2022**, *27*(17), 5556.
[http://dx.doi.org/10.3390/molecules27175556] [PMID: 36080322]

[30]   Uchida, Y.; Misava, S.; Nakahara, T. Factors affecting the formation of succinyl trehalose lipids by *Rhodococcus erythropolis* grown SD-74 on n-alkanes. *Agric. Biol. Chem.,* **1989**, *53*, 765-769.

[31]   Uchida, Y.; Tsuchiya, R.; Chino, M. Extracellular accumulation of mono & di succinyl trehalose lipids by strain *Rhodococcus erythropolis* grown on n-alkanes. *Agric. Biol. Chem.,* **1989**, *53*, 757-763.

[32]   Shaikh, S.R.; Fessler, M.B.; Gowdy, K.M. Role for phospholipid acyl chains and cholesterol in pulmonary infections and inflammation. *J. Leukoc. Biol.,* **2016**, *100*(5), 985-997.
[http://dx.doi.org/10.1189/jlb.4VMR0316-103R] [PMID: 27286794]

[33]   Rodrigues.L, Banat.I.M, Teixeira.J, Oliveira.R, Biosurfactants: applications in medicine.. *J. Antimicrob. Chemother.,* **2006**, *57*(4), 609-618.
[PMID: 16469849]

[34]   Grangemard, I.; Wallach, J.; Maget-Dana, R.; Peypoux, F. Lichenysin: A more efficient cation chelator than surfactin. *Appl. Biochem. Biotechnol.,* **2001**, *90*(3), 199-210.
[http://dx.doi.org/10.1385/ABAB:90:3:199] [PMID: 11318033]

CHAPTER 5

Application of Biosurfactant in Agriculture

Swapnashree Satapathy[1,*], **Ananya Kuanar**[1] and **Arun Kumar Pradhan**[1]

[1] Centre for Biotechnology, Siksha 'O' Anusandhan (Deemed to be University), Bhubaneswar, Odisha-751003, India

Abstract: All countries are concerned about meeting the growing demands of the human population in terms of agricultural output in a timely manner. Biosurfactants are substances that bacteria, yeasts and fungus are said to create as green surfactants which are less harmful and environmentally compassionate. Several forms of biosurfactants could be commercialised for use in the pharmaceutical, cosmetics, and food industries. Surface active molecules are frequently utilised in agricultural soil remediation to improve soil quality. In recent years, the chemical compounds have gotten a lot of attention because they're seen as a viable and environmentally benign alternative to traditional remedial solutions. The bio molecules which showed the potential to replace the harsh surfactants currently utilised in the multibillion-dollar pesticide industry. The microbial population is screened for biosurfactant synthesis using traditional methods. The modern civilization is confronted with a number of issues in terms of enforcing environmental protection, implementation and addressing climate change for future generation. As a result, studies on environmental safety and human are being carried out in order to improve the efficiency of sustainable environmental restoration methods. Cosmetics, medicines, food, petroleum, agriculture, textiles and wastewater treatment are just a few of the areas where biosurfactants have been shown to be successful and efficient. Improved plant pathogen management, antibacterial activity, antibiofilm activity, seed protection and fertility, wound healing and dermatological care, drug delivery systems, and anticancer treatments are just a few of the applications for microbial produced biosurfactants. This study emphasizes the widespread utilize of harsh surfactants in the agrochemical industry and agricultural soil. More research is needed to determine the possible relevance of biosurfactants produced from environmental isolates in plant growth improvement and other agricultural applications.

Keywords: Agriculture, Antibiofilm, Antimicrobial activity, Biosurfactant, Plant pathogen, Remediation.

---

* **Corresponding author Swapnashree Satapathy:** Centre for Biotechnology, Siksha 'O' Anusandhan (Deemed to be University), Bhubaneswar, Odisha-751003, India; E-mail: satapathyswapnashree2@gmail.com

Arun Kumar Pradhan and Manoranjan Arakha (Eds.)

INTRODUCTION

Biosurfactants are amphiphilic compounds formed on living surfaces, released extracellular non-polar and polar moieties such as microbial cell surfaces or allowed fluid phases to mix while lowering surface and interfacial tension [1]. They have high surface activity, specificity, biodegradability and are naturally benign or least hazardous and may be recycled [2-4]. Pharmaceuticals, cosmetics and personal care items are just a few of the businesses that use them [5]. However, because of their refractory and continuous nature, which is chemically manufactured, most of these substances may pose environmental and toxicological risks. Zhang *et al.* proposed that, biosurfactants are also utilized to increase the steadiness of micro bubble technology, illness detection, molecular imaging, cost-effective water purification, sewage treatment, medication and gene delivery system [6]. Several studies have found that utilizing biosurfactants in the soil remediation process improves agricultural soil health. Pesticides that have collected in agricultural soil can also be broken down with the help of biosurfactants [7-9]. Many researchers have focused on developing more environmentally acceptable strategies for producing various types of biosurfactants from microorganisms as a result of recent biotechnology breakthroughs [10]. There have been reports of surfactin-assisted pesticide biodegradation and glycolipid-assisted chlorinated hydrocarbon degradation, validating the biosurfactant biodegradation accelerator capability. But, *Lactobacillus pentosus* biosurfactant decreased octane hydrocarbon from soil by 58.6 percent to 62.86 percent. *Burkholderia* species, which produce biosurfactants and were identified from oil-contaminated soil, have been discovered to be a viable option for pesticide pollution bioremediation [8].

CLASSIFICATION OF BIOSURFACTANT AND ITS BIOLOGICAL ACTIVITY

Based on microbiological origin as well as their composition, biosurfactants are divided into low molecular weight; glycolipids, lipopeptides (LPs) and phospholipids are of major importance, lipopolysaccharides are high molecular weight biosurfactants. There are various different types of biosurfactant groups, as follows:

1. High and low-molecular-weight biosurfactants include glycolipids, lipopolysaccharides, lipopeptides (LPs) phospholipids. Lipopeptides are anticancer, antibacterial, antiviral, specific toxins, immune-modulators and enzyme inhibitors made up of lipid moieties. According to this the bacterial hydrophobicity, the lipopeptide profile differ substantially between iturin A, strains.This strain is lipopeptide in nature and produced by all *Bacillus subtilis*

strains. Polymixin B and daptomycin both microbial produced LP antibiotics, are the most well-studied LPs. Surfactin (SUR), fengycin and iturin have a wide range of applications and the most well-known LPs, [11].

2. Trehalose lipids derived from Mycobacterium and related bacteria. Glycolipids are the most frequent type of biosurfactant. Sophorolipids derived from yeasts being the most successful in terms of surface-active qualities. Rhamnolipids derived from *Pseudomonas* species. Furthermore, the effects of various types of neurite initiation in PC12 cells and microbial extracellular glycolipids were studied [12]. Differentiations of mouse malignant melanoma cells, growth arrest and death have all been linked to glycolipids [13].

3. Rhamnolipids (Fig. **1**) are amphipathic molecules that include both polar and non-polar moieties, allowing them to interfacial tensions and lower surface. They have antimicrobial properties due to their permeabilizing effect, change bacterial cell hydrophobicity and compromise cell surface charge which causes disruption of the bacterial cell plasma membrane [14]. They can also prevent and inhibit the production of biofilms which are more vulnerable to antimicrobial treatments (Cunha CD *et al.*, 2004).

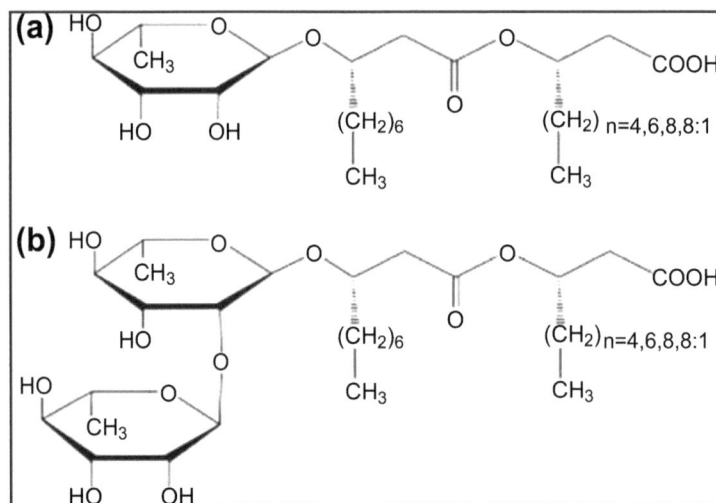

Fig. (1). Mono- and Di-rhamnolipids..

4. Yeasts manufacture sophorolipids. (Fig. **2**) They have a glycosidic connection that connects with two numbers of carbohydrate called sophorose, which has a long-chain hydroxyl fatty acid [15]. Now, it is observed that the spectrum of biosurfactant produced by microorganism. They have extents of bioactivity and quite diverse types to employ highly pure individual of a specific congeners.

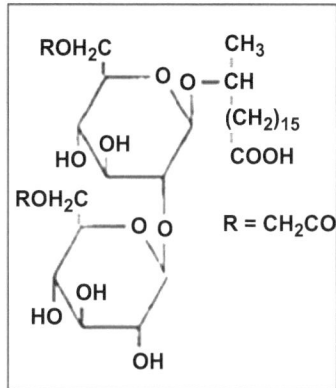

Fig. (2). Sophorolipids.

5. Trehalose lipids (Fig. **3**) are related with most species of *Corynebacterium, Nocardia* and *Mycobacterium*; these are mainly produced by *Rhodococci* and also have physiochemical and biological properties [16]. Trehalolipids from different organisms differ in the the number of carbon atoms, size, the degree of unsaturation and structure of mycolic acid.

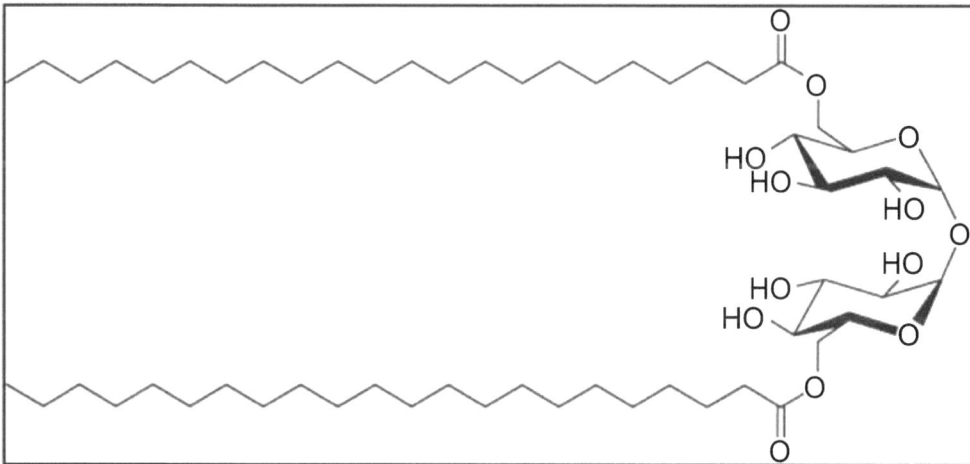

Fig. (3). Trehalose lipids.

6. The most powerful lipopeptide biosurfactants are composed of seven amino acid ring structure which are coupled to a fatty acid chain *via* lactones linkage (Amani H *et al.*, 2010). The inactivation of herpes and retrovirus were also observed with surfactin (Fig. **4**).

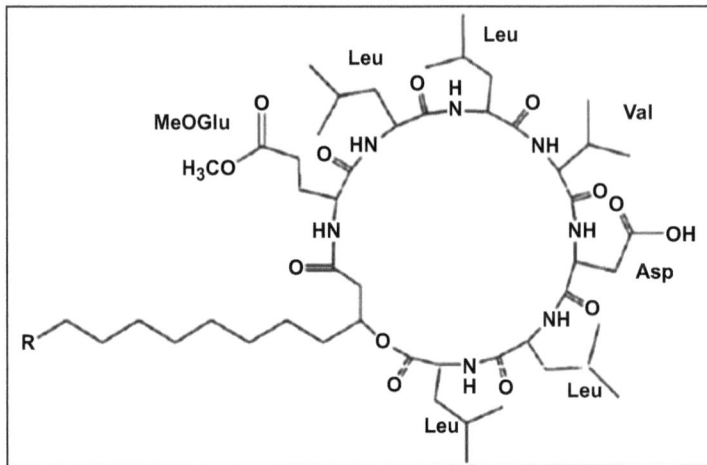

Fig. (4). Surfactin.

7. Lichenysin: Several biosurfactants incorporate with *Bacillus licheniformis* that are synergistic and exhibit extreme temperature, salt, and pH sensitivity. They're also comparable to surfactin in terms of auxiliary and physio-synthetic features.

8. Phospholipids, Neutral lipids and Fatty acids: During growth and development on n-alkanes, some microorganisms, such as yeast and bacteria produces enormous amounts of phospholipids and fatty acids surfactants. These are the most common biosurfactants used in therapeutics. These are: a. Fatty acid (made by *Corynebacterium lepus*); b. Fatty acid (produced by *Corynebacterium lepus*); c. Fatty acid (generated *Nocardia erythropolis* produces neutral lipids. *Thiobacillus thio-oxidans* produces phospholipids.

METHODS AND TECHNIQUES USED FOR BIOSURFACTANT PRODUCTION AND ITS CHARACTERIZATION

The traditional approaches for screening microorganisms for biosurfactant synthesis have been successfully implemented [17, 18]. The most recommended places for isolating biosurfactant-producing microorganisms are ecological niches polluted with hydrocarbons. A metagenomic strategy for screening unculturable microorganisms for biosurfactant synthesis. Metagenomics is the study of microbial communities' genomics without regard to their culture. Genomics (the full investigation of an organism's genetic material) were used to create the word and Meta-analysis (the statistical technique of statistically combining various analyses) and [19, 20]. As a result, this approach is an effective tool for discovering novel chemicals in uncultured bacteria found in natural environments. Another advantage is that when entire soil DNA is sequenced and cloned, it can be used for research. Researchers can use metagenomics to find genes or operons that code for pathways that could drive the production of complicated chemicals like biosurfactants. The genes that code for enzymes or proteins in the biosurfactant synthesis pathway are often clustered in one chromosomal region. A gene cluster of about 3,000–7,000 kb contains a gene relevant to bacterial surfactant production. As a result, novel biosurfactants can be extracted from uncultured bacteria discovered in polluted rhizosphere and agricultural soil using a metagenomic method. A metagenomic approach is essential for biosurfactant synthesis because the majority of studies on economically significant biosurfactants come from pathogenic bacterial strains and there is a greater probability of identifying novel biosurfactants using this technique.

High pressure liquid chromatography, thin layer chromatography and phase separation technology are all approaches for purifying biosurfactants [21, 22]. But the nuclear magnetic resonance, gas chromatography, mass spectrometry, Infrared and fast atom bombardment mass spectrometry are used to characterize the biomolecule [23]. Automation and miniaturization are being used to develop high throughput systems for screening biosurfactant producers (Walter *et al.*, 2010). Biosurfactants have recently been detected and separated using MALDI-TOF mass spectrometry [24].

FACTORS AFFECTING BIOSURFACTANT PRODUCTION

The nature of the nitrogen source, carbon source and nutritional constraints, chemical and physical parameters such as temperature, aeration, divalent cations and pH influence not only the amount of biosurfactant produced but also its composition and emulsifying activity [24]. The following are some of the elements that influence biosurfactant production:

Carbon Sources

The nature of the carbon substrate has an impact on the quality and quantity of biosurfactant production. Carbon substrates for biosurfactant generation have been reported to include diesel, crude oil, glucose, sucrose and glycerol [17].

Nitrogen Sources

It plays a vital role for biosurfactant production medium and for microbial development, protein and enzyme synthesis which are necessary for this sources.

Salt Concentration

It has a similar influence on biosurfactant generation in a specific medium because the cellular activities of microorganisms are impacted by it.

Environmental Factors

These are crucial in influencing the biosurfactants production and characteristics.

Aeration and Agitation

These are major influences on biosurfactant synthesis because they help both oxygen flows from the gas phase to the aqueous phase.

Physical Factors

There are several factors such as pH, temperature, divalent cations, aeration determine not only the amount of biosurfactant generated but also the emulsifying activity [25].

POTENTIAL OF BIOSURFACTANTS IN PESTICIDE INDUSTRIES

In addition, these surfactants have defensive properties and are utilized as pesticides in modern agriculture [26]. Adjuvants are necessary surfactants such as fungicides, insecticides and herbicides. Surfactants of various sorts such as cationic, anionic, nonionic and amphoteric are being used in the pesticide production industry. However, it is vital to keep in mind that soil has an impact on the plant's texture, colour, and growth. These Pesticides which are dangerous to humans and the environment are also leached from the soil water from the ground [27]. These can even be seen on the outside of fruits and vegetables.

Many firms can create good agricultural formulations by combining biosurfactants in various combinations with polymers; nevertheless, agrochemical manufacturers must establish effective formulation technology. Many plant-associated microbes and rhizosphere produces biosurfactant, suggesting that it may play a role in plant-microbe interactions and it could be used in agriculture [28]. Traditional methods for screening microbes for biosurfactant synthesis have proven to be effective. The most recommended places for isolating biosurfactant-producing microorganisms are ecological niches polluted with hydrocarbons.

ANTIFUNGAL AND ANTIMICROBIAL PROPERTIES OF BIOSUR-FACTANTS

In view of the increase in antibiotic resistance, it's become clear that finding new antimicrobials technique are to be regenerated currently for antibiotics preparation which are essential for human being. Inhibition of fungal and bacterial growth is one of the best properties of biosurfactants [29-31]. In *Bacillus pumilus*, rhamnolipid has been demonstrated to divide biofilm activity [32]. Biosurfactants from *Staphylococcus saprophyticus* SBPS 15 were discovered to be antibacterial against *E. coli, Vibrio cholerae, Klebsiella pneumonia, Bacillus subtilis* and *Staphylococcus aureus. Listeria monocytogenes* can be controlled with the biosurfactant SUR. *Listeria monocytogenes* and some Gram-positive bacteria including *M. flavus* and *B. pumulis* can be controlled by biosurfactant SUR in food [33]. Negatively charged cell membranes can be damaged and penetrated by LPs. The cell surface contact has been suggested that the polar element aiming to preset a charge imbalance. Biosurfactant plays a significant role in antiadhesive properties which are especially trying to avoid biofilm formation [34]. Both harmful and nonpathogenic microorganisms rely on biofilm development to survive (Kim K *et al.*, 1998). As previously stated, the focus of study has shifted away from biosurfactants' possible antibacterial properties. In the context of antibiotic resistance, these may include antibacterial adjuvant and inhibitory actions against diverse microorganisms [35, 36].

IMPROVEMENT OF SOIL QUALITY

In case of clean up combined technology and soil washing technology, Biosurfactants are used in various technologies to be successfully removed hydrocarbon and metal [37-39]. The occurrence of organic (living) and inorganic (non-living) contaminants produces abiotic stress to the farmed crop plants and decreased the productivity in agricultural land. The value of such soil tainted by heavy metals and hydrocarbons must be improved through bioremediation. Pollutant desorption by hydrophobic pollutants is a key phenomenon that is strongly bound to soil particles. But, biosurfactant made from microorganisms can be used to effectively remove hydrocarbons [40].

A recent study looked into the role of biosurfactant-producing bacteria in heavy metal bioremediation and hydrocarbon contamination.The following are the details: Several studies using *Acinetobacter sp*; *Bacillus sp., Pseudomonas* sp. have been published. Heavy metals such as Li, Mn, Ba, Cu and Zn (ions), Mg, Cd, Ca, Ni have been reported to be removed from contaminated soil utilising biosurfactants such as rhamnolipid and surfactin [41, 42]. As a result, they have the potential to be the most advantageous for bioremediation as overproducers of harmful synthetic surfactants [43]. In this review, several supplements such as cyclodextrin were also mentioned as aiding in the breakdown of hydrocarbons by soil bacteria [44]. Surfactants, on the other hand, are needed at higher concentrations and have a negative influence on the environment due to biodegradation by microorganisms [45].

PLANT PATHOGEN ELIMINATION

Antibacterial activity of bacteria-derived biosurfactants against plant diseases makes them a feasible biocontrol molecule for achieving plant pathogen power.

Rhizobacteria produce biosurfactants with recognized antagonistic properties [46]. Parasitism, antibiosis, competition, induced systemic resistance, and hypovirulence are some of the biocontrol methods employed by plant growth-promoting bacteria when chemical surfactants and biosurfactants are utilized in agriculture (Singh *et al.*, 2009). Surfactants are employed in agriculture in large quantities to increase microbial products and the antagonistic activities of microorganisms (Kim *et al.*,1998). Surfactants have been shown to improve the insecticidal activities of other systems in a number of *in vitro* and *in situ* experiments [47, 48].

ADVANTAGES OF AGRICULTURAL MICROBES

Many researchers have established that biosurfactants are more efficient than synthetic surfactants at removing organic insoluble contaminants from soil [47]. Iron is required by *Pseudomonas* sp. for improved biosurfactant synthesis and poly aroma bioavailability. Poly aromatic hydrocarbons and pentachlorophenol have been discovered to be eliminated from soil by rhamnolipids [41]. Biosurfactants produced at a high cost are employed in the bioremediation of petroleum-contaminated soil or crude oil . However, at greater concentrations, it can cause root tissue necrosis and foliage, both of which are detrimental to plant growth.

Another benefit of metagenomics is that it allows for the sequencing and cloning of the entire set of soil DNA, allowing for the capture of genes or operons encoding specific mechanisms that could control the manufacture of complex compounds.The genes which rules for enzymes/ proteins concerned in the biosurfactant production pathway are typically grouped at one chromosomal section. A gene cluster of about 3,000–7,000 kb contains a gene relevant to bacterial surfactant production. As a result, the metagenomic technique can be used to obtain novel biosurfactants from bacteria that have never been grown contaminated agricultural soil and rhizosphere. Before, restriction enzymes are put into an appropriate expression vector, it should be used to fragment the entire DNA retrieved. All metagenomic clones needs to be tested for biosurfactant synthesis by using standard and/or high throughput methods but the DNA construct should be converted into an *E.coli* host.

ADVANTAGES OF BIOSURFACTANTS

According to physical, chemical manufactured equivalents, biosurfactants have several varieties of advantages respectively. They are listed below as follows:

Raw Material Availability

Biosurfactants can be made from relatively inexpensive raw materials that are readily available in huge numbers [49].

Surface and Interface Activity

According to Mulligan, an effective surfactant may reduce water surface tension from 75 to 35 mN/m [41].

Other Advantages

These are biocompatibility and digestibility which allows their application in cosmetic and pharmaceutical as functional food additives.

DISADVANTAGES OF BIOSURFACTANTS

Despite the numerous benefits, there are some drawbacks as well:

Maintenance of Aseptic Condition

Because sterile media is required for biosurfactant production, large-scale biosurfactant manufacture is a complex and costly procedure.

Poor Yield from Raw Substrate Material

Raw substrate materials generate less yield than treated substrate materials during the biosurfactant synthesis process. The manufacture of biosurfactant on raw substrate takes longer.

Problems in Product Recovery and Purification

Because numerous consecutive purification procedures are necessary for the metabolic broth, obtaining products with a high degree of purity is difficult.

Problem in Control of the Process Like Foaming

The creation of foam is complicating the process of increasing output. It is necessary to utilize a diluted medium to solve the problem.

Lack of Knowledge Regarding the Biosurfactant Producing Microbes

Microbial species that produce a lot of food are quite rare. Surfactant yields are not capable of being produced in significant quantities by known microbial species, and it also necessitates a complicated growth medium. Exploration of possible microorganisms for biosurfactant synthesis is a current demand in this domain. Whether these biomolecules are created as secondary metabolites or in conjunction with microbial growth, the regulation of biosurfactant synthesis is still unknown.

Adverse Effects of Pesticides

The harmful effects of pesticides and surfactants, which were formerly connected with pesticides, must be replaced. As a result, these toxic surfactants are not used in a multi-million dollar pesticide contamination [50]. *Pseudomonas sp.* and *Burkholderia sp.* bacteria found in rice fields have been identified to break down surfactants, according to a study [51].

AGRICULTURE RELATED APPLICATIONS OF BIOSURFACTANTS

In today's world, To meet upon the raising agricultural production the world's ever-growing food demands is a top priority for all countries. In April 2009, emphasizing the significance of restoring native soil systems for increased crop productivity, The Economic and Social Commission for Asia and the Pacific (ESCAP) of the United Nations published a theme study titled "Sustainable Agriculture and Food Security in Asia and the Pacific". Despite the fact that biosurfactant-producing bacteria, particularly *Pseudomonas* sp. *Acinetobacter* sp. *Candida* sp. and *Bacillus* sp. have been the subject of several patents (Simpson DR). A number of rhizobacteria-related processes that aid in plant growth promotion. According to Singh *et al.*, biosurfactants generated from microbial sources have both hydrophobic (non-polar) and hydrophilic (polar) properties. Because soil-dwelling bacteria use them as a carbon source, these biosurfactants can also replace the harsh surfactants currently employed in pesticide businesses [52]. The following are some other health-related applications:

APPLICATION IN WOUND HEALING

In vitro wound healing and antioxidant activity of biosurfactants derived from *B. subtilis* SPB1 LP on experimental rat excision wounds were examined . SPB1 LPs completely re-epithelized wounds with flawless epidermal re-growth in biopsies.

APPLICATION IN DERMATOLOGICAL CARE

In contrast, Biosurfactants are made up of proteins and lipids that are compatible with the skin's membrane. While the majority of biosurfactant research has focused on extracellular biosurfactants produced by microorganisms, the cell-bound biosurfactants such as those produced by proboscis bacteria have received far less attention. Synthetic antibacterial preservatives are found in the majority of personal care products and their interaction with collagen, keratin and elastin can cause skin irritation and allergic reactions, as well as encourage the removal of lipids from the skin's surface and change the skin cells themselves.

FUTURE ASPECTS AND CONCLUSION

Finally, biosurfactants have been proven to be an environmentally benign product that may be used in medical applications such as healthcare and disease treatment. It is widely used in biodegradation of oils, environmental pollution control, oil fields, cosmetics, the oil industry, oil pollution removal, agriculture and agrochemicals. It's simple to make utilizing low-cost agricultural wastes and oil as a substrate. As a carbon source, biosurfactant manufacturing can be achieved using waste oils such as edible or motor oil. When compared to a single microbe, bacterial consortia can be more efficient at producing biosurfactant. These biosurfactants can also be utilized as biocontrol agents because they exhibit antibacterial properties. Hours are required to investigate prospective microorganisms for its production, as well as the standardization of the best substrate and acceptable methods. Agricultural wastes can be utilized to make biosurfactants as well. The biosurfactant's emulsifying activity is another branch with a wide range of applications. They can be utilized as emulsion forming agents for hydrocarbons and oils, resulting in stable emulsions. Despite the fact that biosurfactant has so many advantages, its industrial application is currently limited due to the high cost of its production process. Because of their non-toxic and environmentally benign nature, biosurfactants have sparked a lot of interest for current and future applications.

Surfactants are employed in a variety of ways in agriculture and the agrochemical sector. The use of agricultural waste for biosurfactant overproduction also requires greater consideration. The chemical compositions of biosurfactants that have been described as effective biocontrol agents can be changed by altering the manufacturing process. The significant frequency of biosurfactants and biosurfactant-producing bacteria in the rhizosphere is evidence of their importance in sustainable agriculture. Biosurfactant producers are largely *Pseudomonas* and *Bacillus* species in the literature, indicating that only a few genera have been investigated.This could result in the identification of new green surfactants. As a result, it is clear that a collaborative effort including experts from many domains such as biochemistry, microbiology and molecular biology. It has also endowed humans with a variety of useful or future qualities. Biosurfactants have long been known for their antibacterial properties, as well as evidence of their benefits in wound healing, dermatological and agricultural application is growing.

REFERENCES

[1]   Cunha, C.D.; Rosario, D.o.; Rosado, A.S.; Leite, S.G. Serratia sp. SVGG16: a promising biosurfactant producer isolated from tropical soil during growth with ethanol-blended gasoline. In: *Proc Biochem*; , **2004**; pp. 2277-2282.

[2]   Xu, Q.; Nakajima, M.; Liu, Z.; Shiina, T. Biosurfactants for microbubble preparation and application. *Int. J. Mol. Sci.,* **2011**, *12*(1), 462-475.
[http://dx.doi.org/10.3390/ijms12010462] [PMID: 21339998]

[3]   Lima, T.M.; Procópio, L.C.; Brandão, F.D.; Carvalho, A.M.; Tótola, M.R.; Borges, A.C. Biodegradability of bacterial surfactants. In: *Biodegrad*; , **2011**; pp. 585-592.

[4]   Koglin, A.; Doetsch, V.; Bernhard, F. Molecular engineering aspects for the production of new and modified biosurfactants. In: *Biosurfactants*; , **2010**; pp. 158-169.
[http://dx.doi.org/10.1007/978-1-4419-5979-9_12]

[5]   Das, P.; Mukherjee, S.; Sivapathasekaran, C.; Sen, R. Microbial surfactants of marine origin: potentials and prospects. In: *Biosurfactants*; , **2010**; pp. 88-101.

[6]   Zhang, F.; Gu, W.; Xu, P.; Tang, S.; Xie, K.; Huang, X.; Huang, Q. Effects of alkyl polyglycoside (APG) on composting of agricultural wastes. *Waste Manag.,* **2011**, *31*(6), 1333-1338.
[http://dx.doi.org/10.1016/j.wasman.2011.02.002] [PMID: 21376559]

[7]   Singh, P.B; Sharma, S.; Saini, H.; Chadha, B.S. **2009**.

[8]   Wattanaphon, H.T.; Kerdsin, A.; Thammacharoen, C.; Sangvanich, P.; Vangnai, A.S. A biosurfactant from *Burkholderia cenocepacia* BSP3 and its enhancement of pesticide solubilization. *J. Appl. Microbiol.,* **2008**, *105*(2), 416-423.
[http://dx.doi.org/10.1111/j.1365-2672.2008.03755.x] [PMID: 18298537]

[9]   Neilson, J.W.; Artiola, J.F.; Maier, R.M. *Characterization of lead removal from contaminated soils by nontoxic soil-washing agents.* **2003**.
[http://dx.doi.org/10.2134/jeq2003.8990]

[10]   Lotfabad, T.B.; Abassi, H.; Ahmadkhaniha, R.; Roostaazad, R.; Masoomi, F.; Zahiri, H.S.; Ahmadian, G.; Vali, H.; Noghabi, K.A. Structural characterization of a rhamnolipid-type biosurfactant produced by *Pseudomonas* aeruginosa MR01: enhancement of di-rhamnolipid proportion using gamma irradiation. In: *Col Surf Biointer*; , **2010**; pp. 397-405.

[11]   Fracchia, L.; Banat, J.J.; Cavallo, M.; Banat, I.M. Potential therapeutic applications of microbial surface-activecompounds. *AIMS Bioeng.,* **2015**, *2*(3), 144-162.
[http://dx.doi.org/10.3934/bioeng.2015.3.144]

[12]   Isoda, H.; Shinmoto, H.; Matsumura, M.; Nakahara, T. The neurite-initiating effect of microbial extracellular glycolipids in PC12 cells. In: *Cytotech*; , **1999**; p. 165.

[13]   Makkar, R.S.; Rockne, K.J. Comparison of synthetic surfactants and biosurfactants in enhancing biodegradation of polycyclic aromatic hydrocarbons. In: *Env Toxi Chem Int J*; , **2003**; pp. 2280-2292.

[14]   Sotirova, A.; Spasova, D.; Vasileva-Tonkova, E.; Galabova, D. Effects of rhamnolipid-biosurfactant on cell surface of *Pseudomonas aeruginosa. Microbiol. Res.,* **2009**, *164*(3), 297-303.
[http://dx.doi.org/10.1016/j.micres.2007.01.005] [PMID: 17416508]

[15]   De Oliveira, M.R.; Magri, A.; Baldo, C.; Camilios-Neto, D.; Minucelli, T.; Celligoi, M.A. Sophorolipids A promising biosurfactant and it's applications. *Int. J. Adv. Biotechnol. Res.,* **2015**, *6*(2), 161-174.

[16]   Lima, T.M.; Procópio, L.C.; Brandão, F.D.; Carvalho, A.M.; Tótola, M.R.; Borges, A.C. Biodegradability of bacterial surfactants. In: *Biodegrad*; , **2011**; pp. 585-592.

[17]   Walter, V.; Syldatk, C.; Hausmann, R. Screening concepts for the isolation of biosurfactant producing microorganisms. In: *Biosurfactants*; , **2010**; pp. 1-3.
[http://dx.doi.org/10.1007/978-1-4419-5979-9_1]

[18]   Satpute, S.K.; Bhuyan, S.S.; Pardesi, K.R.; Mujumdar, S.S.; Dhakephalkar, P.K.; Shete, A.M.; Chopade, B.A. Molecular genetics of biosurfactant synthesis in microorganisms. In: *Biosurfactants*; , **2010**; pp. 14-41.
[http://dx.doi.org/10.1007/978-1-4419-5979-9_2]

[19]   Rondon, M.R.; August, P.R.; Bettermann, A.D.; Brady, S.F.; Grossman, T.H.; Liles, M.R.; Loiacono, K.A.; Lynch, B.A.; MacNeil, I.A.; Minor, C.; Tiong, C.L. *Cloning the soil metagenome: a strategy for accessing the genetic and functional diversity of uncultured microorganisms.* , **2000**.

[20]   Schloss, P.D.; Handelsman, J. Biotechnological prospects from metagenomics. *Curr. Opin. Biotechnol.,* **2003**, *14*(3), 303-310.
[http://dx.doi.org/10.1016/S0958-1669(03)00067-3] [PMID: 12849784]

[21]   Baker, S.C.; Chen, C.Y. Enrichment and purification of lipopeptide biosurfactants. In: *Biosurfactants*; , **2010**; pp. 281-288.

[22]   Heyd, M.; Kohnert, A.; Tan, T.H.; Nusser, M.; Kirschhöfer, F.; Brenner-Weiss, G.; Franzreb, M.; Berensmeier, S. *Development and trends of biosurfactant analysis and purification using rhamnolipids as an example.* , **2008**.
[http://dx.doi.org/10.1007/s00216-007-1828-4]

[23]   Petrović, M.; Barceló, D. Analysis and fate of surfactants in sludge and sludge-amended soils. , **2004**.
[http://dx.doi.org/10.1016/j.trac.2004.07.015]

[24]   Kurtzman, C.P.; Price, N.P.; Ray, K.J.; Kuo, T.M. , **2010**.

[25]   Mulqueen, P. *Recent advances in agrochemical formulation.* , **2003**.
[http://dx.doi.org/10.1016/S0001-8686(03)00106-4]

[26]   Rostás, M.; Blassmann, K. Insects had it first: surfactants as a defence against predators. In: *Proceed Royal Soci Biol Sci*; , **2009**; pp. 633-638.

[27]   Blackwell, P.S. Management of water repellency in Australia, and risks associated with preferential flow, pesticide concentration and leaching. *J. Hydrol. (Amst.),* **2000**, *231-232*, 384-395.
[http://dx.doi.org/10.1016/S0022-1694(00)00210-9]

[28]   Singh, P.B.; Sharma, S.; Saini, H.S.; Chadha, B.S. , **2009**.

[29]   Kim K, Jung SY, Lee DK, Jung JK, Park JK, Kim DK, Lee CH, et al. Suppression of inflammatory responses by surfactin, a selective inhibitor of platelet cytosolic phospholipase A2. *Biochem pharmaco,* **1998**; *55*(7): 975-85.

[30]   Díaz De Rienzo, M.A.; Stevenson, P.S.; Marchant, R.; Banat, I.M. Pseudomonas aeruginosa biofilm disruption using microbial surfactants. In: *J App Microbiol*; , **2016**; pp. 868-876.

[31]   Lotfabad, T.B.; Abassi, H.; Ahmadkhaniha, R.; Roostaazad, R.; Masoomi, F.; Zahiri, H.S.; Ahmadian, G.; Vali, H.; Noghabi, K.A. Structural characterization of a rhamnolipid-type biosurfactant produced by *Pseudomonas* aeruginosa MR01: enhancement of di-rhamnolipid proportion using gamma irradiation. In: *Col Surf Biointer*; , **2010**; pp. 397-405.

[32]   Dusane, D.H.; Nancharaiah, Y.V.; Zinjarde, S.S.; Venugopalan, V.P. Rhamnolipid mediated disruption of marine *Bacillus pumilus* biofilms. *Colloids Surf. B Biointerfaces,* **2010**, *81*(1), 242-248.
[http://dx.doi.org/10.1016/j.colsurfb.2010.07.013] [PMID: 20688490]

[33]   Sabaté, D.C.; Audisio, M.C. Inhibitory activity of surfactin, produced by different *Bacillus subtilis* subsp. subtilis strains, against Listeria monocytogenes sensitive and bacteriocin-resistant strains. *Microbiol. Res.,* **2013**, *168*(3), 125-129.
[http://dx.doi.org/10.1016/j.micres.2012.11.004] [PMID: 23265790]

[34]   Galie, S.; García-Gutiérrez, C.; Miguélez, E.M.; Villar, C.J.; Lombó, F. *Biofilms in the food industry: health aspects and control methods.* , **2018**.

[35]   Borsanyiova, M.; Patil, A.; Mukherji, R.; Prabhune, A.; Bopegamage, S. Biological activity of

sophorolipids and their possible use as antiviral agents. *Folia Microbiol. (Praha)*, **2016**, *61*(1), 85-89.
[http://dx.doi.org/10.1007/s12223-015-0413-z] [PMID: 26126789]

[36]   Fracchia, L.; Cavallo, M.; Martinotti, M.G.; Banat, I.M. *Biosurfactants and bioemulsifiers biomedical and related applications–present status and future potentials.* , **2012**.
[http://dx.doi.org/10.5772/23821]

[37]   Coppotelli, B.M.; Ibarrolaza, A.; Dias, R.L.; Del Panno, M.T.; Berthe-Corti, L.; Morelli, I.S. Study of the degradation activity and the strategies to promote the bioavailability of phenanthrene by Sphingomonas paucimobilis strain 20006FA. *Microb. Ecol.*, **2010**, *59*(2), 266-276.
[http://dx.doi.org/10.1007/s00248-009-9563-3] [PMID: 19609598]

[38]   Pacwa-Płociniczak, M.; Płaza, G.A.; Piotrowska-Seget, Z.; Cameotra, S.S. Environmental applications of biosurfactants: recent advances. *Int. J. Mol. Sci.*, **2011**, *12*(1), 633-654.
[http://dx.doi.org/10.3390/ijms12010633] [PMID: 21340005]

[39]   Partovinia, A.; Naeimpoor, F.; Hejazi, P. Carbon content reduction in a model reluctant clayey soil: Slurry phase n-hexadecane bioremediation. *J. Hazard. Mater.*, **2010**, *181*(1-3), 133-139.
[http://dx.doi.org/10.1016/j.jhazmat.2010.04.106] [PMID: 20570040]

[40]   Sun, X.; Wu, L.; Luo, Y. Application of organic agents in remediation of heavy metals-contaminated soil. In: *J Appl Eco*; , **2006**; pp. 1123-1128.

[41]   Mulligan, C.N.; Wang, S. Remediation of a heavy metal-contaminated soil by a rhamnolipid foam. *Eng. Geol.*, **2006**, *85*(1-2), 75-81.
[http://dx.doi.org/10.1016/j.enggeo.2005.09.029]

[42]   Neilson, J.W.; Artiola, J.F.; Maier, R.M. *Characterization of lead removal from contaminated soils by nontoxic soil-washing agents.* **2003**.
[http://dx.doi.org/10.2134/jeq2003.8990]

[43]   Rosenberg, E.; Ron, E.Z. Surface active polymers from the genus Acinetobacter. Biopolymers from renewable resources. , **1998**.

[44]   Bardi, L.; Mattei, A.; Steffan, S.; Marzona, M. *Hydrocarbon degradation by a soil microbial population with β-cyclodextrin as surfactant to enhance bioavailability.* , **2000**.
[http://dx.doi.org/10.1016/S0141-0229(00)00275-1]

[45]   Colores, G.M.; Macur, R.E.; Ward, D.M.; Inskeep, W.P. Molecular analysis of surfactant-driven microbial population shifts in hydrocarbon-contaminated soil. *Appl. Environ. Microbiol.*, **2000**, *66*(7), 2959-2964.
[http://dx.doi.org/10.1128/AEM.66.7.2959-2964.2000] [PMID: 10877792]

[46]   Nihorimbere, V.; Ongena, M.; Smargiassi, M.; Thonart, P. Beneficial effect of the rhizosphere microbial community for plant growth and health. In: *Biotech Agro Soc Env*; , **2011**; p. 15.

[47]   Gronwald, J.W.; Plaisance, K.L.; Ide, D.A.; Wyse, D.L. Assessment of *Pseudomonas syringae* pv. tagetis as a biocontrol agent for Canada thistle. *Weed Sci.*, **2002**, *50*(3), 397-404.
[http://dx.doi.org/10.1614/0043-1745(2002)050[0397:AOPSPT]2.0.CO;2]

[48]   Jazzar, C.; Hammad, E.A. The efficacy of enhanced aqueous extracts of Melia azedarach leaves and fruits integrated with the Camptotylus reuteri releases against the sweetpotato whitefly nymphs. *Bull. Insectol.*, **2003**, *56*(2), 269-276.

[49]   Kosaric, N. Biosurfactants and their application for soil bioremediation. In: *Food Tech Biotech*; , **2001**; pp. 295-304.

[50]   Hopkinson, M.J.; Collins, H.M.; Goss, G.R. Pesticide formulations and application systems: ASTM Committee E-35 on Pesticides. In: *ASTM International*; , **1997**; pp. 1-331.

[51]   Nishio, E.; Ichiki, Y.; Tamura, H.; Morita, S.; Watanabe, K.; Yoshikawa, H. Isolation of bacterial strains that produce the endocrine disruptor, octylphenol diethoxylates, in paddy fields. *Biosci. Biotechnol. Biochem.*, **2002**, *66*(9), 1792-1798.

[http://dx.doi.org/10.1271/bbb.66.1792] [PMID: 12400675]

[52]   Lima, T.M.; Procópio, L.C.; Brandão, F.D.; Carvalho, A.M.; Tótola, M.R.; Borges, A.C. Biodegradability of bacterial surfactants. In: *Biodegrad*; , **2011**; pp. 585-592.

CHAPTER 6

Use of Biosurfactants in Food Processing Technology

Gargi Balabantaray[1], Bhabani Shankar Das[2] and Pradeepta Sekhar Patro[1,*]

[1] *Department of Immunology and Rheumatology, Institute of Medical Science, Sum Hospital, Siksha 'O' Anusandhan (Deemed to be University), Bhubaneswar, Odisha-751003, India*

[2] *Centre for Biotechnology, School of Pharmaceutical Sciences, Siksha 'O' Anusandhan (Deemed to be University), Bhubaneswar, Odisha-751003, India*

Abstract: Biological surfactants are a class of amphipathic biomolecules that contain a diverse range of constituents derived from different biological sources and have been analysed for their ability to lower surface tension. Their distinct properties with cumulative applications have expanded in different fields starting from human health to detergent industry. According to estimates, the global market for biosurfactants will grow from $4.18 billion in 2022 to $6.04 billion by 2029. Biosurfactants outperform artificial surfactants due to their unique attributes. This provides opportunities for commercial utilization of biosurfactants. Thus, the present chapter aims to describe the various biosurfactants present in the market along with their potential application in food industries.

Keywords: Bioemulsifiers, Enhanced oil recovery, Lactic acid bacteria.

INTRODUCTION

Biosurfactants are surface-active amphiphilic biomolecules that are abundantly produced by aerobically growing bacteria, fungi, and yeast. Biosurfactants are made up of a hydrophilic region with polar properties and a hydrophobic region with non-polar properties. The polar region is either anionic or cationic, whereas the non-polar region is either non-ionic or amphoteric, consisting of a hydrophobic chain illustrated in Fig. (**1**). This distinguishing feature of biosurfactants allows them to aggregate and accumulate at the interface of two fluid phases, such as oil in water with two different polarities, lowering interfacial and surface tension and forming emulsions in which hydrocarbons can solubilize. As a result, the monomers can form micelles or aggregate into micellar tubes, bilayers, and vesicles. Biosurfactant production by microorganisms or bacteria

*Corresponding author Pradeepta Sekhar Patro: Department of Immunology and Rheumatology, Institute of Medical Science, Sum Hospital, Siksha 'O' Anusandhan (Deemed to be University), Bhubaneswar, Odisha-751003, India; E-mail: patrodrpradeepta@gmail.com

Arun Kumar Pradhan and Manoranjan Arakha (Eds.)

depends on fermentation conditions, the environment, and carbon and nitrogen availability [1, 2]. Biosurfactants have large, complex structures, high biodegradability, low toxicity, high surface activity, and extreme stability [2]. There are numerous applications for biosurfactants, from everyday chores like laundry and personal hygiene products to the medical sector, where biosurfactants have antimicrobial, antitumor, and anti-inflammatory properties because of their bioactivity. Biosurfactants are also used as food additives, antioxidant agents, improving the texture of certain foods *etc.*, thus making them potential biomolecules to be used in food industries [3].

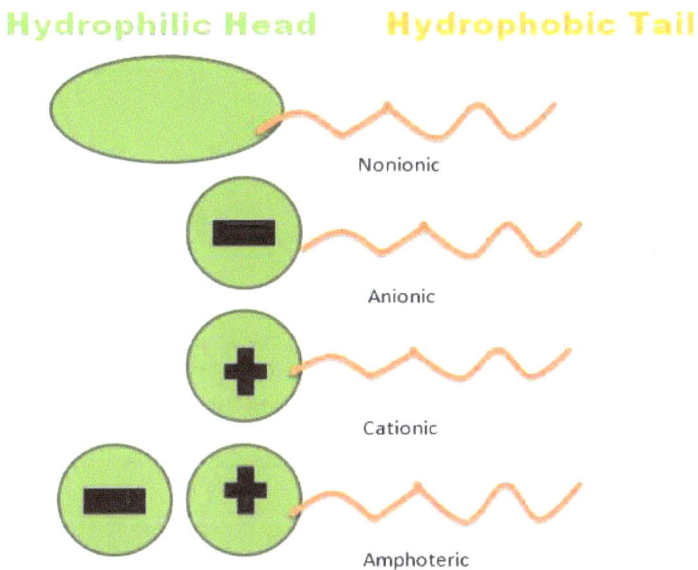

Hydrophilic Head **Hydrophobic Tail**

Nonionic

Anionic

Cationic

Amphoteric

Fig. (1). General structure of a surfactant according to their confirmation.

PROPERTIES OF BIOSURFACTANTS

Biosurfactants have unique and distinguishing properties when compared to synthetic biosurfactants and have sparked widespread commercial interest due to the ever-expanding spectrum of available substances. The following are the distinguishing characteristics of biosurfactants, along with a brief description of each property.

Surface and Interface Activity

Biosurfactants reduce strain and interfacial pressure at lower concentrations, resulting in greater effectiveness and efficiency when compared to conventional surfactants. A lipopeptide surfactant synthesized by *Bacillus subtilis* is able to decrease both surface and interfacial tension of water to 25 mN m^{-1} and water/hexadecane to below 1 mN m^{-1}, respectively [4]. Following the trend, rhamnolipids made by *P. aeruginosa* lowered the surface and interfacial tension of water to 26 mN m^{-1} and water/hexadecane to below 1mN m^{-1}, respectively [5] Biosurfactants are much more capable and effective because their CMC is lower than synthetic surfactants, so less is needed for maximum surface strain reduction [6].

Biodegradability

Compared to synthetic surfactants, microbial compounds are easily degraded and suitable for bioremediation and biosorption [7]. Increasing environmental problems force us to use biosurfactants. Synthetic surfactants pose environmental issues, so biodegradable biosurfactants from marine microorganisms were used to remove phenanthrene from aquatic surfaces [8].

Temperature and pH Tolerance

The lipopeptide from *Bacillus licheniformis* is stable at 75 degrees for 140 hours and within a pH range of 5 to 12 [9, 10]. This is only one example of the wide variety of biosurfactants that may be employed in harsh environments. While conventional surfactants are rendered ineffective at 2% NaCl concentrations, biosurfactants can withstand concentrations of up to 10% salt. Novel microbial items must be separated because industrial procedures involve extremes of heat, pH, and weight.

Low Toxicity

Biosurfactants are low toxic or non-harmful, making them suitable for food, cosmetic, and pharmaceutical industries. A study showed that a chemically-derived surfactant Corexit is 10 times more toxic than rhamnolipids [11]. The toxicity and mutagenicity profile of *P.aeruginosa* biosurfactants was studied and when they were compared to synthetic surfactants, they were found to be nontoxic and non-mutagenic [9, 12].

Emulsion Forming and Breaking

Biosurfactants may destabilize emulsions. An emulsion is a diversified framework that includes one immiscible fluid scattered in 0.1mm-diameter beads [13]. Biosurfactants' minimal stability keeps emulsions stable for years. Liposan can emulsify edible oils without reducing the surface tension. The oil droplets are coated with polymeric surfactants, resulting in stable emulsions [14].

Anti-adhesive Properties

Depending on the type of surface, biosurfactants can form the first layer and serve as wetting agents. A biofilm is a layer of microbes or organic matter that adheres to a surface [15]. The first step in biofilm formation is host adherence to a surface, which is influenced by the type of microorganism, the surface's hydrophobicity, electrical charge, environmental factors, and microorganisms' ability to release extracellular polymers, which aid in cell adhesion. Biosurfactants can change the hydrophobicity of a surface, affecting microorganism adhesion [16].

Anti-microbial Activity

Biosurfactants like surfactin and rhamnolipids have antimicrobial properties. Biosurfactants are used to deliver drugs to infection sites, exhibit potential healing activity and can act as vaccine adjuvants [17, 18].

MECHANISM OF INTERACTION

A biosurfactant is an organic compound that is both hydrophobic and hydrophilic, having tensioactive features. It has a hydrophilic part, which is made up of alcohol, carboxylic acid, phosphate group, cyclic peptide, amino acid, and carbohydrates, and a hydrophobic part, which is made up of α and β alkyl-hydroxy fatty acid, hydroxyl fatty acid, and long-chain fatty acid [19]. Biosurfactants support micelle formation, which reduces both surface and interfacial tension between the two non-miscible liquids. The biosurfactant-induced formation of micelles increases the dissolvability and accessibility of biological compounds, thereby promoting the formation of microemulsions which is depicted in Fig. (**2**) [20, 21]. An emulsion is made by breaking up droplets of a single liquid phase into a solid, fine-coloured mixture that is separated by a single layer of aggregates of the tensioactive agent. Direct emulsions are formed when oil is mixed in water, where the continuous phase is water. A reverse microemulsion is formed where the oil serves as the continuous phase, when water is dispersed in oil [22, 23].

CLASSIFICATION OF BIOSURFACTANTS

Biosurfactants are usually characterized into groups based on their chemical composition and where they come from. This is different from chemically produced surfactants, which are usually put into groups based on the nature of their polar groups. Biosurfactants are divided into two groups: (i) low molecular weight polymers, which are capable of decreasing surface and interfacial tension, and (ii) high-molecular weight polymers, which are better at stabilizing emulsions Table **1**.

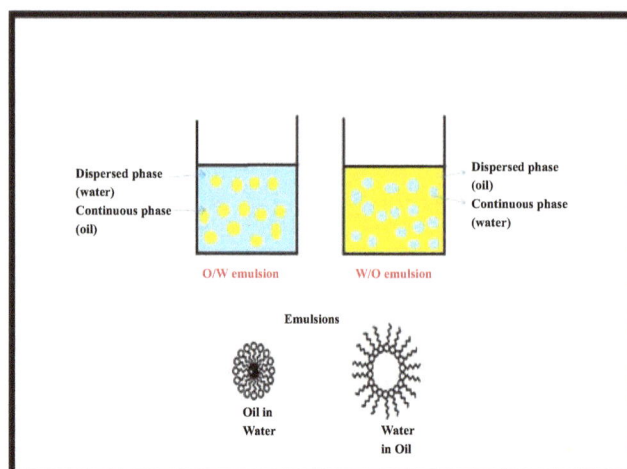

Fig. (2). Surfactants are distinguished by their capacity to facilitate the creation of emulsions of two non-miscible liquids like oil and water.

Table 1. Classification of biosurfactants.

Groups	Biosurfactant	Sub-Class	Producer Microorganisms	Applications	References
Low molecular weight	Glycolipids	Rhamnolipids	*Pseudomonas aeurginosa Serratia rubidea*	Bioremediation Cosmetic industry Antimicrobial and antiviral activity Dissolution of hydrocarbons Gene delivery	[25, 26]
		Sophorolipids	*Torulopsisbombicola*		[26]
		Trehalolipids	*Rhodococcuserythropolis Mycobacterium* sp.		[24, 25]
		Mannosyerythritol lipids	*Candida antartica*		[32, 43]
	Lipopeptides and lipoprotein	Surfactin	*B.subtilis, B.pumilus*	Antimicrobial and antifungal activity Oil extraction Chelating agents	[21, 26]
		Viscosin	*Pseudomonas flourescens, Leuconostocmesenteriods*		[13, 26]
		Serrawettin	*Serratia marcenscens*		[25]
		Subtilisin	*Bacillus subtilis*		[24, 28]
		Lichenysin	*Bacillus licheniformis*		[26, 35]
		Gramicidin	*Bacillus brevis*		[13, 45]
		Polymixin	*Bacillus polymyxia*		[44]
	Fatty acid, neutral lipids and phospholipids	Corynomycolic acid	*Corneybacteriumlepus*	Bioemulsifiers Therapeutic application	[34, 74]
		Spiculisporic acid	*Penicillium spiculisporum*		[32, 44]
		Phosphatidylethanolamine	*Rhodococcuserythropolis*		[25, 26, 75]

(Table 1) cont.....

Groups	Biosurfactant	Sub-Class	Producer Microorganisms	Applications	References
High molecular weight	Polymeric surfactant	Emulsan	*Acinetobacter calcoaceticus*	Biodegradation of aromatic compounds, EOR, Coating agents	[24, 32]
		Biodispersan	*Acinetobacter calcoaceticus*		[24, 44]
		Liposan	*Candida tropicalis*		[24, 26]
		Alasan	*Acinetobacter radioresistens*		[26, 28, 74]
		Lipomannan	*C.tropicalis*		[25, 35]
	Particulate biosurfactant	Vesicles	*Acinetobacter calcoaceticus*	Stimulates alkanes take up by microbial cells	[25, 26, 24]
		Whole cells	*Pseudomonas marginalis Cyanobacteria*		[44, 45]

*EOR – enhanced oil recovery

APPLICATIONS OF BIOSURFACTANTS

Due to their structural and functional diversity, biosurfactants are used in a wide range of industrial processes, with more yet to be discovered. Because of this, many people think that biosurfactants will gain prominence as "functional materials" having multiple features in the coming years. Due to their specific properties, biosurfactants have many potential uses in various sectors, some of which are briefly described in Table **2**.

Table 2. Potential applications of biosurfactants.

Area	Use	Functions	References
Medicine and Health	Wound healing, adjuvants for vaccines.	Antimicrobial, anti-inflammatory and antioxidant agents.	[75, 76]
Agriculture	Animal waste management, agro-industrial waste, residue remediation, phosphate fertilizers.	Wetting and dispersing agents, emulsification of pesticides, spraying powdered pesticides.	[77]
Environmental	Monitoring soil health, and bioavailability of substrates.	Bioremediation, wastewater treatment, and oil spill remediation.	[78, 79]
Metals	Cutting, forming, casting, plating.	Mining, inhibitors in rolling oils, electric cleaning, and electrolytic plating.	[80]
Paint and coating	Pigment preparation, waxes and polishes.	Thickening, washing and defoaming agent, and colouring labelling.	[81]

(Table 2) cont.....

Area	Use	Functions	References
Food and beverages	Food processing, bakery and ice-cream, crystallization of cooking oil and fats	Pesticide and wax coating removal, flower oil solubilisation, consistency and texture control, staling delay.	[27]
Textiles	Preparation of fibres, dyeing and printing, finishing of textiles	Dye levelling, emulsifier in raw wool scoaring, lubricant and antistat in hydrophobic filaments.	[83, 44]
Cleaning	Descaling, Janitorial supplies, soft goods	Wetting agents and corrosion inhibitors, as well as detergents and sanitizers.	[81, 83]
Petroleum products	Drilling fluids, refined products, workover of producing wells	Emulsify oil, disperse solids, emulsify and disperse sludge, and demulsify crude petroleum.	[84]

Emulsifiers

The agent used to blend and combine numerous food constituents to make it a marketable product in the food business is known as an emulsifier. Because of its amphiphilic nature, biosurfactants are effective in increasing the polarity of two non-miscible liquids by lowering the surface tension [27]. There are many food ingredients that have an important role in texture and solubility due to the use of emulsification. Controlling globular clustering and stabilising aerated systems are the primary functions of emulsifying agents. The taste, texture, and nature of food are all impacted when synthetic or chemically produced compounds are used. Biosurfactants are used as an anti-spattering agent and fat stabiliser in food industries [28]. A biosurfactant directs the accumulation of fat globules and improves food consistency and mean life. It also modifies the rheologic features of wheat dough and fat-based products [29]. Lecithin, obtained from proteins like egg and soya, and a variety of other emulsifiers, primarily from artificial sources, are the two most routinely used commercially available emulsifiers in the food and beverage industry [30]. Food-grade emulsifiers are always permitted in most dairy-based products, such as milk, curd, cheese, and creams. Rather than being used as nutritional additives, these compounds are frequently added to a variety of other foods, such as salad dressings, mayonnaise, and desserts, to enhance the taste, appearance, and in storage of foods. Monoglycerides, lecithin, glycolipids, and fatty alcohols with low molecular weight effectively lessen surface and interfacial tension, whereas most of the molecules with high molecular weight are made up of proteins and polysaccharides which aid in emulsion stabilisation [31, 32]. Under these conditions, electrostatic interactions increase the power to penetrate. Gels may form from the particle and droplet aggregation structures in a wide variety of food systems. Several different interactions, such as repulsive forces and van der Waals forces, lead to the clustering of surfactant and polymer molecules. These processes work splendidly with fatty meals. A decrease in surface tension improves the consistency and makes it easier to create emulsions between two immiscible phases [33]. Emulsifiers are added to food to either change or preserve the product's original characteristics after it has been through many procedures, including preparation, processing, dressings, manufacturing, storage, packing, handling, and shipping. A change is observed in terms of chemical properties such as pH, temperature and biological properties to make it safe for consumption and physical properties to maintain the consistency and appearance [34]. Carboxymethyl cellulose and Glyceryl monostearate are frequently used in large quantities as potential emulsifiers [35]. Rhamnolipids obtained from *P. aeruginosa* 47 T2 have been shown to emulsify isopropyl myristate and different types of oils such as olive oil, soyabean oil and Casablanca oil [36].

Lecithin from soybeans or egg deutoplasm, as well as plant gums including xanthan gum, tragacanth, fenugreek gum, guar gum, and Arabic gum that is secreted by the Gram-negative bacterium *Xanthomonas campestris*, are all utilised as natural emulsifiers in the food industry. Phospholipids and amphilic milk proteins also serve the same purpose. Carbon derivatives like sugar esters of fatty acids, mono- and di-glycerides of fat acids, carboxymethyl cellulose and ethyl cellulose are all examples of artificial emulsifiers utilised in the food business [37]. *B.subtilis* LB5a produced biosurfactants made up of lipopeptides that can mix well with soybean oil, coconut fat, and linseed oil to make stable emulsions. This indicates that they could be used in the food industry as a stabiliser for emulsions [38]. Since, *Kluyveromyces marxian* FRR 1586 thrives in an inexpensive lactose-enriched medium, it has the potential to be used as a natural emulsifier in pre-packaged and canned foods, along with preserved meat products. This is because it is able to generate a conjugated protein, which leads to stable oil in water emulsions with maize oil throughout a wide range of pH values (from 3 to 11) and salinities (2-50 g NaCl) [39].

Trichosporon montevidensis CLOA70, *Trichosporon loubiere* CLV20, and *Geotrichumsp* CLOA40 are only a few examples of the glycolipid surfactants generated by different yeasts that have the ability to effectively emulsify vegetable oils at extreme heat, salinities, and pH values. Due to their excellent stability under different conditions and emulsifying properties that are on par with or even superior to those of other commonly used food emulsifiers, these biosurfactants are increasingly being incorporated into food formulations as possible emulsion-stabilizing agents [40].

Food Additives

Food processing is not just about making it safe to eat in the future; it is also about making it look, smell, and taste as fresh as possible. Food processing ensures the quality of the final product by using specific ingredients. It is possible to alter the physical, chemical, biological, or sensory characteristics of a food product through the use of food additives that have no nutritional value. This includes properties like thickening, gelling and stabilisation, as well as the ability to emulsify and disperse. Food preservatives are essential to our food supply because unappetizing food would not be eaten. The food industry uses a wide variety of additives, and today's consumers are very particular about their internal and external health. As a result, a wide range of food additives such as pentosanases, hydrocolloids and enzymes such as amylases, lipases, and hemicellulases are being used to enhance the appearance of food [41]. Botulism is a potentially fatal toxin that originates in bacteria. Antioxidants are commonly used as preservatives to prevent oil and fat oxidation and bad flavour development. Microorganisms, such as bacteria, yeast, fungi, moulds, and actinomycetes, can cause spoilage in food products. Microbial growth in food is greatly aided by exposure to air. Because of this, it is difficult to maintain the desired quality of the food. Additionally, ensuring the safety of food used for both human and animal consumption is a top priority for every step in the production process. The utilization and nutritional quality of most of the food products are impacted by the presence of multiple micronutrients, vitamin supplements, fibers, sugars, fats, and proteins. In order to improve the nutritional value of food products, it may be necessary to add additional nutritional value components; however, food quality and taste must be preserved during these alterations [42]. It is common for food products to have flavouring spices and sweeteners added in order to enhance their taste, while colourants are commonly used to enhance their appearance and appeal to the public. For the desired homogeneity, rheology, aesthetics, textural characteristics, acidity and alkalinity, and other properties of food, emulsifiers, stabilisers and thickeners are used in addition to the regular ingredients. Rhamnolipids enhance the characteristics of cream, margarine, biscuits, and frozen confectioneries, as well as of some fat-based products such as bread loaf, firm and soft rolls, sandwiches, and sponge cake, by improving dough and batter stability and protecting the food from microbial contamination [43]. The US Environmental Protection Agency (EPA) has authorised the use of rhamnolipids in the food industry. Rhamnolipids inhibit haemophilic spores in ultrahigh-temperature soymilk, thereby extending its shelf life. Salads, along with cottage cheese with added niacin and rhamnolipids, prevent the growth of spores, bacteria, and mould, thereby extending their shelf life [44]. In the food industry, synthetic emulsifiers such as carboxymethylcellulose are common. Despite being highly effective, these additives have been restricted because consumers want less "artificial" or chemically synthesised ingredients and more natural ingredients [35].

Biosurfactants in Bakery and Ice-cream Industry

Biosurfactants are utilised in confectionary industries to preserve freshness, regulate consistency, and dissolve flavour oils. In confectionary formulations, biosurfactants enhance consistency by slowing down coagulation, dissolving flavor oils, stabilizing fats, and preventing splattering [45]. Starch-based products have better texture and shelf life as a result of their ability to prevent fat globules from aggregating, and fat-based dairy products have improved consistency and texture as a result of their ability to maintain their shape. These additives are used in the baking industry to alter the dough's rheological properties, increase its water-holding capacity, and enhance its handling properties. In addition, they are used to prevent crystallization in salad oils and as anti-spattering agents in margarine and cooking oils [35]. They are also used as anti-adhering agents, which are used in candies in order to modify their viscidness, thereby improving their texture. They are also used to stabilize the defrosting process in cake toppings and are used as beverage whiteners. Emulsifiers give dairy products, especially low-fat ones, their characteristic texture and creaminess [46]. Rhamnolipids, which stabilize fats, solubilize flavour oils, and retard aging, can be used to regulate ice cream and bakery formulations by maintaining their consistency [34]. Recent years have seen an increase in the popularity of flour-based items like bread, cakes, and cookies as targets for new, healthier formulations that are aimed at addressing specific nutritional criteria connected with market desires. In case of salad dressings, the goal of using biosurfactants in baked goods is to reduce the amount of already commercialized additives while also improving the rheology of the final product. The inclusion of the *S. cerevisiae* URM 6670-produced glycolipid biosurfactant had no discernible effect on the baked items' physical properties or textural profile. This demonstrates the biomolecule's potential for usage in food formulations, where it may emulsify and provide vital fatty acids for human health [47]. Hydrophobins are potential globulins produced by fungi and are used as foam and emulsion stabilizers in aerated foods. These proteins act as adsorption and adhesion agents and are very resistant to denaturation due to their surface-covering properties [48]. When compared to a commercial emulsifier like soy lecithin, adding a lipopeptide biosurfactant synthesized by *B. subtilis SPB1* to dough formulations resulted in more compact dough with considerably better textural qualities and gas retention capacity [49]. As a substitute emulsifier in the production of cupcakes, *Nesterenkonia alba MSA31* was investigated and found to be rather effective. When compared to muffins manufactured with traditional emulsifiers, muffins containing this biosurfactant have lower stiffness, crunchiness and higher smoothness, glossiness, and compactness [50]. Rhamnolipids synthesized from *Pseudomonas aeruginosa* undergo a hydrolysis reaction to produce L-rhamnose, which is an excellent flavouring additive like Furaneol. Furthermore, rhamnolipids are used as a source of L-rhamnose, which has important applications in the production of high-quality aromatic compounds [51, 52].

Anti-adhesive Agents

Biosurfactants may be used to protect the inner parts of food processing equipment, pipelines, and other solid surfaces to hinder the development of harmful bacteria and other microbes. Biosurfactants may be used to prolong the shelf life of different items due to their anti-microbial and potential anti-adhesive qualities. The ability of microbes to stick to the surfaces of food items and machinery may be limited by using these naturally occurring chemicals, which helps to ensure the safety and quality of the items produced [53]. In order to eliminate the biofilm, the interfacial tension between the solid surface and the biofilm must be reduced, where a surface-active chemical interacts with the biofilm matrix to accomplish this. Both the food and biomedical industries have made use of biosurfactant's adsorption ability to reduce microbial adherence and fight colonization by harmful bacteria on solid surfaces [35]. The anti-adhesive properties of biosurfactants synthesized from *Lactobacillus paracasei* have shown to be effective against numerous bacteria and fungi, which include *Staphylococcus aureus, Staphylococcus epidermis, Escherichia coli, Candida albicans*, and *Pseudomonas aeruginosa* [54]. An earlier study demonstrated that rhamnolipids, when present in a concentration of 100 mM, have the ability to inhibit the formation of biofilm by bacteria. *B. pumilus* [55]. Rufisan, a biosurfactant produced by *Candida lipolytica* UCP0988, demonstrated antiadhesive activity against bacteria, yeast, and filamentous fungi at concentrations ranging from 3 to 50 mg/ml [56]. Surface-active compounds change a substrate's hydrophobicity by interfering with microbe adhesion and desorption. It has been hypothesised that biosurfactants with anti-adhesion activity against bacteria might be employed as coating agents for utensils and surfaces in contact with food, or to decrease the rate or incidence of antifouling [57]. Some time ago, it was thought that there might be ways to reduce microbial fouling in dairy processing, which is caused when microbes stick to heat exchangers. These attachments are troublesome because they can easily contaminate pasteurized milk. Biosurfactants, which are produced by cells attaching to surfaces, operate as an anti-adhesive biological layer, lowering microbial fouling during dairy processing [58]. It has also been reported that *L.acidophilus* Biosurfactant inhibits the development and adhesion of uropathogenic *Enterococci* that are known to invade the human body. One strain of the probiotic found in dairy products, *i.e.*, *L. acidophilus*, is well-suited for oral administration, suggesting its widespread use in different industries with favourable benefits. Antimicrobial and anti-adhesive qualities are particularly exciting, because they are used as potent tools in the fight to keep food secure from contamination [59]. Biosurfactants have the potency to be used at various stages throughout the food industry due to their diverse and distinct properties, as shown in Fig. (**3**).

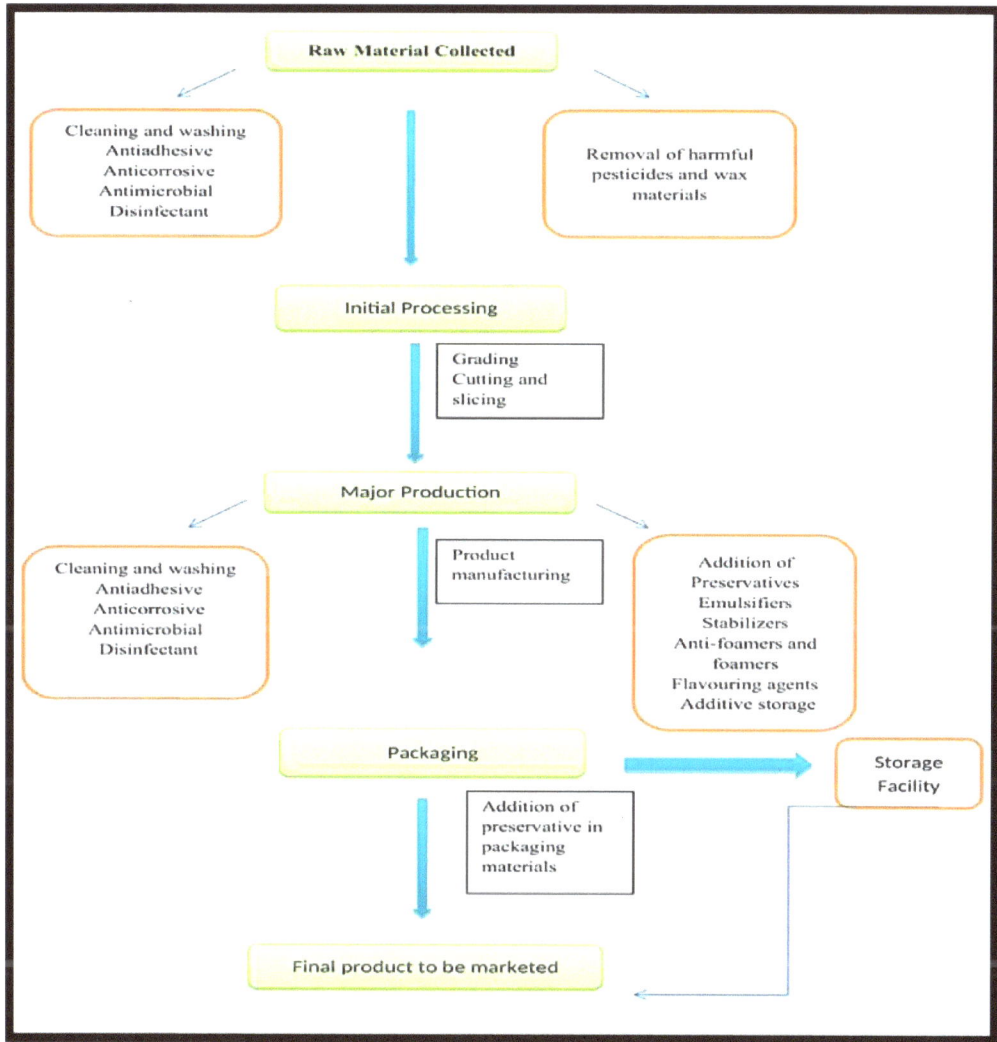

Fig. (3). A schematic representation of biosurfactants used in different stages of the food sector.

Biosurfactants with Anti-microbial Activity

Lipopeptides are the most widely reported class of biosurfactants with antimicrobial activity. Bacillus is the genus responsible for the most well-known lipopeptide biosurfactants. Surfactin, the first biosurfactant with this antimicrobial property, is produced by *B. subtilis*. Because of its antimicrobial properties, it has been reported as a safe alternative to synthetic components [60]. Biosurfactant molecules such as lipopeptides and sophorolipids have been shown to have antimicrobial properties against *Propionibacterium* acne, and a study found that these biosurfactant molecules can penetrate into the membrane of gram-positive bacteria, leading to the leakage of cytoplasmic material and cell lysis [61]. Biosurfactants derived from *L. paracasei* sp. A20 have antimicrobial activity against *Candida albicans*, *Staphylococcus aureus*, and *Staphylococcus epidermidis*.

Sophorolipid has antifungal activity against the fungal pathogen *Botrytis cinerea*. Due to its amphiphilic nature, lipopeptide biosurfactant attaches to the lipoteichoic as well as lipopolysaccharide acid membranes of fungal and bacterial cell walls to dissolve them, acting in a detergent-like manner [61, 62] Biosurfactants made by lactic acid bacteria (LAB) or bacteria found in dairy products have structures ranging from glycoproteins and glycopeptides to glycolipids and lipoproteins and exhibit antibacterial action against pathogenic bacteria, yeasts, and filamentous fungus. We assume that biosurfactants made from *Lactobacillus acidophilus* (LAB) and other bacteria found in dairy products will be safe for mammals because these organisms have long been accepted as safe for human consumption (GRAS). These compounds can be used as food preservatives or as cleaning agents to wipe down surfaces and equipment that interacts with edible food to prevent contamination. As a result, food shelf life can be extended and food safety can be ensured [54, 60]. The physicochemical characteristics can vary depending on the structural changes. Sophorolipid congeners exhibit potential antimicrobial activity against bacteria *E.coli* O157:H7, in the presence and absence of ethanol. The research found that after two hours of treatment in 20% ethanol, the population of *E. coli* O157:H7 was undetectable when *Lactonic sophorolipids* with a hydrophobic domain of stearic or oleic acid were present at a concentration of 10 gl^{-1}. Sophorolipids that are internally esterified with a free acid, on the other hand, have a low antibacterial activity [63]. Food makers typically utilize low-pH preservatives to avoid food spoiling since they have been validated as alternative preservatives due to their strong antibacterial activity and resistance to proteases.

Antioxidant and Anticorrosive Agents

Corrosion has become a major issue for all types of industrial operations, particularly in the food business. Corrosion results in the unevenness of the surface, allowing microbial attachment and causing unsanitary conditions during food manufacturing. Furthermore, it shows the adhesion of *L. monocytogenes* to the stainless steel [64]. The role of bacteria as a source of corrosion in the food processing industry is generally disregarded; however, some research supports the use of biosurfactants to prevent this problem. A study reported the anti-corrosive properties of biosurfactants, suggesting that it could be used as a surface coating agent to stop corrosion by using chromium oxide and decreasing iron oxide (Corrosion/anticorrosion changes the alloy's chromium concentration). This would also make it harder for bacteria to grow on the surface [65]. Antioxidant activity is highly valued in the food industry because antioxidant-rich foods lessen the risk of heart disease, prevent degenerative diseases, and extend the usable life of products by slowing or obstructing oxidation events. As a result, food compositions with these traits benefit sales, as customers' desire for healthy foods grows [66]. There are various tests available to measure a compound's antioxidant activity. A biosurfactant could be more impactful at securing free radicals or reducing complexes depending on the technique used, and its application is based on the outcomes of such tests [67]. Unsaturated fatty acids in the biosurfactant molecule may also be responsible for its high antioxidant activity. Consequently, the chemical structure plays an important role in achieving the desired outcomes [68]. Mannosylerythritol lipids (MELs) are biosurfactants with versatile interfacial and biochemical properties.

Research utilising a free-radical and superoxide anion-scavenging test showed antioxidant activity *in vitro*. Based on their findings, they hypothesised that MEL-C, due to its potent antioxidant and protective properties on cells, may be employed as an anti-aging skin care product [69]. Previously, it was demonstrated that a polysaccharide emulsifier from *Klebsiella* sp. effectively inhibits the auto-oxidation of soybean oil. The emulsifier inhibited the peroxidation of soybean oil by encapsulating the oil and isolating it from the surrounding medium [70]. In three independent antioxidant assays, researchers found that a biosurfactant from *C. utilis* exhibited values of more than 70% at a concentration of 5 mgml^{-1} for both complex-reducing activity and radical-sequestration capacity [67]. Furthermore, a linear relationship was discovered between biosurfactant concentration and antioxidant activity. Biosurfactants from Lactobacillus species have the highest radical-sequestering activity at the same concentration (5 mgml^{-1}) when subjected to the 2, 2-diphenyl-1-picrylhydrazyl reaction (DPPH) [71].

Biosurfactants in Biofilm Formation

The cells of microorganisms adhere to two surfaces, forming biofilms. They consist of bacteria and a combination of surface-produced extracellular material from microbes and any other materials that become trapped in the resulting matrix. It is common knowledge that bacterial biofilms on surfaces are major contributors to contamination in the food industry, which in turn can lead to the spread of disease and the spoilage of perishable goods. Therefore, one of the most important things you can do to ensure your products are safe for consumers to use is to reduce the number of bacteria that stick to the surfaces they come into contact with. A previous study reported that preconditioning metal surface areas with a *P. fluorescens* anionic biosurfactant to reduce Listeria adhesion favoured disinfectants' bactericidal activities [72]. Bio-conditioning areas with bacterial surfactants may reduce adherence. By delaying the adherence of both expanding and non-growing microbes to preconditioned surfaces made of stainless-steel polypropylene and polystyrene, surfactin was able to decrease the number of adhering cells. Microbial wetting of the surface *via* surface active substances has been reported as the key adhesion criteria for a complex set of heterogeneous biofilms observed in nature. Biosurfactant enriched biofilms which are extensively used in microbial-enhanced oil recovery. According to these characteristics, biosurfactants are promising candidates for use in efforts to postpone or avoid microbial colonization of industrial plants where food is prepared [73, 74].

CONCLUSION

In today's world, consumers have a strong desire for completely natural foods, thereby encouraging researchers to seek out more suitable organic ingredients for use in food formulations. Biosurfactants are preferred because of their distinct characteristics of being low toxic and biodegradable. Regardless of the fact that biosurfactants increase the physical and aesthetic aspects of food while preserving its nutritional content, their use is still restricted. This is owing to problems in optimizing numerous processes in biosurfactant manufacturing, including large-scale purifying operations. Furthermore, the expense of applying biosurfactants is still extremely high, limiting their use. Other considerations, such as safety, must be addressed prior to the use of these biomolecules in food industries, as certain biosurfactants are created by aggressive bacteria. The only way to deal with this issue is to look out for non-pathogenic bacterial strains producing potential biosurfactants. Despite these challenges, biosurfactants have the clear ability to replace synthetic surfactants while benefiting public health and the environment. In the coming years, these molecules will definitely be used as constituents in food formulas and surface-altering preservatives in the food processing industry.

ACKNOWLEDGEMENTS

The authors acknowledge the support provided by Siksha 'O' Anusandhan (Deemed to be University), Bhubaneswar, for encouraging us to publish scientific literature.

REFERENCES

[1]   Volkering, F.; Breure, A.M.; Rulkens, W.H. Microbiological aspects of surfactant use for biological soil remediation. *Biodegradation,* **1997-1998**, *8*(6), 401-417.
[http://dx.doi.org/10.1023/A:1008291130109] [PMID: 15765586]

[2]   Rahman, P.K.S.M.; Gakpe, E. Production, characterisation and applications of biosurfactants-Review. *Biotechnology,* **2008**, *7*(2), 360-370.
[http://dx.doi.org/10.3923/biotech.2008.360.370]

[3]   Roy, A. Review on the biosurfactants: Properties, types and its applications. *J Fundam Renew Energy Appl.,* **2017**, *8*, 1-4.

[4]   Syldatk, C.; Lang, S.; Wagner, F.; Wray, V.; Witte, L. Chemical and physical characterization of four interfacial-active rhamnolipids from *Pseudomonas* spec. DSM 2874 grown on n-alkanes. *Z. Naturforsch. C J. Biosci.,* **1985**, *40*(1-2), 51-60.
[http://dx.doi.org/10.1515/znc-1985-1-212] [PMID: 3993180]

[5]   Chakrabarti, S. Bacterial biosurfactant: Characterization, antimicrobial and metal remediation properties (Doctoral dissertation).

[6]   Cooper, D.G.; Macdonald, C.R.; Duff, S.J.B.; Kosaric, N. Enhanced production of surfactin from *Bacillus subtilis* by continuous product removal and metal cation additions. *Appl. Environ. Microbiol.,* **1981**, *42*(3), 408-412.
[http://dx.doi.org/10.1128/aem.42.3.408-412.1981] [PMID: 16345840]

[7]   Mulligan, C.N.; Yong, R.N.; Gibbs, B.F. Remediation technologies for metal-contaminated soils and groundwater: an evaluation. *Eng. Geol.,* **2001**, *60*(1-4), 193-207.
[http://dx.doi.org/10.1016/S0013-7952(00)00101-0]

[8]   Gharaei-Fa, E. Biosurfactants in pharmaceutical industry: a mini-review. *Am. J. Drug Disc. Develop.,* **2010**, *1*(1), 58-69.
[http://dx.doi.org/10.3923/ajdd.2011.58.69]

[9]   Das, K.; Mukherjee, A.K. Crude petroleum-oil biodegradation efficiency of *Bacillus subtilis* and *Pseudomonas aeruginosa* strains isolated from a petroleum-oil contaminated soil from North-East India. *Bioresour. Technol.,* **2007**, *98*(7), 1339-1345.
[http://dx.doi.org/10.1016/j.biortech.2006.05.032] [PMID: 16828284]

[10]   McInerney, M.J.; Javaheri, M.; Nagle, D.P., Jr Properties of the biosurfactant produced by *Bacillus licheniformis* strain JF-2. *J. Ind. Microbiol.,* **1990**, *5*(2-3), 95-101.
[http://dx.doi.org/10.1007/BF01573858] [PMID: 1366681]

[11]   Poremba, K.; Gunkel, W.; Lang, S.; Wagner, F. Toxicity testing of synthetic and biogenic surfactants on marine microorganisms. *Environ. Toxicol. Water Qual.,* **1991**, *6*(2), 157-163.
[http://dx.doi.org/10.1002/tox.2530060205]

[12]   Flasz, A.; Rocha, C.A.; Mosquera, B.; Sajo, C. A comparative study of the toxicity of a synthetic surfactant and one produced by *Pseudomonas aeruginosa* ATCC 55925. *Med. Sci. Res.,* **1998**, *26*(3), 181-185.

[13]   Velikonja, JO; Kosaric, NA Biosurfactants in food applications. Biosurfactants: production, properties, applications. **1993**; 48.

[14]   Hu, Y.; Ju, L.K. Purification of lactonic sophorolipids by crystallization. *J. Biotechnol.,* **2001**, *87*(3), 263-272.
[http://dx.doi.org/10.1016/S0168-1656(01)00248-6] [PMID: 11334668]

[15]   Jadhav, M.; Kalme, S.; Tamboli, D.; Govindwar, S. Rhamnolipid from *Pseudomonas desmolyticum* NCIM-2112 and its role in the degradation of Brown 3REL. *J. Basic Microbiol.,* **2011**, *51*(4), 385-396.
[http://dx.doi.org/10.1002/jobm.201000364] [PMID: 21656804]

[16]   Zottola, E.A.; Sasahara, K.C. Microbial biofilms in the food processing industry—Should they be a concern? *Int. J. Food Microbiol.,* **1994**, *23*(2), 125-148.
[http://dx.doi.org/10.1016/0168-1605(94)90047-7] [PMID: 7848776]

[17]   Bhawsar, S.D.; Path, S.D.; Chopade, B.A. Antimicrobial activity of purified emulsifier of *Acinetobacter genospecies* isolated from rhizosphere of wheat. *Agric. Sci. Dig.,* **2011**, *31*(4), 239-246.

[18]   Sana, S.; Datta, S.; Biswas, D.; Auddy, B.; Gupta, M.; Chattopadhyay, H. Excision wound healing activity of a common biosurfactant produced by *Pseudomonas* sp. *Wound Medicine,* **2018**, *23*, 47-52.
[http://dx.doi.org/10.1016/j.wndm.2018.09.006]

[19]   Karlapudi, A.P.; Venkateswarulu, T.C.; Tammineedi, J.; Kanumuri, L.; Ravuru, B.K.; Dirisala, V.; Kodali, V.P. Role of biosurfactants in bioremediation of oil pollution : A review. *Petroleum,* **2018**, *4*(3), 241-249.
[http://dx.doi.org/10.1016/j.petlm.2018.03.007]

[20]   Oyeleke, S.B.; Oyewole, O.A.; Aliyu, G.M.; Shaba, A.M.; Ikekwem, C.C.; Ayisa, T.T. Production and characterization of biosurfactants by candida boleticola H09 and Rhodotorulabogoriensis H15 for crude oil recovery and cleaning of oil contaminated fabrics. *Int. J. Life Sci.,* **2017**, *10*(15), 109-123.

[21]   Rocha e Silva, F.C.P.; Rocha e Silva, N.M.P.; Luna, J.M.; Rufino, R.D.; Santos, V.A.; Sarubbo, L.A. Dissolved air flotation combined to biosurfactants: A clean and efficient alternative to treat industrial oily water. *Rev. Environ. Sci. Biotechnol.,* **2018**, *17*(4), 591-602.
[http://dx.doi.org/10.1007/s11157-018-9477-y]

[22]   Rocha e Silva, N.M.P.; Meira, H.M.; Almeida, F.C.G.; Soares da Silva, R.C.F.; Almeida, D.G.; Luna, J.M.; Rufino, R.D.; Santos, V.A.; Sarubbo, L.A. Natural surfactants and their applications for heavy oil removal in industry. *Separ. Purif. Rev.,* **2019**, *48*(4), 267-281.
[http://dx.doi.org/10.1080/15422119.2018.1474477]

[23]   Milani, J.M. Some new aspects of colloidal systems in foods. *London: IntechOpen.* **2019**: pp. 98.
[http://dx.doi.org/10.5772/intechopen.75145]

[24]   Gürkök, S.; Özdal, M. Microbial Biosurfactants: Properties, Types, and Production. *Anatol J Biol.,* **2021**, *2*(2), 7-12.

[25]   Desai, J.D.; Banat, I.M. Microbial production of surfactants and their commercial potential. *Microbiol. Mol. Biol. Rev.,* **1997**, *61*(1), 47-64.
[PMID: 9106364]

[26]   Deleu, M.; Paquot, M. From renewable vegetables resources to microorganisms: new trends in surfactants. *C. R. Chim.,* **2004**, *7*(6-7), 641-646.
[http://dx.doi.org/10.1016/j.crci.2004.04.002]

[27]   Smyth, T.; Perfumo, A.; Marchant, R.; Banat, I. *Isolation and Analysis of Low Molecular WeightMicrobial Glycolipids. InHandbook of Hydrocarbon and Lipid Microbiology*; Springer, **2010**, pp. 3705-3723.

[28]   Campos, J.M.; Montenegro Stamford, T.L.; Sarubbo, L.A.; de Luna, J.M.; Rufino, R.D.; Banat, I.M. Microbial biosurfactants as additives for food industries. *Biotechnol. Prog.,* **2013**, *29*(5), 1097-1108.
[http://dx.doi.org/10.1002/btpr.1796] [PMID: 23956227]

[29]   McClements, D.J.; Gumus, C.E. Natural emulsifiers : Biosurfactants, phospholipids, biopolymers, and colloidal particles: Molecular and physicochemical basis of functional performance. *Adv. Colloid*

Interface Sci., **2016**, *234*, 3-26.
[http://dx.doi.org/10.1016/j.cis.2016.03.002] [PMID: 27181392]

[30] Flasz, A.; Rocha, C.A.; Mosquera, B.; Sajo, C. A comparative study of the toxicity of a synthetic surfactant and one produced by *Pseudomonas aeruginosa* ATCC 55925. *Med. Sci. Res.*, **1998**, *26*(3), 181-185.

[31] Kosaric, N.; Sukan, F.V. Biosurfactants: production and utilization—processes, technologies, and economics. *CRC press,* **2014**, pp. 389.
[http://dx.doi.org/10.1201/b17599]

[32] Kosaric, N. Biosurfactants and their application for soil bioremediation. *Food Technol. Biotechnol.,* **2001**, *39*(4), 295-304.

[33] Satpute, S.K.; Banat, I.M.; Dhakephalkar, P.K.; Banpurkar, A.G.; Chopade, B.A. Biosurfactants, bioemulsifiers and exopolysaccharides from marine microorganisms. *Biotechnol. Adv.,* **2010**, *28*(4), 436-450.
[http://dx.doi.org/10.1016/j.biotechadv.2010.02.006] [PMID: 20172021]

[34] Hasenhuettl, G.L. *Overview of food emulsifiers inFood emulsifiers and their applications*; Springer, **2008**, pp. 1-9.
[http://dx.doi.org/10.1007/978-0-387-75284-6]

[35] Kralova, I.; Sjöblom, J. Surfactants used in food industry: A review. *J. Dispers. Sci. Technol.,* **2009**, *30*(9), 1363-1383.
[http://dx.doi.org/10.1080/01932690902735561]

[36] Haba, E.; Bouhdid, S.; Torrego-Solana, N.; Marqués, A.M.; Espuny, M.J.; García-Celma, M.J.; Manresa, A. Rhamnolipids as emulsifying agents for essential oil formulations: Antimicrobial effect against *Candida albicans* and methicillin-resistant *Staphylococcus aureus*. *Int. J. Pharm.,* **2014**, *476*(1-2), 134-141.
[http://dx.doi.org/10.1016/j.ijpharm.2014.09.039] [PMID: 25269010]

[37] McClements, D.J.; Bai, L.; Chung, C. Recent advances in the utilization of natural emulsifiers to form and stabilize emulsions. *Annu. Rev. Food Sci. Technol.,* **2017**, *8*(1), 205-236.
[http://dx.doi.org/10.1146/annurev-food-030216-030154] [PMID: 28125353]

[38] Nitschke, M.; Pastore, G.M. Production and properties of a surfactant obtained from *Bacillus subtilis* grown on cassava wastewater. *Bioresour. Technol.,* **2006**, *97*(2), 336-341.
[http://dx.doi.org/10.1016/j.biortech.2005.02.044] [PMID: 16171690]

[39] Lukondeh, T.; Ashbolt, N.J.; Rogers, P.L. Evaluation of Kluyveromyces marxianus FII 510700 grown on a lactose-based medium as a source of a natural bioemulsifier. *J. Ind. Microbiol. Biotechnol.,* **2003**, *30*(12), 715-720.
[http://dx.doi.org/10.1007/s10295-003-0105-6] [PMID: 14689315]

[40] Monteiro, D.R.; Gorup, L.F.; Takamiya, A.S.; Ruvollo-Filho, A.C.; Camargo, E.R.; Barbosa, D.B. The growing importance of materials that prevent microbial adhesion: antimicrobial effect of medical devices containing silver. *Int. J. Antimicrob. Agents,* **2009**, *34*(2), 103-110.
[http://dx.doi.org/10.1016/j.ijantimicag.2009.01.017] [PMID: 19339161]

[41] Mnif, I.; Ghribi, D. Glycolipid biosurfactants: main properties and potential applications in agriculture and food industry. *J. Sci. Food Agric.,* **2016**, *96*(13), 4310-4320.
[http://dx.doi.org/10.1002/jsfa.7759] [PMID: 27098847]

[42] Banat, I.M.; Thavasi, R. *Microbial biosurfactants and their environmental and industrial applications*; CRC Press, **2019**.
[http://dx.doi.org/10.1201/b21950]

[43] Nitschke, M.; Costa, S.G.V.A.O. Biosurfactants in food industry. *Trends Food Sci. Technol.,* **2007**, *18*(5), 252-259.
[http://dx.doi.org/10.1016/j.tifs.2007.01.002]

[44] Pardhi, D.S.; Bhatt, R.; Panchal, R.R.; Raval, V.H.; Rajput, K.N. *Rhamnolipid Biosurfactants Structure, Biosynthesis, Production, and Applications. InMicrobial Surfactants*; CRC Press, **2021**.

[45] Muthusamy, K.; Gopalakrishnan, S.; Ravi, T.K.; Sivachidambaram, P. Biosurfactants: Properties, commercial production and application. *Curr. Sci.*, **2008**, 736-747.

[46] Rosenberg, E.; Ron, E.Z. High- and low-molecular-mass microbial surfactants. *Appl. Microbiol. Biotechnol.*, **1999**, *52*(2), 154-162.
[http://dx.doi.org/10.1007/s002530051502] [PMID: 10499255]

[47] Silva, I.A.; Veras, B.O.; Ribeiro, B.G.; Aguiar, J.S.; Campos Guerra, J.M.; Luna, J.M.; Sarubbo, L.A. Production of cupcake-like dessert containing microbial biosurfactant as an emulsifier. *PeerJ*, **2020**, *8*, e9064.
[http://dx.doi.org/10.7717/peerj.9064] [PMID: 32351793]

[48] Penfold, J.; Thomas, R.K. Adsorption properties of plant based bio-surfactants: Insights from neutron scattering techniques. *Adv. Colloid Interface Sci.*, **2019**, *274*, 102041.
[http://dx.doi.org/10.1016/j.cis.2019.102041] [PMID: 31655367]

[49] Mnif, I.; Besbes, S.; Ellouze-Ghorbel, R.; Ellouze-Chaabouni, S.; Ghribi, D. Improvement of bread dough quality by *Bacillus subtilis* SPB1 biosurfactant addition: optimized extraction using response surface methodology. *J. Sci. Food Agric.*, **2013**, *93*(12), 3055-3064.
[http://dx.doi.org/10.1002/jsfa.6139] [PMID: 23512731]

[50] Kiran, G.S.; Priyadharsini, S.; Sajayan, A.; Priyadharsini, G.B.; Poulose, N.; Selvin, J. Production of lipopeptide biosurfactant by a marine Nesterenkonia sp. and its application in food industry. *Front. Microbiol.*, **2017**, *8*, 1138.
[http://dx.doi.org/10.3389/fmicb.2017.01138] [PMID: 28702002]

[51] Linhardt, R.J.; Bakhit, R.; Daniels, L.; Mayerl, F.; Pickenhagen, W. Microbially produced rhamnolipid as a source of rhamnose. *Biotechnol. Bioeng.*, **1989**, *33*(3), 365-368.
[http://dx.doi.org/10.1002/bit.260330316] [PMID: 18587925]

[52] Federici, F.; Fava, F.; Kalogerakis, N.; Mantzavinos, D. Valorisation of agro‐industrial by‐products, effluents and waste: concept, opportunities and the case of olive mill wastewaters. *J. Chem. Technol. Biotechnol.*, **2009**, *84*(6), 895-900.
[http://dx.doi.org/10.1002/jctb.2165]

[53] McLandsborough, L.; Rodriguez, A.; Pérez-Conesa, D.; Weiss, J. Biofilms: At the interface between biophysics and microbiology. *Food Biophys.*, **2006**, *1*(2), 94-114.
[http://dx.doi.org/10.1007/s11483-005-9004-x]

[54] Gudiña, E.J.; Teixeira, J.A.; Rodrigues, L.R. Isolation and functional characterization of a biosurfactant produced by *Lactobacillus paracasei*. *Colloids Surf. B Biointerfaces*, **2010**, *76*(1), 298-304.
[http://dx.doi.org/10.1016/j.colsurfb.2009.11.008] [PMID: 20004557]

[55] Dusane, D.H.; Nancharaiah, Y.V.; Zinjarde, S.S.; Venugopalan, V.P. Rhamnolipid mediated disruption of marine *Bacillus pumilus* biofilms. *Colloids Surf. B Biointerfaces*, **2010**, *81*(1), 242-248.
[http://dx.doi.org/10.1016/j.colsurfb.2010.07.013] [PMID: 20688490]

[56] Rufino, R.D.; Luna, J.M.; Sarubbo, L.A.; Rodrigues, L.R.M.; Teixeira, J.A.C.; Campos-Takaki, G.M. Antimicrobial and anti-adhesive potential of a biosurfactant Rufisan produced by *Candida lipolytica* UCP 0988. *Colloids Surf. B Biointerfaces*, **2011**, *84*(1), 1-5.
[http://dx.doi.org/10.1016/j.colsurfb.2010.10.045] [PMID: 21247740]

[57] Bouman, S.; Lund, D.B.; Driessen, F.M.; Schmidt, D.G. Growth of thermoresistant *Streptococci* and deposition of milk constituents on plates of heat exchangers during long operating times. *J. Food Prot.*, **1982**, *45*(9), 806-813.
[http://dx.doi.org/10.4315/0362-028X-45.9.806] [PMID: 30866295]

[58] Lalande, M.; Rene, F.; Tissier, J.P. Fouling and its control in heat exchangers in the dairy industry.

Biofouling, **1989**, *1*(3), 233-250.
[http://dx.doi.org/10.1080/08927018909378111]

[59] Reid, G. *In vitro* testing of *Lactobacillus acidophilus* NCFMTM as a possible probiotic for the urogenital tract. *Int. Dairy J.,* **2000**, *10*(5-6), 415-419.
[http://dx.doi.org/10.1016/S0958-6946(00)00059-5]

[60] Rodrigues, L.; Banat, I.M.; Teixeira, J.; Oliveira, R. Biosurfactants: Potential applications in medicine. *J. Antimicrob. Chemother.,* **2006**, *57*(4), 609-618.
[http://dx.doi.org/10.1093/jac/dkl024] [PMID: 16469849]

[61] Kim, K.; Dalsoo, Y.; Youngbum, K.; Baekseok, L.; Doonhoon, S.; Eun-Ki, K.I. Characteristics of sophorolipid as an antimicrobial agent. *J. Microbiol. Biotechnol.,* **2002**, *12*(2), 235-241.

[62] Naruse, N.; Tenmyo, O.; Kobaru, S.; Kamei, H.; Miyaki, T.; Konishi, M.; Oki, T. Pumilacidin, a complex of new antiviral antibiotics. Production, isolation, chemical properties, structure and biological activity. *J. Antibiot.,* **1990**, *43*(3), 267-280.
[http://dx.doi.org/10.7164/antibiotics.43.267] [PMID: 2157695]

[63] Zhang, X.; Ashby, R.D.; Solaiman, D.K.Y.; Liu, Y.; Fan, X. Antimicrobial activity and inactivation mechanism of lactonic and free acid sophorolipids against *Escherichia coli* O157:H7. *Biocatal. Agric. Biotechnol.,* **2017**, *11*, 176-182.
[http://dx.doi.org/10.1016/j.bcab.2017.07.002]

[64] Mai, T.L.; Sofyan, N.I.; Fergus, J.W.; Gale, W.F.; Conner, D.E. Attachment of Listeria monocytogenes to an austenitic stainless steel after welding and accelerated corrosion treatments. *J. Food Prot.,* **2006**, *69*(7), 1527-1532.
[http://dx.doi.org/10.4315/0362-028X-69.7.1527] [PMID: 16865881]

[65] Araujo, L.V.; Freire, D.M.; Nitschke, M. Biosurfactants: Anticorrosive, antibiofilm and antimicrobial properties. *Quim. Nova,* **2013**, *36*, 848-858.
[http://dx.doi.org/10.1590/S0100-40422013000600019]

[66] Amiri, A.; Mousakhani-Ganjeh, A.; Amiri, Z.; Guo, Y.; Pratap Singh, A.; Esmaeilzadeh Kenari, R. Fabrication of cumin loaded-chitosan particles: Characterized by molecular, morphological, thermal, antioxidant and anticancer properties as well as its utilization in food system. *Food Chem.,* **2020**, *310*, 125821.
[http://dx.doi.org/10.1016/j.foodchem.2019.125821] [PMID: 31753687]

[67] Ribeiro, B.G.; de Veras, B.O.; dos Santos Aguiar, J.; Medeiros Campos Guerra, J.; Sarubbo, L.A. Biosurfactant produced by Candida utilis UFPEDA1009 with potential application in cookie formulation. *Electron. J. Biotechnol.,* **2020**, *46*, 14-21.
[http://dx.doi.org/10.1016/j.ejbt.2020.05.001]

[68] Kiran, G.S.; Priyadharsini, S.; Sajayan, A.; Priyadharsini, G.B.; Poulose, N.; Selvin, J. Production of lipopeptide biosurfactant by a marine Nesterenkonia sp. and its application in food industry. *Front. Microbiol.,* **2017**, *8*, 1138.
[http://dx.doi.org/10.3389/fmicb.2017.01138] [PMID: 28702002]

[69] Takahashi, M.; Morita, T.; Fukuoka, T.; Imura, T.; Kitamoto, D. Glycolipid biosurfactants, mannosylerythritol lipids, show antioxidant and protective effects against H_2O_2-induced oxidative stress in cultured human skin fibroblasts. *J. Oleo Sci.,* **2012**, *61*(8), 457-464.
[http://dx.doi.org/10.5650/jos.61.457] [PMID: 22864517]

[70] Kawaguchi, K.; Satomi, K.; Yokoyama, M.; Ishida, Y. Antioxidative properties of an extracellular polysaccharide produced by a bacterium *Klebsiella* sp. isolated from river water. *Nippon Suisan Gakkaishi,* **1996**, *62*(1), 123-128.
[http://dx.doi.org/10.2331/suisan.62.123]

[71] Merghni, A.; Dallel, I.; Noumi, E.; Kadmi, Y.; Hentati, H.; Tobji, S.; Ben Amor, A.; Mastouri, M. Antioxidant and antiproliferative potential of biosurfactants isolated from *Lactobacillus casei* and their anti-biofilm effect in oral *Staphylococcus aureus* strains. *Microb. Pathog.,* **2017**, *104*, 84-89.

[http://dx.doi.org/10.1016/j.micpath.2017.01.017] [PMID: 28087493]

[72] Meylheuc, T.; Renault, M.; Bellonfontaine, M. Adsorption of a biosurfactant on surfaces to enhance the disinfection of surfaces contaminated with Listeria monocytogenes. *Int. J. Food Microbiol.,* **2006**, *109*(1-2), 71-78.
[http://dx.doi.org/10.1016/j.ijfoodmicro.2006.01.013] [PMID: 16488496]

[73] Quinn, G.A.; Maloy, A.P.; Banat, M.M.; Banat, I.M. A comparison of effects of broad-spectrum antibiotics and biosurfactants on established bacterial biofilms. *Curr. Microbiol.,* **2013**, *67*(5), 614-623.
[http://dx.doi.org/10.1007/s00284-013-0412-8] [PMID: 23783562]

[74] Meylheuc, T.; Herry, J.M.; Bellon-Fontaine, M.N. *Biosurfactants, surface-active biomolecules with wide potential applications*; Sciences des Aliments: France, **2001**.

CHAPTER 7

Biosurfactants and Their Application in Remediation of Environmental Contaminants

Meena Choudhary[1,2], Monali Muduli[1,2] and Sanak Ray[1,2,*]

[1] *Analytical and Environmental Science Division & Centralized Instrument Facility, CSIR-Central Salt & Marine Chemicals Research Institute, G.B. Marg, Bhavnagar-364002, India*

[2] *Academy of Scientific and Innovative Research (AcSIR), Ghaziabad-201002, India*

Abstract: The demand for bio-surfactants is growing daily over synthetic surfactants due to their less harmful effects on the environment, biodegradability, and nontoxic effects on public health. Biosurfactants play a significant role in foam generation, emulsification, oil dispersion, and detergency due to their amphipathic structure with the hydrophilic and hydrophobic sites. In recent years, tremendous development in research has resulted in different methods to produce several types of biosurfactants from microorganisms. Several biosurfactants are grown commercially and applied in the pharmaceutical and cosmetic sectors, food, petroleum, and agricultural sectors to mitigate environmental contaminants. The current chapter discusses the potentiality of biosurfactants to degrade environmental pollutants in various fields.

Keywords: Amphipathic, Biosurfactant, Emulsification, Environmental contaminants, Interfacial tension.

INTRODUCTION

Surfactants, surface active agents, are amphipathic molecules with polar and nonpolar structures. They align themselves at the air-water interface in such a way (hydrophobic part in air and hydrophilic part in water) that they would be able to reduce the surface tension. This interfacial tension reduction of surfactants helps them to become effective emulsifiers, cleaning agents, and oil dispersers. Recently, the whole world has been focusing on developing a sustainable society. Thus, surfactants produced by microorganisms have been highlighted more than chemically synthesized surfactants due to their less harmful impact on humans and their environment. The most promising natural surfactants are those derived

---

** **Corresponding author Sanak Ray:** Analytical and Environmental Science Division & Centralized Instrument Facility, CSIR-Central Salt & Marine Chemicals Research Institute, G.B. Marg, Bhavnagar-364002, India & Academy of Scientific and Innovative Research (AcSIR), Ghaziabad-201002, India; Tel: 91-278-2567760; E-mail: sanakray@csmcri.res.in*

Arun Kumar Pradhan and Manoranjan Arakha (Eds.)

from microorganisms, also referred to as "microbial surfactants" or "biosurfactants" [1, 2]. They are surface-acting substances that have the potential to enhance surface-surface interactions by producing micelles from their natural sources, including plants, microorganisms (bacteria, yeast, and archaea), and animals [3].

Biosurfactants adsorb at the interface of the two phases by minimizing surface tensions. They reduce the repulsive force of the two phases and allow them to mix quickly [4].

Amphipathic molecules with both hydrophilic and hydrophobic components typically make up biosurfactants. The hydrophilic chemicals typically consist of amphoteric or positively and negatively charged ions. At the same time, the hydrophobic molecules comprise (water resisting, such as unsaturated or saturated hydrocarbon chains or fatty acids). Similar to their chemical equivalents, biosurfactants can be categorized according to their molecular weight (low or high), critical micelle concentration (C.M.C.), the type of microbe they create, and their action method. The most frequently reported low-molecular-weight substances include phospholipids, glycolipids, and lipopeptides; however, high-molecular-weight biosurfactants comprise polysaccharides and a complex blend of biopolymers [5, 3].

Biosurfactants, one of the latest researched microbially produced/synthesized biomolecules, offer several potential uses. The advantages, like long storage time, biodegradability nature, and feasibility over changing abiotic factors like pH, temperature, and salt concentrations, have made biosurfactants suitable for various applications [2, 5 - 7].

Many microbiological strains of bacteria, fungi, and yeast have been identified for effective biosurfactant generation. The kind of microbe, medium supplements, substrate characteristics, and other internal and external parameters during microbial culture growth all impact the quality and the amount of biosurfactants [8, 9]. The initial stage in the synthesis of biosurfactants is the selection of the microbial strain. However, when nutritional conditions are restricted during the exponential, stationary growth phase, microbial strain synthesizes biosurfactants intracellularly or extracellularly [10]. The source of the microorganisms that degrade a specific pollutant also affects the nature of biosurfactants. This idea probably stems from the fact that an isolated bacterium can utilize a contaminant as a source of energy or a food supply, whereas other bacteria or microorganisms that do not produce surfactants cannot [11].

Bio-surfactants from lower to higher molecular weight can be derived from genera like *Acinetobacter, Pseudomonas, Brevibacterium, Bacillus, Rhodococcus, Clostridium, Leuconostoc, Citrobacter, Thiobacillus, Candida, Corynebacterium, Penicillium, Aspergillus, Ustilago, Saccharomyces, Enterobacter,* and *Lactobacillus* [5, 12, 13].

MECHANISM OF BIOSURFACTANT

Low concentrations of the biosurfactant CMC trigger the mobilization process. In systems involving soil and water and air and water, biosurfactants at such concentrations lower surface and interfacial tension. Interaction between the soil/oil system and biosurfactants increases the contact angle. It decreases the capillary force that holds the soil and oil simultaneously due to reduced interfacial strength. The solubilization method then happens above the biosurfactant CMC. Biosurfactant molecules produce micelles when present in these amounts, greatly enhancing the solubility of oil. In turn, the solubilization process occurs above the biosurfactant CMC. Biosurfactant molecules gather to form micelles at these concentrations, significantly improving the solubility of oil. The hydrophilic ends of biosurfactant molecules are open to the aqueous phase outside, while the hydrophobic ends are connected inside the micelle. As a result, a micelle's inside produced a favorable setting for hydrophobic organic compounds. Solubilization refers to incorporating these molecules into a micelle (Fig. **1**) [14].

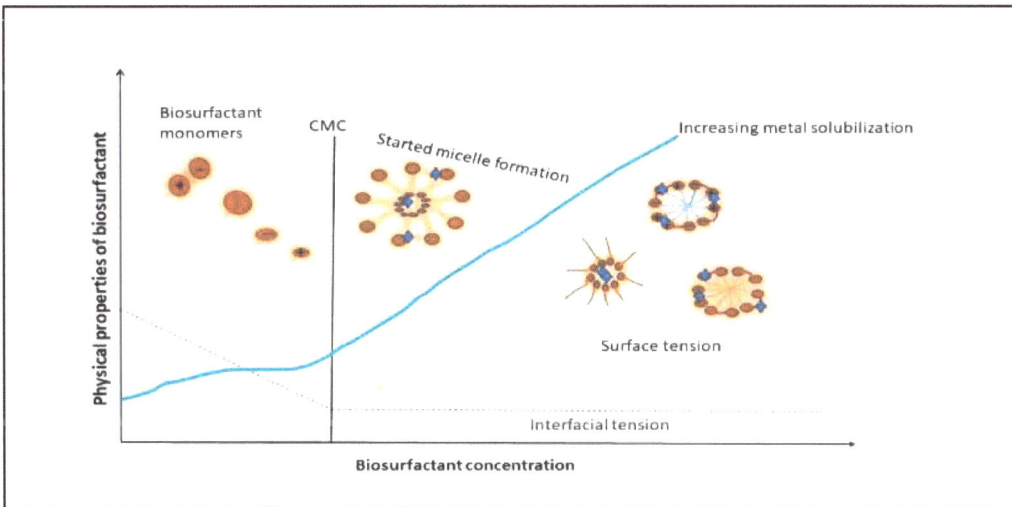

Fig. (1). Mechanism of Biosurfactant.

CLASSIFICATION OF BIOSURFACTANT

The type of biosurfactant could be determined based on the biochemical nature, structure, and source. There are five biosurfactant classes: glycolipids, lipopeptides, fatty acids, polymeric, and particulate biosurfactants Table **1** [15, 16].

Table 1. Classification of biosurfactants.

Class of Biosurfactant	Nature	Microorganisms and Associated Biosurfactant
Glycolipids	The hydrocarbons employed as the substrate determine the degree of polarity.	*Candida*: Sophorolipids *Pseudomonas aeruginosa*: Rhamnolipids. *Mycobacterium*: Trehalose
Lipopeptides	One of the most potent biosurfactants.	*Bacillus subtilis*: Surfactin, Iturin, Fengycin *Bacillus licheniformis*: Lichenysin
Fatty Acid biosurfactant	It is saturated in fatty acids (C_{12}-C_{14})	*Arthobacter* *P. aeruginosa*
Polymeric biosurfactant	High viscosity, resistant to shear and tensile strength	*Candida lipolytica*: Liposan *Acinetobacter:* Alasan
Particulate biosurfactant	Extracellular membrane vesicles of microbial cells	*Acinetobacter* sp.

Glycolipids

Glycolipids are composed of long-chain aliphatic or hydroxy aliphatic acids and carbohydrates such as mono-, di-, tri-, and tetrasaccharides like glucose, mannose, galactose, glucuronic acid, rhamnose, and galactose sulfate. Trehalose lipids, rhamnolipids, sophorolipids, diglycosyl diglycerides, and mannosyl erythritol lipids are some glycolipids that are the most effective [1]. Many distinct hexose lipids are also included.

As sugar-based bio emulsifiers, mannosylerythritol lipid A, glycoglycerolipid [17], and other glycolipids have been documented [1].

Trehalose Lipids

There are numerous documented structural varieties of microbial trehalose lipid biosurfactants. Most *Mycobacterium, Nocardia*, and *Corynebacterium* species are linked *via* the disaccharide trehalose to mycolic acids at positions C-6 and C-6'. Mycolic acids are long-chain, α-branched- β- hydroxy fatty acids. The size, shape, and quantity of carbon atoms in mycolic acid and the degree of unsaturation vary amongst trehalolipids from various organisms [1].

Rhamnolipids

One of the characteristics of some *Pseudomonas* species is their high production of biosurfactants, composed of one or two rhamnose molecules connected to one or two hydroxy decanoic acid molecules. *Pseudomonas aeruginosa* was found to produce glycolipid type R-1 in 1965, according to Edward and Hayashi [18]. This type of lipid contains two rhamnose and two hydroxy decanoic units.

Sophorolipids

The primary producers of sophorolipids are yeasts like *Torulopsis bombicola*, *Wickerhamiella domericqiae*, and *T. apicola*, which link a dimeric carbohydrate called sophorose to a long-chain hydroxy fatty acid [19, 20]. When sophorolipids are present at concentrations between 40 and 100 mg/l in micelles, they can minimize the interfacial tension of water from 72.8 mN/m to 40 mN/m [1].

Mannosylerythritol Lipids

Yeast species include *Candida Antarctica* [21] and *Candida* sp. SY 16 [22] produces this glycolipid biosurfactant, which is made up of the sugar mannosylerythritol. It was discovered that hexanoic, dodecanoic, tetra decanoic, or tetra decanoic acids make up the majority of the fatty acid content of biosurfactants; at critical micelle concentrations of 10 mg/l, mannosylerythritol lipids produced by *Candida* sp. SY 16 reduced the surface tension of water to 29 dyne/cm, whereas the minimal interfacial tension was 0.1 dyne/cm against kerosene [22].

Lipopeptides

Surfactin

One of the most efficient biosurfactants is surfactin; a cyclic lipopeptide first discovered in the bacteria *B. subtilis* ATCC-21332. Surfactin is the term given to it because of its high surfactant activity. With a critical micelle value of 0.017 g/l, surfactin can decrease surface tension from 72 to 27.9 mN/m. Surfactin exhibits potent antiviral, antimycoplasma, antitumoral, anticoagulant, and enzyme inhibitory properties, according to recent investigations [23, 24].

Iturin

Iturin A, the most well-known and first iturin group chemical, was extracted from a *Bacillus subtilis* strain collected from the land of Ituri, Zaire, and its structure was clarified [25]. Five additional lipopeptides, including iturin A_L, mycosubtilin, bacillomycin L, D, F, and L_C (or bacillopeptin), were later isolated from different strains of *Bacillus subtilis*. Because these lipopeptides share a similar chemical makeup, they are collectively referred to as "iturins," which is the general term for this class of lipopeptides [26].

Fengycin

Fengycin is a lipodecapeptide with a side chain of β-hydroxy fatty acid and consists of C15 to C17 variations with a distinctive Ala-Val dimorphy at placesix in the peptide ring [1]. By employing the electrospray ionization mass spectrometry (ESI-MS) approach, Wang *et al* . [27] have shown how to identify the fengycin homologues produced by *B. subtilis*.

Lichenysin

Bacillus licheniformis produces lichenysin, which resembles surfactin in terms of structure and physiochemical characteristics. Lichenysin is said to be far more effective at promoting the dispersion of colloidal 3-silicon carbide and aluminum nitride slurries than chemical agents and to be steady for various environmental factors pH, temperature, and NaCl concentration [1].

Fatty Acid Biosurfactant

Saturated fatty acids in the C12–C14 range and complex fatty acids with hydroxyl groups and alkyl branching make up the fatty acid biosurfactants. When growing on n-alkanes, some bacteria and yeast create significant quantities of phospholipid and fatty acids surfactants. *Acinetobacter* sp forms phosphatidyl ethanolamine-filled vesicles that develop transparent micro-emulsions of alkanes in water. 1-N [28].

Polymeric Biosurfactant

Polymeric biosurfactants are high molecular weight biopolymers with high viscosity, tensile strength, and shear resistance. Various kinds of polymeric biosurfactants are illustrated by the following [1].

Emulsan

Emulsan, an effective extracellular polymeric bio-emulsifier produced by *Acinetobacter calcoaceticus* RAG-1, is classified as a polyanionic amphipathic heteropolysaccharide. The repeating units of the trisaccharide are made up of N-acetyl-d-galactosamine, N-acetylgalactosamine uronic acid, and an unexplained N-acetylamino sugar up the heteropolysaccharide backbone. Apoemulsan, a substance produced when the protein fraction is removed, has a substantially weaker emulsifying ability on hydrophobic substrates like n-hexadecane. Cell surface esterase is a crucial protein connected to the emulsan complex [1].

Biodispersan

Biodispersan is an extracellular, nondialyzable surface-active dispersant produced by *A. calcoaceticus* A-2. The surface-active ingredient of biodispersan is an anionic heteropolysaccharide with an average molecular weight of 51,400 that contains four reducing sugars: glucosamine, 6-methylaminohexose, galactosamine, uronic acids, and an unnamed amino sugar. According to Elkeles and his colleagues, strain *A. calcoaceticus* A-2 mutants with deficient protein secretion could help make biodispersan [1].

Alasan

Polyaromatic hydrocarbons have been solubilized and degraded by Alasan produced by *A. radioresistens* KA-53. A protein with a 35.77 kD mass known as AlnA is the surface-active component of alasan that resemble *Escherichia coli* external membrane protein A (OmpA), which does not show emulsifying action. The three proteins having a molecular mass of 16, 31, and 45 kD were isolated from A radioresistant. The emulsifying action of protein 45kD is 11% higher than 16 and 31kD [1].

Liposan

Liposan, an extracellular water-soluble emulsifier produced by *C. lipolytica*, is made up of 83% (w/v) carbohydrates (composed of heteropolysaccharides like glucose, galactose, galactosamine, and galacturonic acid) and 17% (w/v) protein [1].

Particulate Biosurfactant

Extracellular membrane vesicles of microbial cells, which aid in emulsifying hydrocarbons, are some examples of particulate biosurfactants. Protein, phospholipids, and lipo-polysaccharides make up the vesicles of the *Acinetobacter* sp. Strain HO1-N has a diameter of twenty to fifty nm and a buoyant density of 1.158 cubic cm [28].

FACTORS THAT INFLUENCE THE BIOSURFACTANT PRODUCTION

Microorganism plays a vital role in the production of biosurfactants. Besides microbial strain, factors like culture incubation condition, media/ substrate types, environmental parameters like temperature, pH, and salt concentration, and mechanical factors like agitation speed and aeration affect biosurfactant production [29].

Media/Substrate

Carbon Source

Carbon substrate, both qualitatively and quantitatively, affects bio-surfactant production. It has been reported that diesel, crude oil, sucrose, glycerol, and glucose are the primary sources of carbon substrate [30].

Nitrogen Source

Nitrogen influences microbial growth as it is one of the essential components of protein and enzymes. Nitrogen compounds like yeast extract, meat extract, sodium nitrate, ammonium nitrate, ammonium sulfate, and urea peptone have been utilized for biosurfactant synthesis. Yeast extract is one of the most popular nitrogen sources for making biosurfactants. It has been reported that *Arthrobacter paraffineus* prefers ammonium salts and urea as nitrogen sources, whereas *P. aeruginosa* uses nitrate to support maximal surfactant synthesis [30].

Salt Concentration

As salt content affects the cellular activities of microorganisms, it also has a comparable impact on the generation of biosurfactants in a particular medium. However, contradictory observations were noted for some biosurfactant products, which were not impacted by values up to 10% (weight/volume), despite the presence of minor CMC reductions [30].

Environmental Parameters

The synthesis of biosurfactants can be affected by a slight change in environmental factors like pH and temperature. Thus, it is essential to standardize the bioprocess to produce significant amounts of biosurfactants. According to the reports, most biosurfactant productions occur at temperatures of 25-300°C and pH 8 [30].

Mechanical Factors

The movement of O_2 from the gas to the liquid phase is facilitated by both aeration and agitation, which are significant features that affect the synthesis of biosurfactants. It has been proposed that synthesizing bio-emulsifiers can improve the solubilization of water-insoluble substrates and ease nutrient delivery to bacteria. It was studied that when the airflow rate was one vvm and the dissolved oxygen concentration was kept at 50% of saturation, the most excellent manufacturing value of the surfactant (45.5g/l) was attained [31].

APPLICATION OF BIOSURFACTANT

Bio-surfactants can be applied in industry, cosmetics, pharmaceuticals, detergent, food, and agriculture for processes like lubrication, wetting, softening, fixing dyes, making emulsions, stabilizing dispersions, and foaming. In addition, bio-surfactant can potentially remediate environmental pollutants like heavy metals, hydrocarbons, and crude oil from contaminated soil, wasteland, aquatic bodies, and oil refineries. Biosurfactants are eco-friendly as well as nontoxic by nature. They can perform at extreme pH, high temperature, and high salt concentration conditions. They can be synthesized from waste product, making them pocket-friendly and encouraging the world to apply them to environmental pollutant degradation [14].

Role of Bio-surfactants in the Remediation of Agricultural Contaminants

All nations are very concerned about how to increase agricultural output to fulfill the demands of the expanding human population. The use of environmentally friendly substances is currently required for sustainable agriculture. Biosurfactants derived from environmental isolates have a promising future in the agriculture sector. Biosurfactants have more advantages than chemically produced surfactants; they can be applied in agricultural fields to enhance soil quality, accelerate the biodegradation of pollutants, and indirectly promote plant growth because they have antimicrobial activity and increase plant-microbe interactions that are beneficial for plants. Since these natural surfactants are found to be employed as a carbon source by soil-dwelling bacteria, they potentially replace the harsh surfactants presently used in the pesticide industries [32, 33].

Optimal soil micro- or macronutrient concentrations are necessary for the growth and production of the crop ecosystem. The elements found in minimum concentration in the soil directly influence the physiological functions of the plant. These nutrients' deficit or surplus availability reduces plant growth and causes various diseases. The effectiveness of trace elements in the soil is negatively impacted by ongoing climate change, an increase in global temperature, variations in soil pH, an increase in salinity, or the deposition of environmental contaminants, ultimately leading to lower crop production and poor food quality [34].

Biosurfactants like rhamnolipid (produced by *Pseudomonads)* and lipopeptide are associated with the soil amendment process to enhance crop production. Surface water, groundwater, and waste streams contaminated with hydrophobic organic compounds, such as polycyclic aromatic hydrocarbons, could be treated through bio-surfactants [15, 35] and metals [15]. The excessive use of fungicides with metal salts, sewage, and sludge in agricultural lands causes heavy metal pollution. Although these heavy metals serve as crucial micronutrients and help in various physiological processes of plant metabolism, they can damage plant growth at greater doses, resulting in root tissue necrosis and leaf purpling [32]. The biosurfactants produced by *Pseudomonas* sp., *Bacillus* sp., and *Acinetobacter* sp. can remove heavy metals from polluted soil and potentially speed up the degradation of pesticides [36]. Moreover, biosurfactants like rhamnolipid and surfactin use foaming-surfactant technology to remove heavy metals, including Ni, Cd, Mg, Mn, Ca, Ba, Li, Cu, and Zn (ions) from the soil [37, 38].

Role of Biosurfactant in Remediation of Oil Pollution

The biosurfactants can be applied extensively in the petroleum industry to prevent clogging pipelines, manage oil spills, control oil pollution, and clean up oil sludge. The demand for soil washing and microbial-enhanced oil recovery (MEOR) technologies in the petroleum industry is increasing daily [1, 4, 15, 39, 40].

Microbial Enhanced Oil Recovery (MEOR)

Microbial Enhanced Oil Recovery (MEOR) is an organic oil recovery process that uses microbes and their metabolites like acids, biopolymers, enzymes, gases, solvents, biomass, biosurfactants, *etc.* to extract residual crude oil, the recovery of which is impossible during the first two initial phases (primary and secondary) [41]. Biosurfactant-enhanced oil recovery is one of the most innovative and promising techniques for recovering a significant amount of residual oil. The high viscosity of crude petroleum oil and the low permeability of the reservoir make high interfacial tension between water and petroleum crude. It causes difficulty extracting oil from the pool [4].

The facilitation of oil drive through porous media by changing the interfacial properties (decrease of interfacial tension of oil-water-minerals, increase of the reservoir driving force, and speed up of fluidity) of the oil-water-mineral is the first principle behind the MEOR. In the MEOR process, the microorganisms consume oil (hydrocarbons) as their food source and produce bio-products like acids, biopolymers, enzymes, gases, solvents, biomass, and surfactants. These metabolites assist in changing the physicochemical properties of crude oil to stimulate oil-water-rock interactions to improve remaining oil recovery [40]. The MEOR methods can be applied on and of site also. During *in-situ* (onsite) MEOR treatment, the native or pre-cultured bacteria insert the required nutrient into the contaminated area and produce surfactants that help degrade the oil pollution. The cost of *in-situ* MEOR is not high as the bacteria produce the surfactant on site to treat the contaminated area. However, *ex-situ*-generated biosurfactants can also be employed to promote microbial development in oil reserves [42].

Alvarez *et al* . [43] performed MEOR simulations using the biosurfactant made by *B. amyloliquefaciens* and attained a petroleum hydrocarbon recovery rate of more than 90%. The core flood model used rhamnolipid generated by *P. aeruginosa* 5514 for MEOR [4]. It has been studied that *ex-situ* bioaugmentation using rhamnolipid could increase oil recovery by 8.82% as a percentage of residual oil saturation (R.O.S.). They reported that rhamnolipid mobilized oil in the micro model and recommended rhamnolipid usage in MEOR. However, the use of biosurfactants in MEOR might be debatable since the quantities needed to recover oil leftovers trapped in the porous rock may make the procedure uneconomical [15].

Oil Cleanup of Storage Tank

Daily, a massive volume of crude oil is transported, transferred to refineries, and shifted into storage tanks. Periodic cleaning is needed for these tanks' maintenance. However, waste and heavy oil fractions deposited at the lowest and on the peripheral area of storage tanks are highly viscous and turn into solid stacks, occupying a significant amount of space in the tanks as well as lowering their capacity and affecting the quality of the products that cannot be removed with the conventional pumping. This substance must be manually cleaned and washed off with solvents, which is a risky, labor-intensive, expensive, and time-consuming process that may involve hot water spraying and solvent liquefaction [39, 44, 45].

As an alternative cleaning method, microbial biosurfactants can reduce the viscosity of sludge and oil deposits by forming an oil-in-water emulsion that makes waste pumping easier. Additionally, when the emulsion breaks, this method enables crude oil recovery [39]. The *Actinomycete Gordonia* sp. produced new biosurfactant *JE1058BS* was investigated to treat oil tank bottom sludge [45]. The biosurfactant proved more effective at dispersion than a chemical or plant-derived surfactant, and it remained stable for at least three weeks [46]. They assessed the impact of *P.aeruginosa SH 29* supernatant on cleaning oil-polluted vessels. The oil was removed from the vessels' bottoms and walls within 15 minutes and drifted on the supernatant as a separate phase. Scientists/authors suggest that the biosurfactant in *P. aeruginosa* S.H. 29's sterilized supernatant can be used to clean crude oil storage tanks and other vessels.

Role of Bio-surfactants in Hydrocarbon Degradation

Several sources of hydrocarbon soil pollution include drilling, leaking pipelines, storage tanks, transportation, *etc.*, which cause long-lasting harmful effects on the environment. Heavy metals and hydrocarbons resist removal due to their high hydrophobicity, especially when adsorbing soil particles [39, 47]. The U.S. Environmental Protection Agency offers a variety of physical, chemical, and biological solutions for cleaning the soil contaminated with hydrocarbons. The implementation of biosurfactants appears as a harmless substitution for increasing the solubility of hydrophobic compounds by enabling the desorption and solubilization of hydrocarbons and making it easier for microbial cells to assimilate these compounds [48]. Biosurfactants use two different processes to biodegrade oil-derived hydrocarbons. The first includes increasing the hydrophobic substrate's bioavailability to microorganisms, which directs to minimizing the surface tension of the medium around the bacterium and the interfacial tension between its cell wall and hydrocarbon molecules. The other process involves the biosurfactant interacting with the cell surface, which leads to changes in the membrane, promoting hydrocarbon adhesion (increasing hydrophobicity) and lowering the lipopolysaccharide index of the cell wall without harming the membrane. As a result, biosurfactants prevent the development of hydrogen bridges and permit hydrophobic-hydrophilic interactions, which result in molecular rearrangements, lower the liquid's surface tension by increasing its surface area, and promote bioavailability and subsequent biodegradability [49]. Biosurfactants facilitate the biodegradation, solubilization, mobilization, or emulsification of hydrocarbons to remove them from the environment [36]. The biosurfactant produced by *B. brevis* was used by Mouafi *et al* . [50] to disperse and emulsify motor oil in water successfully. The ability of several biosurfactants to remove petroleum-derived compounds from contaminated soil and water has been investigated. Rhamnolipids have been effectively employed in environmental pollutant degradation processes [10, 39]. Other surfactants released by species of *Pseudomonas* [46], *Bacillus* [39], and *Candida* [39, 51] have also been efficaciously used in the treatment of contaminated soil. The deficit or surplus availability of these nutrients can damage plant health and affect plant growth and development. An emulsifier generated from *P. aeruginosa* SB30 can disperse the oil into fine droplets. Therefore, it can be applied to remove oil from the contaminated beach [14].

Clean-up Combined Technology

Clean-up combined technology is pocket-friendly to treat soil contaminated with oil pollutants. This technology consists of two steps; in the first step, washing of the contaminated soil occurs where the migrated portion is treated by using the bio-surfactant; in the second stage, the amount that remains in the ground is reduced by using biodegradation or phytoremediation process. A pilot scale was designed in an area of $340m^2$ to treat the contaminated soil of size $1000m^3$. Initially, the oil pollutants were in the range of $180-270$ $g \cdot kg^{-1}$ of ground in the contaminated soil; after washing, the concentration of contaminants was decreased to $34-59$ $g \cdot kg^{-1}$ of soil, and then the treatment of soil by application of bio-surfactant reduced the pollutant concentration to $3.2-7.3$ $g \cdot kg^{-1}$ of the soil [36].

Soil Washing Technology

Instead of affecting bacterial metabolism or altering the characteristics of their cell surfaces, the chemical-physical qualities of the biosurfactant define soil-washing technology. However, the procedures could make it more bioavailable for bioremediation. In a process known as washing, aqueous solutions of biosurfactants can also be applied to produce chemicals with limited solubility from soil and other media [14, 36].

When employing a soil-washing procedure to remove crude oil from polluted soil, Urum *et al*. [52] tested the effectiveness of several surfactant solutions. They showed that rhamnolipid biosurfactants (44%) and synthetic surfactants (46%), such as sodium dodecyl sulfate (SDS), eliminated crude oil more effectively than natural surfactants, such as saponins (27%).

Franzetti *et al*. [53] assessed the use of surface-active compounds made by *Gordonia* sp. strain BS29 in two soil treatment techniques: washing of soils polluted with crude oil, PAHs, and heavy metals and bioremediation of soils contaminated with aliphatic and aromatic hydrocarbons (microcosm bioremediation experiment) (batch experiment). This research is the first investigation into how *Gordonia sp* surface-active's chemicals might be used in methods for soil contamination cleanup.

Role of Biosurfactants in Heavy Metal Degradation

Heavy metal contamination of soil habitats poses severe risks to people and other living things in the ecosystem. Heavy metals are exceedingly hazardous; thus, it has been discovered that even small amounts in the soil can have adverse effects. Rich metal-contaminated soil can now be cleaned up using a variety of approaches. Using plants (phytoremediation) or microorganisms (bioremediation), biological processes can remove metals from the soil. To reduce metal contamination, microbes have been used for many years. The mobility and toxicity of heavy metals alter because they cannot degrade biologically and can only be changed from one chemical state to another. There are many ways that microorganisms can affect metals. Alkylation and redox reactions are two ways that some metals can be changed. Both intracellular, metabolism-dependent (active), and metabolism-independent (passive) absorption of metals are possible for bacteria. Through pH changes, the production or release of chemicals that alter metal mobility or microorganisms can indirectly affect metal mobility [14].

The capacity of biosurfactants to make complexes with metals is the primary factor contributing to their efficacy for bioremediation of heavy metal-contaminated soil. Using ionic bonds, the anionic biosurfactants build non-ionic complexes with metals [36, 54]. These bindings are more resilient than the metal's bonds to the soil. Because the interfacial tension is reduced, metal-biosurfactant complexes are desorbed from the soil matrix to the solution. The cationic biosurfactants can replace the same charged metal ions by competing for some negatively charged surfaces, but not all of them (ion exchange). The biosurfactant micelles are also capable of removing metal ions from soil surfaces. Metals can be bound by the polar head groups of micelles, which mobilize the metals in water.

Additionally, Das *et al* . [55] proposed employing a biosurfactant made by a marine bacterium to remove heavy metals from solutions. There has previously been information regarding the beneficial effects of marine biosurfactants in the remediation of polyaromatic hydrocarbons [56], but nothing about the biosurfactant's products in the remediation of heavy metals. The investigation results showed that the tested anionic biosurfactant could bind metal ions and that the amount of the lead and cadmium metals removed influenced the amount of the metals and the biosurfactants. Water treatment for heavy metal-containing wastewater may benefit from the capability of biosurfactants of sea origin to bind harmful heavy metals and generate an insoluble precipitate.

CONCLUSION AND FUTURE PERSPECTIVE

The application of biosurfactants derived from various environmental microorganisms has been investigated for the pollutant degradation that destroys the ecosystem. The biosurfactants are organic, non-harmful, and bio-degradable by nature, urging the world to apply them as a treatment to remediate the environmental pollutant. Although biosurfactants can potentially treat environmental contaminates, commercial production of biosurfactants is still minimal, making them expensive and less available. Whatever the results we have found, all are lab-scale studies. Thus, the scale-up of the application and production of bio-surfactant is essential. Not only scale up, but also the standardization of isolation and purification process of biosurfactant should be further studied.

Moreover, to use biosurfactants in phytoremediation on a broad scale, research is needed to determine whether or not they could be poisonous to plants. Even though biosurfactants are supposed to be environmentally favorable, several investigations showed that, in some cases, they might be harmful to the environment. However, the prudent and controlled application of these intriguing surface-active agents will undoubtedly contribute to the improved removal of hazardous environmental pollutants and give us a healthy environment.

ACKNOWLEDGMENTS

The authors are grateful to the Director, CSIR-CSMCRI, Bhavnagar for providing infrastructural facilities. The manuscript has been assigned CSIR-CSMCRI-166/2022 registration. This work was supported by in-house funding (MLP-0045). MM acknowledges the PhD fellowship grant from UGC.

REFERENCES

[1]　Mukherjee, A.K.; Das, K. *Microbial surfactants and their potential applications: An overview*; Biosurfactants, **2010**, pp. 54-64.
[http://dx.doi.org/10.1007/978-1-4419-5979-9_4]

[2]　Singh, P.; Patil, Y.; Rale, V. Biosurfactant production: Emerging trends and promising strategies. *J. Appl. Microbiol.,* **2019**, *126*(1), 2-13.
[http://dx.doi.org/10.1111/jam.14057] [PMID: 30066414]

[3]　Kumar, A.; Singh, S.K.; Kant, C.; Verma, H.; Kumar, D.; Singh, P.P.; Modi, A.; Droby, S.; Kesawat, M.S.; Alavilli, H.; Bhatia, S.K.; Saratale, G.D.; Saratale, R.G.; Chung, S.M.; Kumar, M. Microbial biosurfactant: A new frontier for sustainable agriculture and pharmaceutical industries. *Antioxidants,* **2021**, *10*(9), 1472.
[http://dx.doi.org/10.3390/antiox10091472] [PMID: 34573103]

[4]　Varjani, S.J.; Upasani, V.N. Critical review on biosurfactant analysis, purification and characterization using rhamnolipid as a model biosurfactant. *Bioresour. Technol.,* **2017**, *232*, 389-397.
[http://dx.doi.org/10.1016/j.biortech.2017.02.047] [PMID: 28238638]

[5]　Jimoh, A.A.; Lin, J. Biosurfactant: A new frontier for greener technology and environmental sustainability. In: *Ecotoxicology and Environmental Safety (184)*; Academic Press, **2019**.
[http://dx.doi.org/10.1016/j.ecoenv.2019.109607]

[6]　Anjum, F.; Gautam, G.; Edgard, G.; Negi, S. Biosurfactant production through *Bacillus* sp. MTCC 5877 and its multifarious applications in food industry. *Bioresour. Technol.,* **2016**, *213*, 262-269.

[http://dx.doi.org/10.1016/j.biortech.2016.02.091] [PMID: 27013189]

[7]   Bezza, F.A.; Nkhalambayausi Chirwa, E.M. Biosurfactant from *Paenibacillus dendritiformis* and its application in assisting polycyclic aromatic hydrocarbon (PAH) and motor oil sludge removal from contaminated soil and sand media. *Process Saf. Environ. Prot.,* **2015**, *98*, 354-364.
[http://dx.doi.org/10.1016/j.psep.2015.09.004]

[8]   Prasad, N.K.; Jayakumar, R.; Wagdevi, P.; Ramesh, S. Application of biosurfactants as bioabsorption agents of heavily contaminated soil and water. **2021**. Available from: https://doi.org/https://doi.org/10.1016/B978-0-12-822696-4.00021-8

[9]   Tahir, Z.; Nazir, M.S.; Aslam, A.A.; Bano, S.; Ali, Z.; Akhtar, M.N.; Azam, K.; Abdullah, M.A. Active metabolites and biosurfactants for utilization in environmental remediation and eco-restoration of polluted soils. *Green Sustainable Process for Chemical and Environmental Engineering and Science,* **2021**, 31-51.
[http://dx.doi.org/10.1016/B978-0-12-822696-4.00007-3]

[10]   Santos, D.; Rufino, R.; Luna, J.; Santos, V.; Sarubbo, L. Biosurfactants: Multifunctional biomolecules of the 21st century. *Int. J. Mol. Sci.,* **2016**, *17*(3), 401.
[http://dx.doi.org/10.3390/ijms17030401] [PMID: 26999123]

[11]   Patowary, K.; Patowary, R.; Kalita, M.C.; Deka, S. Characterization of biosurfactant produced during degradation of hydrocarbons using crude oil as sole source of carbon. *Front. Microbiol.,* **2017**, *8*, 279.
[http://dx.doi.org/10.3389/fmicb.2017.00279] [PMID: 28275373]

[12]   Li, J.; Deng, M.; Wang, Y.; Chen, W. Production and characteristics of biosurfactant produced by *Bacillus pseudomycoides* BS6 utilizing soybean oil waste. *Int. Biodeterior. Biodegradation,* **2016**, *112*, 72-79.
[http://dx.doi.org/10.1016/j.ibiod.2016.05.002]

[13]   Shekhar, S.; Sundaramanickam, A.; Balasubramanian, T. Biosurfactant producing microbes and their potential applications: A review. *Crit. Rev. Environ. Sci. Technol.,* **2015**, *45*(14), 1522-1554.
[http://dx.doi.org/10.1080/10643389.2014.955631]

[14]   Unaeze, CH Application of biosurfactants in environmental remediation. *J environ. sci. toxicol. food. technol., 14*(10), 30-42.
[http://dx.doi.org/10.9790/2402-1410043042]

[15]   Sarubbo, L.A.; Silva, M.G.C.; Durval, I.J.B.; Bezerra, K.G.O.; Ribeiro, B.G.; Silva, I.A.; Twigg, M.S.; Banat, I.M. Biosurfactants: Production, properties, applications, trends, and general perspectives. *Biochem. Eng. J.,* **2022**, *181*, 108377.
[http://dx.doi.org/10.1016/j.bej.2022.108377]

[16]   Sobrinho, HB; Luna, JM; Rufino, RD; Porto, AL; Sarubbo, LA Biosurfactants: Classification, properties and environmental applications. *Recent developments in biotechnology.,* **2013**, *11*(14), 1-29.

[17]   Nakata, K. Two glycolipids increase in the bioremediation of halogenated aromatic compounds. *J. Biosci. Bioeng.,* **2000**, *89*(6), 577-581.
[http://dx.doi.org/10.1016/S1389-1723(00)80060-2] [PMID: 16232801]

[18]   Edwards, J.R.; Hayashi, J.A. Structure of a rhamnolipid from *Pseudomonas aeruginosa. Arch. Biochem. Biophys.,* **1965**, *111*(2), 415-421.
[http://dx.doi.org/10.1016/0003-9861(65)90204-3] [PMID: 4285853]

[19]   Tulloch, A.P.; Hill, A.; Spencer, J.F. A new type of macrocyclic lactone from Torulopsis apicola. *Chem. Commun.,* **1967**, (12), 584-586.

[20]   Chen, J.; Song, X.; Zhang, H.; Qu, Y. Production, structure elucidation and anticancer properties of sophorolipid from Wickerhamiella domercqiae. *Enzyme Microb. Technol.,* **2006**, *39*(3), 501-506.
[http://dx.doi.org/10.1016/j.enzmictec.2005.12.022]

[21]   Kitamoto, D.; Yanagishita, H.; Shinbo, T.; Nakane, T.; Kamisawa, C.; Nakahara, T. Surface active properties and antimicrobial activities of mannosylerythritol lipids as biosurfactants produced by

Candida antarctica. J. Biotechnol., **1993**, *29*(1-2), 91-96.
[http://dx.doi.org/10.1016/0168-1656(93)90042-L]

[22]   Kim, H.S.; Yoon, B.D.; Choung, D.H.; Oh, H.M.; Katsuragi, T.; Tani, Y. Characterization of a biosurfactant, mannosylerythritol lipid produced from *Candida* sp. SY16. *Appl. Microbiol. Biotechnol.,* **1999**, *52*(5), 713-721.
[http://dx.doi.org/10.1007/s002530051583] [PMID: 10570818]

[23]   Sen, R.; Swaminathan, T. Characterization of concentration and purification parameters and operating conditions for the small-scale recovery of surfactin. *Process Biochem.,* **2005**, *40*(9), 2953-2958.
[http://dx.doi.org/10.1016/j.procbio.2005.01.014]

[24]   Mukherjee, A.K.; Das, K. Correlation between diverse cyclic lipopeptides production and regulation of growth and substrate utilization by *Bacillus subtilis* strains in a particular habitat. *FEMS Microbiol. Ecol.,* **2005**, *54*(3), 479-489.
[http://dx.doi.org/10.1016/j.femsec.2005.06.003] [PMID: 16332345]

[25]   Peypoux, F.; Besson, F.; Michel, G.; Delcambe, L.; Das, B.C. Structure de l'iturine C de *Bacillus subtilis. Tetrahedron,* **1978**, *34*(8), 1147-1152.
[http://dx.doi.org/10.1016/0040-4020(78)80138-0]

[26]   Kajimura, Y.; Sugiyama, M.; Kaneda, M. Bacillopeptins, new cyclic lipopeptide antibiotics from *Bacillus subtilis* FR-2. *J. Antibiot.,* **1995**, *48*(10), 1095-1103.
[http://dx.doi.org/10.7164/antibiotics.48.1095] [PMID: 7490214]

[27]   Wang, J.; Liu, J.; Wang, X.; Yao, J.; Yu, Z. Application of electrospray ionization mass spectrometry in rapid typing of fengycin homologues produced by *Bacillus subtilis. Lett. Appl. Microbiol.,* **2004**, *39*(1), 98-102.
[http://dx.doi.org/10.1111/j.1472-765X.2004.01547.x] [PMID: 15189295]

[28]   Saravanan, V.; Vijayakuma, S. Biosurfactants-types, sources and applications. *Res. J. Microbiol.,* **2015**, *10*(5), 181-192.
[http://dx.doi.org/10.3923/jm.2015.181.192]

[29]   Md, F. Biosurfactant: Production and Application. *J. Pet. Environ. Biotechnol.,* **2012**, *3*(4), 124.
[http://dx.doi.org/10.4172/2157-7463.1000124]

[30]   Desai, J.D.; Banat, I.M. Microbial production of surfactants and their commercial potential. *Microbiol. Mol. Biol. Rev.,* **1997**, *61*(1), 47-64.
[http://dx.doi.org/10.1128/mmbr.61.1.47-64.1997] [PMID: 9106364]

[31]   Adamczak, M.; Bednarski, W. Influence of medium composition and aeration on the synthesis of biosurfactants produced by *Candida antarctica. Biotechnol. Lett.,* **2000**, *22*(4), 313-316.
[http://dx.doi.org/10.1023/A:1005634802997]

[32]   Sachdev, D.P.; Cameotra, S.S. Biosurfactants in agriculture. *Appl. Microbiol. Biotechnol.,* **2013**, *97*(3), 1005-1016.
[http://dx.doi.org/10.1007/s00253-012-4641-8] [PMID: 23280539]

[33]   Scott, M.J.; Jones, M.N. The biodegradation of surfactants in the environment. *Biochim. Biophys. Acta Biomembr.,* **2000**, *1508*(1-2), 235-251.
[http://dx.doi.org/10.1016/S0304-4157(00)00013-7] [PMID: 11090828]

[34]   Robinson, B.H.; Bañuelos, G.; Conesa, H.M.; Evangelou, M.W.H.; Schulin, R. The phytomanagement of trace elements in soil. *Crit. Rev. Plant Sci.,* **2009**, *28*(4), 240-266.
[http://dx.doi.org/10.1080/07352680903035424]

[35]   Mehetre, G.T.; Dastager, S.G.; Dharne, M.S. Biodegradation of mixed polycyclic aromatic hydrocarbons by pure and mixed cultures of biosurfactant producing thermophilic and thermo-tolerant bacteria. *Sci. Total Environ.,* **2019**, *679*, 52-60.
[http://dx.doi.org/10.1016/j.scitotenv.2019.04.376] [PMID: 31082602]

[36]   Pacwa-Płociniczak, M.; Płaza, G.A.; Piotrowska-Seget, Z.; Cameotra, S.S. Environmental applications

of biosurfactants: recent advances. *Int. J. Mol. Sci.,* **2011**, *12*(1), 633-654.
[http://dx.doi.org/10.3390/ijms12010633] [PMID: 21340005]

[37] Mulligan, C.N.; Wang, S. Remediation of a heavy metal-contaminated soil by a rhamnolipid foam. *Eng. Geol.,* **2006**, *85*(1-2), 75-81.
[http://dx.doi.org/10.1016/j.enggeo.2005.09.029]

[38] Neilson, J.W.; Artiola, J.F.; Maier, R.M. Characterization of lead removal from contaminated soils by nontoxic soil-washing agents. *J. Environ. Qual.,* **2003**, *32*(3), 899-908.
[http://dx.doi.org/10.2134/jeq2003.8990] [PMID: 12809290]

[39] Silva, R.; Almeida, D.; Rufino, R.; Luna, J.; Santos, V.; Sarubbo, L. Applications of biosurfactants in the petroleum industry and the remediation of oil spills. *Int. J. Mol. Sci.,* **2014**, *15*(7), 12523-12542.
[http://dx.doi.org/10.3390/ijms150712523] [PMID: 25029542]

[40] Shibulal, B.; Al-Bahry, S.N.; Al-Wahaibi, Y.M.; Elshafie, A.E.; Al-Bemani, A.S.; Joshi, S.J. Microbial enhanced heavy oil recovery by the aid of inhabitant spore-forming bacteria: an insight review. *ScientificWorldJournal,* **2014**, *2014*, 1-12.
[http://dx.doi.org/10.1155/2014/309159] [PMID: 24550702]

[41] Gao, C.H.; Zekri, A. Applications of microbial-enhanced oil recovery technology in the past decade. *Energy Sources A Recovery Util. Environ. Effects,* **2011**, *33*(10), 972-989.
[http://dx.doi.org/10.1080/15567030903330793]

[42] Geetha, S.J.; Banat, I.M.; Joshi, S.J. Biosurfactants: Production and potential applications in microbial enhanced oil recovery (MEOR). *Biocatal. Agric. Biotechnol.,* **2018**, *14*, 23-32.
[http://dx.doi.org/10.1016/j.bcab.2018.01.010]

[43] Alvarez, V.M.; Jurelevicius, D.; Marques, J.M.; de Souza, P.M.; de Araújo, L.V.; Barros, T.G.; de Souza, R.O.M.A.; Freire, D.M.G.; Seldin, L. *Bacillus amyloliquefaciens* TSBSO 3.8, a biosurfactant-producing strain with biotechnological potential for microbial enhanced oil recovery. *Colloids Surf. B Biointerfaces,* **2015**, *136*, 14-21.
[http://dx.doi.org/10.1016/j.colsurfb.2015.08.046] [PMID: 26350801]

[44] Chrysalidis, A.; Kyzas, G.Z. Applied cleaning methods of oil residues from industrial tanks. *Processes (Basel),* **2020**, *8*(5), 569.
[http://dx.doi.org/10.3390/pr8050569]

[45] Matsui, T.; Namihira, T.; Mitsuta, T.; Saeki, H. Removal of oil tank bottom sludge by novel biosurfactant, JE1058BS. *J. Jpn. Petrol. Inst.,* **2012**, *55*(2), 138-141.
[http://dx.doi.org/10.1627/jpi.55.138]

[46] Diab, A.; Din, G.E. Application of the biosurfactants produced by *Bacillus* spp.(SH 20 and SH 26) and *P. aeruginosa* SH 29 isolated from the rhizosphere soil of an Egyptian salt marsh plant for the cleaning of oil-contaminataed vessels and enhancing the biodegradat. *Afr. J. Environ. Sci. Technol.,* **2013**, *7*(7), 671-679.
[http://dx.doi.org/10.5897/AJEST2013.1451]

[47] Huesemann, M.H. *Biodegradation and bioremediation of petroleum pollutants in soil. InApplied Bioremediation and Phytoremediation*; Springer: Berlin, Heidelberg, **2004**, pp. 13-34.
[http://dx.doi.org/10.1007/978-3-662-05794-0_2]

[48] Kuyukina, M.S.; Ivshina, I.B.; Makarov, S.O.; Litvinenko, L.V.; Cunningham, C.J.; Philp, J.C. Effect of biosurfactants on crude oil desorption and mobilization in a soil system. *Environ. Int.,* **2005**, *31*(2), 155-161.
[http://dx.doi.org/10.1016/j.envint.2004.09.009] [PMID: 15661276]

[49] Aparna, A.; Srinikethan, G.; Hedge, S. Effect of addition of biosurfactant produced by *Pseudomonas* sp. on biodegradation of crude oil. *Int. Proc. Chem. Biol. Environ. Eng.,* **2011**, *6*, 71-75.

[50] Mouafi, F.E.; Abo Elsoud, M.M.; Moharam, M.E. Optimization of biosurfactant production by *Bacillus brevis* using response surface methodology. *Biotechnol. Rep.,* **2016**, *9*, 31-37.

[http://dx.doi.org/10.1016/j.btre.2015.12.003] [PMID: 28352589]

[51]   Rufino, R.D.; Luna, J.M.; Sarubbo, L.A.; Rodrigues, L.R.M.; Teixeira, J.A.C.; Campos-Takaki, G.M. Antimicrobial and anti-adhesive potential of a biosurfactant Rufisan produced by Candida lipolytica UCP 0988. *Colloids Surf. B Biointerfaces,* **2011**, *84*(1), 1-5.
[http://dx.doi.org/10.1016/j.colsurfb.2010.10.045] [PMID: 21247740]

[52]   Urum, K.; Grigson, S.; Pekdemir, T.; McMenamy, S. A comparison of the efficiency of different surfactants for removal of crude oil from contaminated soils. *Chemosphere,* **2006**, *62*(9), 1403-1410.
[http://dx.doi.org/10.1016/j.chemosphere.2005.05.016] [PMID: 16005939]

[53]   Franzetti, A.; Caredda, P.; Ruggeri, C.; Colla, P.L.; Tamburini, E.; Papacchini, M.; Bestetti, G. Potential applications of surface active compounds by Gordonia sp. strain BS29 in soil remediation technologies. *Chemosphere,* **2009**, *75*(6), 801-807.
[http://dx.doi.org/10.1016/j.chemosphere.2008.12.052] [PMID: 19181361]

[54]   Singh, P.; Cameotra, S.S. Enhancement of metal bioremediation by use of microbial surfactants. *Biochem. Biophys. Res. Commun.,* **2004**, *319*(2), 291-297.
[http://dx.doi.org/10.1016/j.bbrc.2004.04.155] [PMID: 15178405]

[55]   Das, P.; Mukherjee, S.; Sen, R. Biosurfactant of marine origin exhibiting heavy metal remediation properties. *Bioresour. Technol.,* **2009**, *100*(20), 4887-4890.
[http://dx.doi.org/10.1016/j.biortech.2009.05.028] [PMID: 19505818]

[56]   Das, P.; Mukherjee, S.; Sen, R. Improved bioavailability and biodegradation of a model polyaromatic hydrocarbon by a biosurfactant producing bacterium of marine origin. *Chemosphere,* **2008**, *72*(9), 1229-1234.
[http://dx.doi.org/10.1016/j.chemosphere.2008.05.015] [PMID: 18565569]

CHAPTER 8

Biosurfactants: New Insights in Bioengineering and Bioremediation of Crude Oil Contamination

Pyari Payal Beura[1] and **Sanjay Kumar Raul**[1,*]

[1] *Department of Biotechnology, Rama Devi Women's University, Vidya Vihar, Bhubaneswar, Odisha-751003, India*

Abstract: Human activities are the principal source of various kinds of hazardous substances in our environment, which have serious consequencesfor the well-being of the environment and people. Using standard means to lessen, degrade, and eliminate these substances is usually causing harmful effects to environment. Pesticides, crude oil sludge, and polycyclic aromatic hydrocarbons (PAHs) are toxic, mutagenic, and carcinogenic in nature. It has recently been shown to be possible to use microorganisms to breakdown and cleanse contaminated soil and water ecosystems, a process known as bio-reclamation. Biosurfactants, which are amphiphillic chemicals generated by bacteria, fungus, and yeast, have immense potential to lower the surface tension of a liquid as well as tension at the interface between 2 liquids or among a liquid and a solid. Additionally, bio surfactants strongly emulsify hydrophobic substances and create stable emulsions. Bio emulsifiers and biosurfactants are especially useful in a wide range of commercial and scientific applications, including pharmaceuticals, better oil recovery, and pollutant biodegradation. Thus, using biosurfactants to reduce crude oil pollution is an environmentally responsible strategy to developing a sustainable ecosystem.

Keywords: Biosurfactant, Biodegradable, Bioemulsifier, Environment, Pharmaceuticals.

INTRODUCTION

Surfactants are considered to be among the most important groups of chemical products since they are widely utilized in daily life and have several applications in everyday life, the agro industry, or healthcare [1]. At the moment, surfactants are almost entirely synthetic [2]. Recently, there is rising interest in biosurfactants, particularly microbial surfactants. Due to harmlessness and biodegradable characteristics, they are first regarded as environmental friendly. In

---

* **Corresponding author Sanjay Kumar Raul:** Department of Biotechnology, Rama Devi Women's University, Vidya Vihar, Bhubaneswar, Odisha-751003, India; E-mail: sanjaykumarraul@rdwu.ac.in

Arun Kumar Pradhan and Manoranjan Arakha (Eds.)

addition, they have unique physical characteristics that render them essential for their potential application in a range of industrial sectors, ranging from biotechnology to environmental remediation. Finally, they may be synthesized from renewable feedstock and have improved foaming powers, more selectivity, and specialized action in challenging environmental conditions of temperature, pH, and salinity [3].

Secondary metabolites known as biosurfactants (surface active compounds) are physiologically created during the stationary stage of the development of microbial cells. Biosurfactants comprises amphiphilic molecules having hydrophilic and hydrophobic motifs [4]. These compounds are made by microbes living on their surfaces, and they consist of separate polar and non-polar components allowing them to transform into micelles that cluster at the interface of several fluids with varying polarities [5]. They function similarly to chemical surfactants like water and oil in that they are made to lower surface pressure and have the distinctive property of lowering surface and interfacial tension. In contrast to hydrophobic domains, which are frequently made up of long-chain fatty acids, hydrophilic domains are frequently made up of sugars, amino acids, and phosphate chains [4]. Water surface tension (ST) can potentially be decreased from 72 to 35 mN/m by a superior surfactant, and the water/hexadecane interfacial tension (IT) is capable of being decreased from 40 to 1 mN/m [1]. Surfactants enable detergency, emulsification, lubrication, solubilization, and phase dispersion by reducing surface and interfacial tension. Surfactants are a highly important chemical element that are used in a wide variety of goods in very big amounts due to their numerous home and commercial uses [6]. The amount of surfactants that cause micelles form and all further surfactants introduced into the system will form micelles is known as the critical micelle concentration (CMC) in colloidal and surface chemistry. A detergent's CMC is a crucial characteristic (Fig. **1**).

Biological origin and chemical make-up are the main factors used to classify biosurfactants. Biosurfactants are technically essential because they enhance the number of surfactant types available and have distinct surface-active properties in comparison to synthetic surfactants.The major types of biosurfactants include glycolipids, lipoproteins or lipopeptides, phospholipids, fatty acids or natural lipids, polymeric surfactants, and particle surfactants. They are often biodegradable as well, which reduces the possibility of contamination.Recyclable uses for bio surfactants include emulsifiers, flocculating agents, demulsifiers, adhesives, detergents, and as amethod used in the tertiary recovery process of oil. All of these characteristics are used by bio systems in an environment. Numerous environmental uses of biosurfactants include the transfer of crude oil, increased oil recovery, and bioremediation of oil spills. Additional uses for biosurfactants include the food, cosmetics, and healthcare sectors as well as the cleansing of harmful compounds with both industrial and agricultural origins (Fig. **2**) [8]. Therefore, biosurfactants are a natural alternative to chemically produced surfactants and are chosen for a variety of industrial applications in the bioremediation, healthcare, cosmetics, oil, and food processing industries.

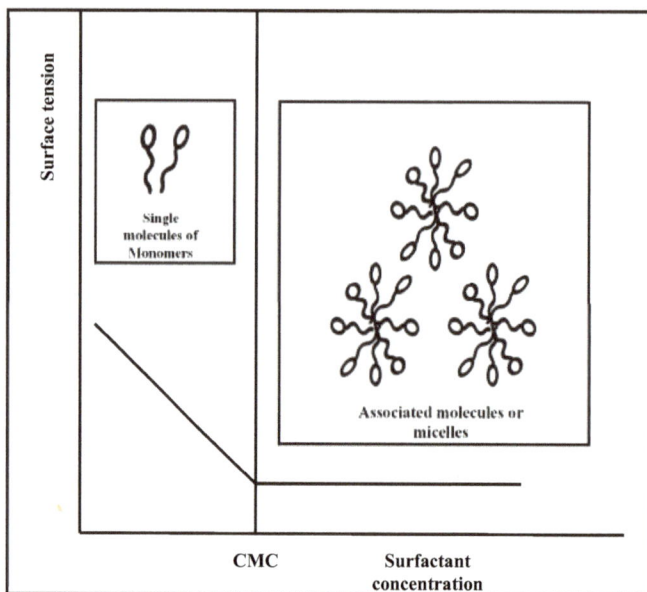

Fig. (1). CMC (critical micelle concentration) as a function of chemical or natural surfactant concentration [7].

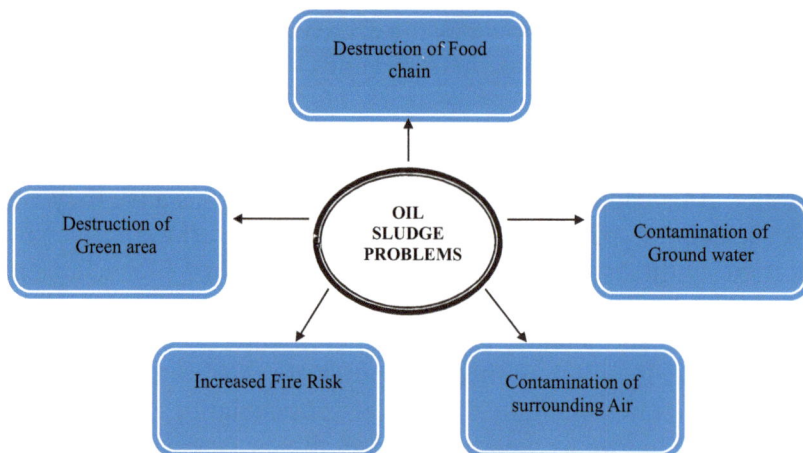

Fig. (2). Various impact of oily sludge.

Surfactants are frequently used throughout mechanical, agricultural, feeding, beauty care items, and medicinal applications; nevertheless, the great majority of these mixtures is created artificially and may generate environmental and toxicological issues due to their refractory and determined character.With recent breakthroughs in biotechnology, focus has been given to the most eco-friendly approach for producing various types of biosurfactants from microorganisms.

Characteristics of Biosurfactants

Biosurfactants are suited for commercial applications due to their superior qualities to those of their chemically manufactured equivalents and their wide substrate availability. The characteristics of microbial surfactants include surface activity, resistance to temperature, pH, and ionic quality changes, biodegradability, low toxicity, emulsifying and demulsifying capability, and antibacterial action [5].

Surface and Interface Activity

Surfactant aids in the reduction of interfacial pressure and surface strain [9]. The surface tension of water may be minimized to 25 mN m-1 and the interfacial strain between water and hexadecane to under 1 mN m-1 by using *B. subtilis*-produced surfactin [5].

Temperature, pH and Ionic Strength Tolerance

A decade ago, the generation of biosurfactants from extremophiles began to get interest due to their potential economic use. Most biosurfactants and their surface-active properties are safe when exposed to environmental variables like temperature and pH.According to McInerney *et al.*, lichenysin from *Bacillus licheniformis* is proved to be tolerant to temperature approximately 50°C, pH between 4.5 to 9.0, and NaCl and Ca levels up to 50 and 25 g/L [5].

Biodegradability

When compared to manufactured surfactants, microbially produced chemicals are quickly destroyed and can be employed for natural applications such as bioremediation/biosorption [5]. Research suggests that the poly-anionic surfactant 'Emulsan' has a substantially lower LC50 against *Photobacterium phosphoreum* than *Pseudomonas rhamnolipids* [9].

Emulsion Breaking and Emulsion Framing

An emulsion may be defined a heterogeneous framework made up of immiscible fluids that are dispersed as beads (larger than 0.1 mm in diameter).Oil-in-water or water-in-oil is the two main types of emulsions found in nature. They have a low degree of stability that can be countered by the addition of other materials. For instance, biosurfactants can maintain stable emulsions for months or even years [5].

Anti-adhesive Agents

Any collection of microorganisms or bacteria or other organic materials that has gathered on a surface can also be considered of as a biofilm. This process is regulated by a variety of factors, such as the nature of microorganism, hydrophobicity and electrical charges of the surface, and the ecological circumstances [5]. By providing an ideal setting for bacterial adherence, biosurfactants make advantage of their wettability feature.

Types of Biosurfactants

Biosurfactants are generally grouped in accordance to their molecular mass as well as chemical constitution.

Classification by Molecular Weight

Low Molecular Weight Biosurfactants

These are substances that reduce surface and interfacial tension at the air/water interfaces. Glycolipids or lipopeptides are the typical examples of low molecular weight biosurfactants. Rhamnolipids, trehalolipids, and sophorolipids, disaccharides (acylated with long-chain fatty acids or hydroxyl fatty acids) are the glycolipids that have been taken as subject of most of the research till date [4].

High-molecular Weight Biosurfactants

They are most frequently known as bioemulsan. They work better at maintaining oil-in-water emulsions. They are low-concentration emulsifiers that are extremely efficient. It also have extensive substrate specificity [4]. In order for microbial cells to effectively absorb alkanes, extracellular membrane vesicles divide hydrocarbons into a micro-emulsions [6].

Classification based on Chemical Structure

Glycolipids

The majority of known biosurfactants are typically glycolipids in origin. These comprise carbohydrates that when mixed with long-chain aliphatic acids or hydroxyl fatty acids, link themselves with either an ester or an ether group. In this taxon, rhamnolipids, trehalolipids, and sophorolipids are the most well-known lipids [6]. The degree of polarity depends on the hydrocarbons employed as the substrate; examples include the sophorolipids generated by species of *Candida* and the rhamnolipids produced by *Pseudomonas aeruginosa* [10]. Cellobiolipids are a type of additional glycolipid generated by bacteria [4].

Lipopeptides and Lipoproteins

Bacillus brevis and *Bacillus polymyxa* generates a huge number of cyclic lipopeptides, including the antibiotics decapeptide (gramicidin) and polymxin, which showed exceptional surface active characteristics. It is made up of a polypeptide chain and a lipid [4].

Fatty Acids, Phospholipids and Neutral Lipids

Many bacteria and yeast release large amounts of fatty acid and phospholipid surfactants while growing on n-alkanes [4]. Alkanes undergo microbial oxidations that result in the production of fatty acids, which have been utilized as surfactants [4].

Phospholipids

When certain hydrocarbon-degrading bacteria or yeast are grown on alkane substrates, the amount of phospholipids increases significantly. Phospholipids are important components of microbial mem- branes. Using a substrate of hexadecane, the bacteria *Acinetobacter*sp. HOI-N generates phospholipids, mostly 'Phosphatidylethanolamine' [4]. The structure is common to many microorganisms for example, biosurfactant from *Corynebacterium lepus* [10].

Polymeric Microbial Surfactants

Emulsan, liposan, mannoprotein, and other polysaccharide-protein complexes are some of the most well-known constituents of the best researched polymeric biosurfactans. Emulsan,which is made by the bacteria *Acinetobacter calcoaceticus*, consists of a heteropolysaccharide framework wherein fatty acids are attached by covalent bonds [6].

Rhamnolipids

The components of rhamnolipids are rhamnose molecules connected to one or two molecules of hydroxydecanoic acid. The OH group of one of the hydroxydecanoic acids is involved in the production of an ester and the glycosidic bond is present with the reducing end of the rhamnose disaccharide.

Connection Between Bioremediation and Biosurfactants

Due to their widespread distribution, low bioavailability, significant soil persistence, and potential for damage to human health, xenobiotic are a good candidate for bioremediation. These xenobiotic interact with non-aqueous phases and soil organic matter because of their high hydrophobicity and solid-water distribution ratios, which makes them potentially unavailable for microbial breakdown. These compounds are known to only be broken down by bacteria when they are dissolved in water. These elements influence a chemical's so-called bioavailability. In essence, research into bioremediation has focused on enhancing the activity of microbes through the addition of nutrients or bio-augmentation, but the failure of these strategies to increase the reduced bioavailability of potentially dangerous compounds is typically attributed to inherent microbial activity during bioremediation [11].

Commercial Application of Biosurfactants

In the sectors of food processing, cosmetics, medicines, and environmental bio-remediation, biosurfactants are important, especially in enhanced oil recovery (EOR) and oil spills clean up.

Biosurfactants in Cosmetic Industry

Multifunctional biosurfactants are used in a wide range of cosmetic applications due to their exceptional surface properties, including detergency, soaking, emulsifying, solubilizing, dissipating and foaming implications [6]. Biosurfactants are recognized to exhibit advantages over synthetic surfactants, such as reduced irritancy or anti-irritating effects, superior moisturizing characteristics, skin compatibility, and are thus in high demand. The three types of biosurfactant glycolipids that are most frequently utilized in cosmetics are rhamnolipids, sophorolipids, and mannosylerythritol lipids. Biological emulsifiers and surfactants called rhamnolipids could take the place of the majority of cosmetic goods' petrochemical-based surfactants. Outstanding moisturizing and skin-friendly properties are possessed by sophorolipids. Due to their high emulsifying properties, they have also been utilized in toothpaste, anti-wrinkle, anti-dandruff and anti-ageing products, deodorants, acne pads and nail care essentials [6].

Biosurfactants in Food Industry

Biosurfactants are environmentally friendly, biodegradable, and/or harmless, have a number of advantageous properties in the food industry, especially when used as emulsifiers, frothing agents, hydrating agents, and solubilizers [6].

Food Emulsifier

Biosurfactants exhibit a variety of characteristics, including emulsion-based compositions with broad potential uses in the food business. An emulsion is a heterogeneous system that comprises both a dispersed and continuous phase and is composed of at least one immiscible liquid that is intimately distributed in another in the form of droplets. Since they improve the consistency and richness of dairy products, emulsifiers are especially useful when added to low-fat foods. Polymeric surfactants, on the other hand, cover the oil droplets and create extremelysteady, non-coaling emulsions. This capability is very helpful for making oil/water emulsions for food and cosmetics [12].

Food Stabilizer

In ice cream and bread recipes, biosurfactants serve as consistency regulators. The inclusion of rhamnolipid surfactants in the processing of food enhances the taste and shelf-life of starch-containing goods, alters rheological characteristics, and increases the stability of wheat dough. When cooking with fats and oil, it also serve as a fat stabilizer as well as an anti-spattering ingredient [6].

Biosurfactant In Pharmaceutical Industry

Genetic Manipulation

The discovery of a safe and effective method for integrating foreign nucleotides within cells of mammals is critical for basic science and therapeutic applications such as gene therapies. Out of the several known techniques of gene transfection, lipofection employing cationic liposomes is regarded as a safe means to transfer foreign genes to the target cells without causing any negative effects [6].

Immune Modulatory Action

When paired with conventional antigens, bacterial lipopeptides make a powerful immunological adjuvant that is non-pyrogenic and non-toxic [6].

Anti-Adhesive Agents

Biosurfactant adhesion to solid surfaces may provide a novel and efficient strategy for preventing the colonization of surfaces by harmful microorganisms.Biosurfactants have been discovered to decrease pathogenic organism adherence to the site of infection. The anti-adhesive properties of these surfactants enable them effective in treating an variety of disorders as well as being used as probiotic and therapeutic agents [6].

HYDROCARBON AND CRUDE OIL CONTAMINATION SITES

The technologies involved in tertiary oil recovery are utilized to extract oil out of the plentiful unconventional oil reserves in order to ensure energy supply to the world as a result of the depletion of conventional oil reserves and a rise in the population's need for energy [13]. As a result, more and more of waste crude oil is being generated in both the upstream and downstream oil industrial sector activities such as crude oil extraction, transportation, storage, and refining. Petroleum pollution, which has severe impacts on all aquatic life, but especially the microbial population, is one of the most common ecological risks. The first stage of this action is the contact-mediated transfer of hydrocarbons from the oil phase to the cell surface, followed by transportation across the cell membrane. Although much research has been done throughout this domain, the mechanisms by which n-alkanes enter bacterial cells and how hydrocarbons are assimilated by microbial cells remain poorly known [14].

Prior study has shown that some bacterial populations are resilient to oil transfer and effectively breakdown oils/hydrocarbons. Oil/hydrocarbon biodegradation procedures generally involve two distinct kinds of interactions. The method being used involves a number of steps of oil adhesion, pseudo-solubilization, and hydrocarbon disintegration to create tiny drops of oil [15].

Biosurfactants may be released from within or outside of cells [16]. Although bacterial biosurfactants have served as the focus of several studies, the extent of their activity depends on their chemical composition. It has been demonstrated that a *Pseudomonas aeruginosa* variation may produce a biosurfactant similar to rhamnolipids that contains both mono and dirhamnolipid [17]. The type of hydrocarbon or oil that is preferentially broken down by rhamnolipids and the microorganisms that make them is clearly correlated with the type of surfactant that is being utilized. Several investigations on the breakdown of phenanthrene by different chemical surfactants have been documented. A study also showed that when phenanthrene was coupled with a strain of bacteria that created a non-ionic surfactant, the degree of phenanthrene the decomposition increased [18]. By the addition of the biosurfactant trehalose-5, 5'- dicorynomycolates, the oil-degrading ability of the chemical surfactant "FinasolOSR-5" was reported to be increased [19].

When bacteria attach to a layer and produce biosurfactants in the manner of biofilms, they change a number of surface qualities, including wettability. *Pseudomonas aeruginosa*, a marine bacterium isolated from oil-tainted sea water, has demonstrated the capacity to degrade hexadecane, octadecane, heptadecane, and also nonadecane following 28 days post incubation. The creation of a biosurfactant demonstrates these bacteria's ability to breakdown. Another research demonstrated that *Pseudomonas aeruginosa* degradeswide range of hydrocarbons including tetradecane, 2-methylnaphthalene, and pristine [20].

OIL SLUDGE PROBLEMS

Petroleum Sludge is a common hazardous waste produced by the oil industry. Nitrogen, phosphorus, potassium, iron, copper, cadmium, magnesium, calcium, phosphate, chromium, sodium, zinc and lead are the elements found in petroleum sludge [21]. Since the problem of oily sludge formation is becoming more prevalent, most treatment options such as ultrasonic treatment, incineration and solvent extraction are either prohibitively expensive or ineffective. While land cultivation and degradation are linked to heavy metal leaching, incineration is also associated with air pollution. Oil sludge is deposited as a result of the principal pollutants produced by oil production businesses, and it becomes tightly bonded to the effluents throughout conditioning and treatment procedures. The risk of ground water pollution rises as the amount of sludge deposited grows because the hydrocarbons slowly disperse into the subsoil after penetrating through the top soil layer [21]. Oil sludge must thus be treated to reduce its toxicity to the environment.

BIOSURFACTANT IN BIOREMEDIATION AND BIODEGRADATION OF CRUDE OIL CONTAMINATION

Bioremediation is the most widely used treatment approach that employs local microbial flora.Several kinds of hydrocarbons can be metabolized by bacteria that producebiosurfactants. Technologies for bioremediation of hydrocarbon-contaminated soil using microbes that produce biosurfactants have already been created and put to use in the Middle East and Canada. Because glycolipid-rich biosurfactants operate as agricultural nutrients, some hydrocarbon-contaminated land areas in the Middle East and Canada were bio remediated with biosurfactant-producing microbes themselves. The chemical makeup of the petroleum oil, the percentage of its structural classes, and the bioavailability of the feedstock are all factors that affect how quickly it breaks down through microbial action [22]. By melting or emulsifying organic hydrocarbon molecules, microorganisms cause their oxidation. While the rate of solubility of the oil is the key limiting factor in its biodegradation, biosurfactants accelerate the frequency at which organic molecule decay by emulsifying them. A significant number of bacteria that break down crude oil make exogenous biosurfactants to speed up the microbial decomposition of each hydrocarbon and assist with oil absorption.

A study by Thavasi *et al.*, suggested that the potency of *Lactobacillus delbrueckii's* biosurfactant to break down petroleum oil is extremely promising

[8]. Even though it was supported that biosurfactant works best when combined with fertilizer, the difference was not significant and biosurfactant alone can promote degradation of crude oil to a great extent. Almeida *et al.*, studied a biosurfactant obtained from *Artocarpus heterophyllus* plant extract was efficient in removing oil from solid surfaces. This biosurfactant effectively decreased the surface tension of water to roughly 26.9 mN/m and was found to be stable in a variety of temperature, salt and pH levels [23]. Different surface tension values [22] from the selected biosurfactant producing strains are discussed in Table **1**.

Table 1. Data regarding surface tension obtained from the chosen biosurfactant-producing variants [22].

Organism	Biosurfactant Type	Surface Tension (mN m-1)
P. aeruginosa	Rhamnolipids	29
Rhodococcus sp.	Trehalolipids.	36
T. bombicola	Sophorolipids	33
B. licheniformis	Peptide-lipid	27
S. marcescens	Serrawettin	33
P. fluorescens	Viscosin	26.5
B. subtilis	Surfactin	27-32
A. calcoaceticus	Emulsan	32
C. tropicalis	Mannan-lipid-protein	30
C. lipolytica	Liposan	29
Microbacterium sp.	Carbohydrate-protein-lipid	27

Surfactin as a pH-Switchable Biodemulsifier

Effective oil separation is the ideal but still challenging answer to the leftover crude oil dilemma. Due to their greater surface/interfacial activity, absence of contaminants, nontoxicity, and high demulsification effectiveness even in harsh circumstances, biodemulsifiers like bacterial cells and extracellular metabolites are preferable to chemical demulsifiers. Surfactin is one of the most powerful biosurfactants generated by *Bacillus subtilis* and it is made up of a hydroxyl fatty acid and a peptide ring with seven amino acids (Fig. **3**) [24].

At surfactin amounts below 20 mg/L, surfactin lowers the water's surface tension limited to than 27 mN/m [25, 26]. Since strong surface activity could guarantee outstanding demulsification efficiency in accordance with the demulsification process, surfactin is anticipated to be an effective bio emulsifier [27]. Owing to a shortage of water (0.5%), surfactin was able to successfully demulsify model emulsions as well as discarded crude oil. Since over ninety-five percent of the oil was recovered, the refining process may begin anew with the oil. Such a high-performance process ought to be connected to the adequate emulsion droplet breaking brought on by the surfactin precipitation that occurs easily when acid is added. Surfactin holds enormous promise for industrial uses for recovering oil form contaminated crude oil, that is a serious issue typically present in most petroleum-related factories, due to its high de-emulsification efficiency, environmentally friendly characteristics, and cost-efficiency (Fig. **4**).

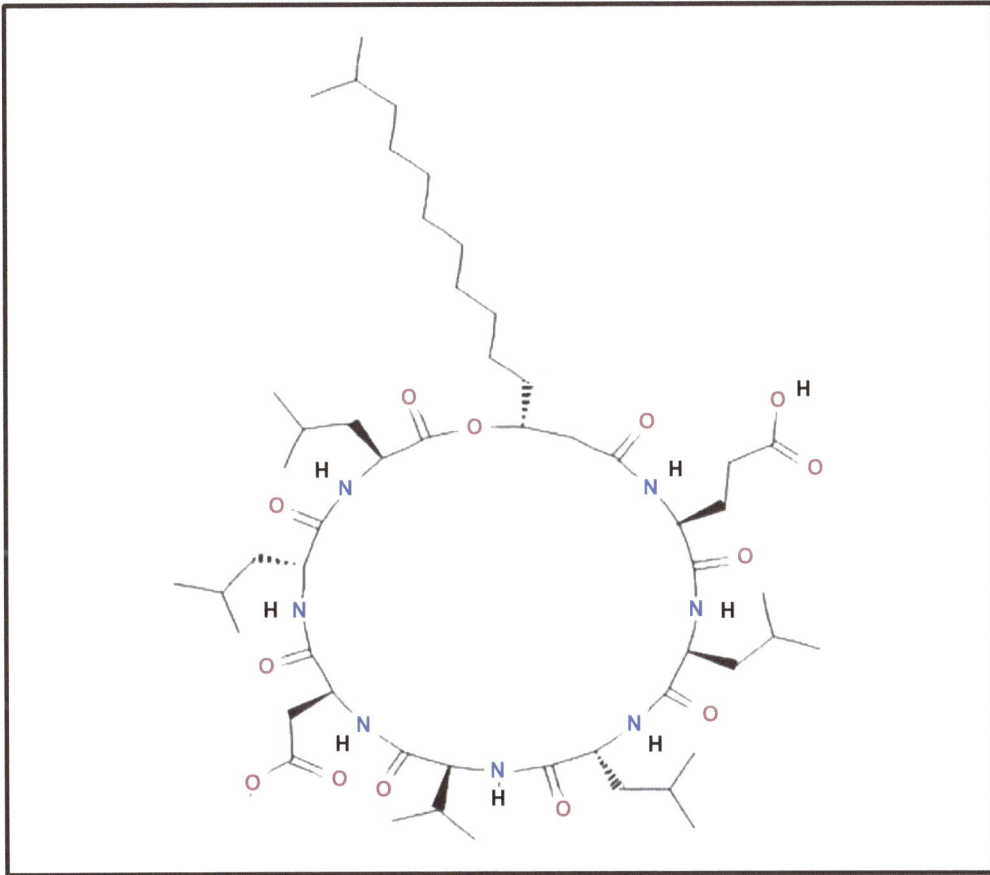

Fig. (3). Primary structure of surfactin [26].

Rhamnolipid Destabilizing Crude Oil

Rhamnolipids have attracted significant attention in a variety of commercial areas, including emulsifier, detergent, and antibiotics, due to their exceptional high surface activity (Fig. **5**) [28]. Over 92% of the water in the discarded crude oil was eliminated by the use of rhamnolipid treatment, and more than 98% of the crude oil was recovered. Following demulsification, the acquired oil phase and the watery phase can undergo further processing. While the obtained oil phase with water content of less than 0.3% can re-enter the refining process, the obtained aqueous phase with oil residual being far less than 0.01% can be processed by active sludge into more of an effluent with soluble COD satisfying the requirements of China's National Discharge Standard (GB8978-1996). Rhamnolipids demonstrated a high degradable capability in a comparative investigation of biosurfactants for washing dirt tainted with crude oil; 80% of the oil was degraded [28].

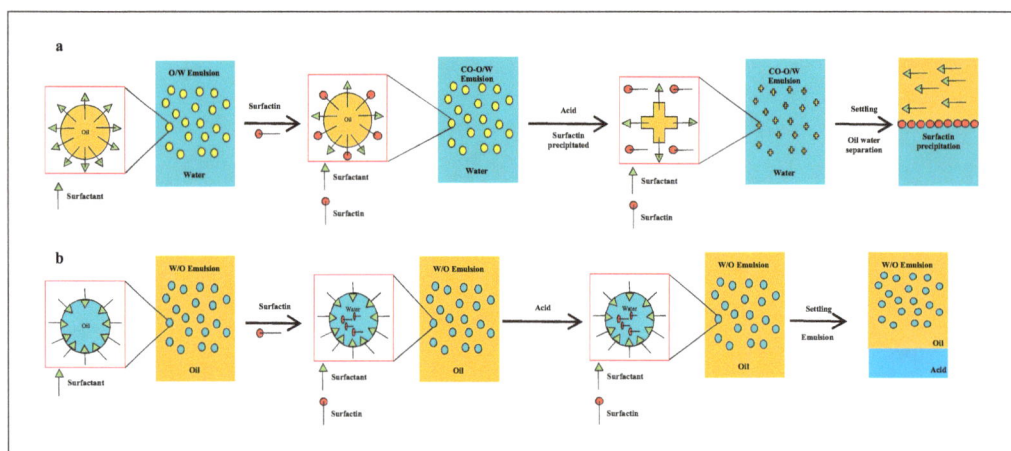

Fig. (4). Schematic representation of the two-step demulsification process on O/W (a) and W/O (b) type emulsions using surfactin as demulsifier [13].

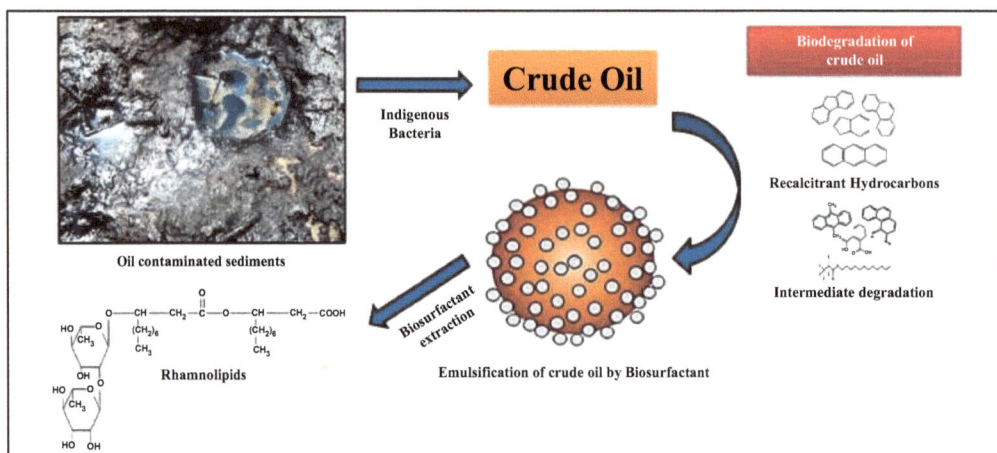

Fig. (5). Indigenous micro-organisms bioremediate petroleum oil using biosurfactants. Taean seashore silt was used to isolate it [28].

Lichenysins as a Potential Biosurfactant

Biosurfactants can be produced in significant quantities on an industrial basis through fermentation; Lichenysinsare most potent anionic cyclic lipoheptapeptide biosurfactants resulting from the well water isolated *B. licheniformis* JF-2. They have the capacity to lower the surface tension of water from 72 to 27 mN/m. Lichenysin A, B, C, D, G and surfactant BL86 are the species-specific variants. Acid precipitated lichenysin B has the lowest recorded interfacial tension against decane of 0.006 mN/m. Surfactant BL86 and lichenysin B have recorded lowest ever CMC of 10 mg/L by any surfactant under optimal conditions [25]. Lichenysin was effective at bringing the surface tension between the interfacial surfaces down to extremely low levels (10-2 mN/m), even at reduced quantities (10-60 mg/l). Its action was unaffected by changes in salt (up to 10% w/v NaCl), pH (6–10), or temperature parameters (140°C).Lichenysin in MEOR has the potential to be used in biotechnology, which has sparked study into the structure-activity link (Fig. **6**).

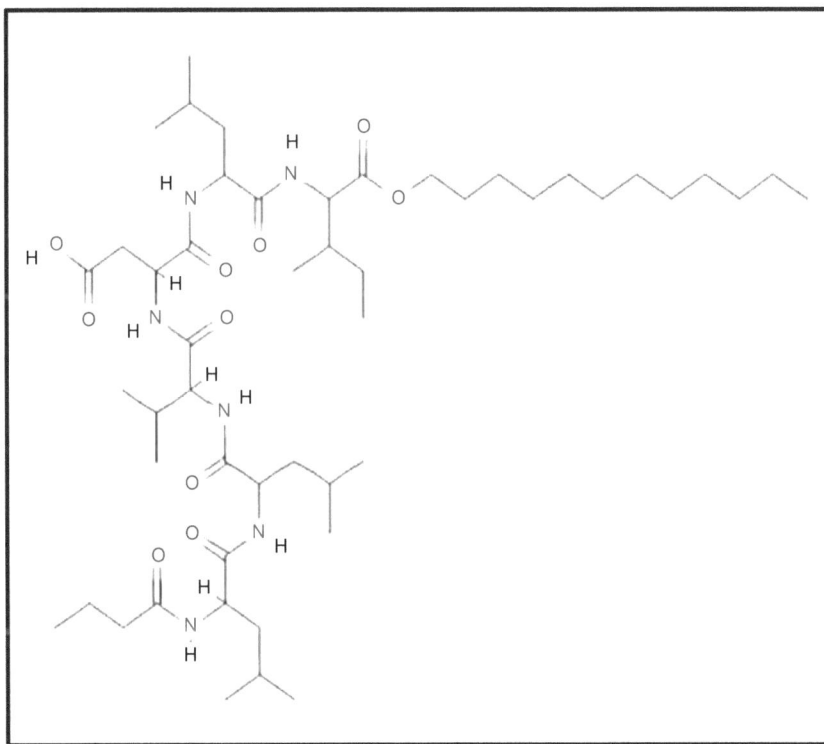

Fig. (6). Lichenysin-A primary structure [26].

RELATIONSHIP OF OIL REMEDIATION WITH MICROBIAL ENHANCED OIL RECOVERY (MEOR)

Oil leaks have a terrible impact on the oceanic ecosystem's sea life. Chemically generated surfactants were rarely used in the remediation process because they had been shown to be highly harmful to marine life. MEOR is a major tertiary recovery method that recovers residual oil by employing microorganisms and/or their metabolic end products. It is widely assumed that current EOR technology can recover around 30% of the oil present in a reservoir [29]. Searching for proteins with surface activity, emulsifying activity and low critical micelle content was one of the obvious options for this goal (CMC). By changing the wettability capacity of the permeable surface, bio surfactant adsorbs the oil. Biosurfactants can be used to clean up crude or tainted oil because they can emulsify hydrocarbons in water to create a variety of combinations and render it permeable in water.

Surfactin, lichenysins and rhamnolipidsare few detergents that have been found to be effective in cleaning up oil pollution. A bacterium that generated a biosurfactant with effective emulsifying qualities on crude oil and paraffin was isolated by Kim *et al.*, from a crude oil sample in 1997 [22]. According to the literature, biosurfactants derived from marine bacteria are capable of destroying oil slicks that float on the surface of water in order to enhance oil dispersion in water by generating a stable emulsion, hence increasing the rate of biodegradation. Because of these qualities, biosurfactants have demonstrated potential in their uses for cleaning up oil spills on shorelines and in the sea.

Both the aliphatic and aromatic components of petroleum oil can be broken down by a property of bacterium consortium known as hydrocarbonoclasticity. In general, biosurfactants made by microbes that break down oil can hasten the absorption of both environmental nutrients and hydrocarbons. Some types of microbes create emulsifying agents or emulsifiers that could aid in the breakdown of hydrocarbons and are therefore used to remove oil [30]. At 0.1 mg/ml concentration, the emulsion formed by *Acineto bactervenetianus* ATCC 31012 eliminates 89 percent of the crude oil that has been reabsorbed to the samples of limestone, and at 0.5 mg/ml concentration, 98 percent removal is accomplished [31].

Aerobic microorganisms in a microbially enhanced oil recovery (MEOR) processrequire strict anaerobic conditions because these conditions promote the synthesis of biosurfactants (Fig. **7**). In recent years, it was demonstrated that two key processes of microbial improved oil recovery (MEOR)were the lowering of interfacial tension and the changing of wetting ability. A well-known anaerobic bacterium *Anaerophagathe rmohalophila* (DSM 12881T) was capable of survivingat temperatures up to50°C and salinity such as 7.5%. It can produce a peptide with very little molecular weight, which is referred to as a surface active molecule. Zhuang *et al.* identified and described a bacterium that breaks down naphthalene-contaminated maritime sediments in 2002 [20].

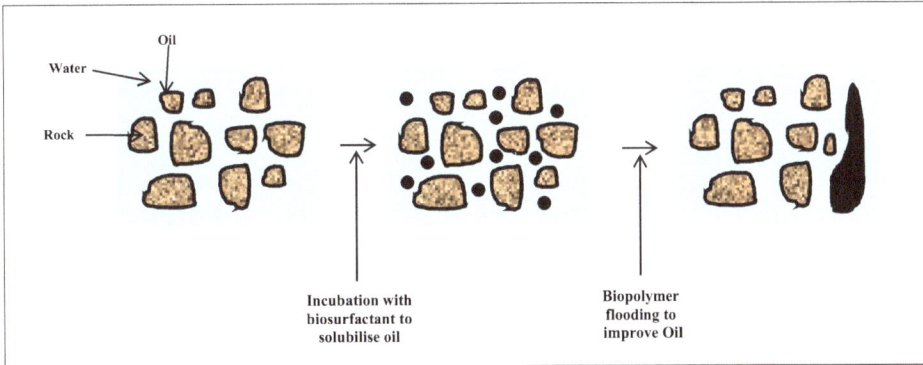

Fig. (7). Biosurfactant-biopolymer driven MEOR [35].

The recovery of oil has frequently used microbial enhanced oil recovery. Certain microbes free the oil deposits locked in reservoirs and rocks so they may use them during their metabolic activity to create a variety of byproducts [32]. It was reported that Biosurfactants were crucial to the oil's successful adsorption onto the rocks [33]. It was found that Surfactin from *Bacillus coagulans* 30 could generate stable emulsions with crude oil at a variety of pH, temperature, and salinity levels. This increased the oil recovery rate from 17 to 31%. The study also demonstrated the effectiveness of microbially enhanced oil recovery using laboratory sand packed columns. Between 85 and 95 mg/l of lipopeptides were generated by *Bacillus* species in the hydrocarbon reservoirs [34].

Approximately ten microbially enhanced oil recovery technologies have been adopted in the United States, Malaysia, China, and Argentina over the previous decade. Maudgalya's research indicated that 20 different biosurfactants were suited to recovering oil, where approximately 26 distinct kinds of biosurfactants wereused in field trials of microbially enhanced oil recovery, with the most encouraging results of microbially enhanced oil recovery to date seen in China's Shengli oil field. It has been demonstrated that microbially enhanced oil research holds great promise for enhancing oil generation and sustainably prolonging the existence of oil fields Table **2** [34].

Table 2. Potential uses of microbially generated biosurfactants as MEOR facilitators [36].

MEOR Agents	Microbes	Product	Possible MEOR Application
Biosurfactants	*Acinetobacter*	Emulsan and alasan	Emulsification and de-emulsification through reduction of interfacial tension
	Bacillus sp.	Surfactin, rhamnolipid, lichenysin	
	Pseudomonas	Rhamnolipid, glycolipids	
	Arthrobacter	-	

Advantage of MEOR

- Cheap nutrition and injected microbes (the injected bacteria are out-dated).
- The novel microbial EOR system only needs to introduce nutrients to encourage the naturally occurring bacteria rather than injecting microbes into the reservoir [37].
- Simple to use in the field and unaffected by energy costs.
- Economically appealing for old oilfields prior to closure.
- Increases oil production.
- Existing facilities require slight modifications.
- Application is easy.
- Set up is less expensive.
- Low energy input requirement for microbes to produce MEOR agents.
- When applied to carbonate oil reservoirs, it is more effective than other EOR techniques.
- Microbial growth increases its activity.
- Even though microbial nutrients decompose naturally, they can be regarded as ecologically benign.

Disadvantages of MEOR

- Layer permeability higher than 20 md, reservoir temperature less than 85 °C, salinity less than 100,000 ppm, and reservoir depth less than 3,500 m are all favourable conditions for microbial development.
- Recent instances and findings of on-going field tracking have demonstrated that there is no corrosion during MEOR. Additionally, the boosted native microorganisms have no impact on the qualities of crude oil, and there is no indication that the number of microbes in the generated liquid is growing [38].

MECHANISM INVOLVED IN BIOSURFACTANTS BASED OIL BIOREMEDIATION

Biosurfactants are involved in bioremediation in at least two ways such as raising the bioavailability of hydrophobic substances and the surface area of hydrophobic water-insoluble substrates.

Increasing the Surface Area of Hydrophobic Water Insoluble Substrates

The surface area at which water and oil interact can restrict the growth rate of bacteria that are growing on hydrocarbons [39]. Rather than growing rapidly, biomass grows arithmetically especially when the surface area becomes limiting. It is indirect confirmation that emulsification is an extracellular agent-driven natural process. As stated, emulsification is a cell-density-dependent phenomenon i.e the greater the number of cells, the higher the concentration of extracellular product.In an open system, such as an oil-polluted body of water, the concentration of cells never reaches a high enough value to efficiently emulsify oil. Furthermore, any emulsified oil would distribute in the water and would be equally accessible to the emulsifier-producing strain and competing bacteria. Theoretical arguments suggest that the emulsifying agents play a natural function in oil degradation but are ineffective in creating macroscopic emulsions in the mass liquid, which is one way to account for the empirical data and these theoretical arguments. Each collection of cells forms its own microenvironment if emulsion happens at, or very near, the cell surface and no mixing takes place at the microscopic level, in which case there should be no general reliance on cell density.

Increasing the Bioavailability of Hydrophobic Water-insoluble Substrates

A possible issue in the bioremediation of contaminated areas is the poor water solubility of many hydrocarbons, particularly the polycyclic aromatic hydrocarbons (PAHs), which restricts their access to microbes. It is believed that surfactants increase the solubility of hydrophobic compounds. Investigations into a number of non-biological lubricants have revealed both detrimental and beneficial effects on biodegradation. A rise in the development of a strain of *Pseudomonas stutzeri* was observed after the inclusion of the surfactant Tergitol NP-10, which accelerated the dissolution of solid-phase phenanthrene [40]. The addition of Tween 80 to two *Sphingomonas* strains had a comparable impact, almost doubling the rate of fluoranthene mineralization. On the other hand, the same surfactant reduced the rate of fluoranthene mineralization by two *Mycobacterium* isolates [41] and no stimulus was found in later studies using different surfactants [42, 43].

Chemical and biological surfactants both work to increase the solubility of hydrophobic substances. In comparison to the majority of synthetic surfactants, they are also selective, ecologically favourable, and typically less durable. Alasan, a high molecular weight bioemulsifier, has recently been demonstrated to greatly speed up the biodegradation of several PAHs [44]. The low water solubility of high molecular weight hydrophobic compounds, which increases their sorption to surfaces and reduces their accessibility to biodegrading microbes, is one of the primary causes of their extended persistence. By either desorbing them from the surface of the substrate or increasing their perceived water solubility, biosurfactants can stimulate development on bound substrates [45]. Mobilizing bonded hydrophobic molecules and opening them up for biodegradation is a key function of surfactants that greatly lower interfacial tension. Biosurfactants can either promote or prevent the biodegradation of hydrocarbons [46].

CONCLUDING REMARKS

Pollutants are constantly entering the ecosystem through natural and human processes, contaminating the soil, sediments, and water (both surface and ground). The impacted locations have been decontaminated and restored using a variety of techniques. The poor solubility of contaminants in watery media, as well as the dearth of microorganism degraders and physicochemical methods, limits the effectiveness of these processes. Recent years have seen an increase in interest in biosurfactants and their HMW/polymeric variants (bioemulsifiers), as well as from businesses looking for chemical components with better functional qualities that can be obtained from sustainable sources.

A range of oil-degrading microorganisms create biosurfactants. These biosurfactants can be high molecular weight and act as biodispersants by stopping the coalescence of oil drops in water, or they can be low molecular weight and work by lowering the oil-water interfacial tension. Although not yet used on a commercial basis, the use of biosurfactants for MEOR to treat oil spills is of great importance to the oil business. Due to a dearth of interdisciplinary study to address many of the limitations or knowledge gaps impeding its development, MEOR has not been widely embraced by industry despite existing for almost 70 years. It is well known that finding and isolating new microbe populations is necessary for study to improve MEOR, especially those active in anaerobic environments and capable of rapidly producing biomass or biosurfactants.

The ability of bacterial biosurfactants to increase the surface area of hydrophobic, water-insoluble materials as well as the solubility and accessibility of hydrocarbons makes them effective in bioremediation. They can be added to bioremediation processes as either pure compounds or as microbes that produce bioemulsifiers. In either instance, they can promote the development of bacteria that break down oil and enhance those bacteria's capacity to use fuels. Because biosurfactants are currently not economically competitive with chemical surfactants, extensive research on the large-scale production of biosurfactants from inexpensive substrates and/or novel bacterial strains is required to lower production costs.

REFERENCES

[1]   Mulligan, C.N. Environmental applications for biosurfactants. *Environ. Pollut.,* **2005**, *133*(2), 183-198.
        [http://dx.doi.org/10.1016/j.envpol.2004.06.009] [PMID: 15519450]

[2]   Desai, J.D.; Banat, I.M. Microbial production of surfactants and their commercial potential. *Microbiol. Mol. Biol. Rev.,* **1997**, *61*(1), 47-64.
        [PMID: 9106364]

[3]   Varjani, S.J.; Upasani, V.N. Critical review on biosurfactant analysis, purification and characterization using rhamnolipid as a model biosurfactant. *Bioresour. Technol.,* **2017**, *232*, 389-397.
        [http://dx.doi.org/10.1016/j.biortech.2017.02.047] [PMID: 28238638]

[4]   Shah, N.; Nikam, R.; Gaikwad, S.; Sapre, V.; Kaur, J. Biosurfactant: Types, detection methods, importance and applications. *Indian J. Microbiol. Res.,* **2016**, *3*(1), 5.
        [http://dx.doi.org/10.5958/2394-5478.2016.00002.9]

[5]   Roy, A. A review on the biosurfactants: Properties, types and its applications. **2017**, *8*(1), 5.

[6]   Shoeb, E.; Akhlaq, F.; Badar, U.; Akhter, J.; Imtiaz, S. Classification and industrial applications of biosurfactants. **2013**, *4*(3), 10.

[7]   Bustamante, M; Durán, N; Diez, MC Biosurfactants are useful tools for the bioremediation of contaminated soil: A review. *J Soil Sci Plant Nutr.,* **2012**, (ahead)

[8]   Saharan, B.; Sahu, R.; Sharma, D. A review on biosurfactants: Fermentation. *Current Developments and Perspectives.,* **2011**, *2011*, 39.

[9]   Karlapudi, A.P.; Venkateswarulu, T.C.; Tammineedi, J.; Kanumuri, L.; Ravuru, B.K.; Dirisala, V.; Kodali, V.P. Role of biosurfactants in bioremediation of oil pollution : A review. *Petroleum,* **2018**, *4*(3), 241-249.
        [http://dx.doi.org/10.1016/j.petlm.2018.03.007]

[10]   Sobrinho, HBS; Luna, JM; Rufino, RD; Porto, ALF; Sarubbo, LA. Biosurfactants: Classification, properties and environmental applications. **2014**, *11*, 303-330.

[11]   Cameotra, S.S.; Makkar, R.S. Biosurfactant enhanced bioremediation of hydrophobic pollutants. *Pure Appl. Chem.,* **2010**, *82*(1), 97-116.
        [http://dx.doi.org/10.1351/PAC-CON-09-02-10]

[12]   Rosenberg, E.; Ron, E.Z. High- and low-molecular mass microbial surfactants. *Appl. Microbiol. Biotechnol.,* **1999**, *52*(2), 154-162.
        [http://dx.doi.org/10.1007/s002530051502] [PMID: 10499255]

[13]   Yang, Z.; Zu, Y.; Zhu, J.; Jin, M.; Cui, T.; Long, X. Application of biosurfactant surfactin as a pH-switchable biodemulsifier for efficient oil recovery from waste crude oil. *Chemosphere,* **2020**, *240*, 124946.
        [http://dx.doi.org/10.1016/j.chemosphere.2019.124946] [PMID: 31726598]

[14]   Ampelli, C.; Centi, G.; Passalacqua, R.; Perathoner, S. Electrolyte-less design of PEC cells for solar fuels: Prospects and open issues in the development of cells and related catalytic electrodes. *Catal. Today,* **2016**, *259*, 246-258.
        [http://dx.doi.org/10.1016/j.cattod.2015.07.020]

[15]   Paleček, E.; Tkáč, J.; Bartošík, M.; Bertók, T.; Ostatná, V.; Paleček, J. Electrochemistry of nonconjugated proteins and glycoproteins. Toward sensors for biomedicine and glycomics. *Chem. Rev.,* **2015**, *115*(5), 2045-2108.
[http://dx.doi.org/10.1021/cr500279h] [PMID: 25659975]

[16]   Antoniou, E.; Fodelianakis, S.; Korkakaki, E.; Kalogerakis, N. Biosurfactant production from marine hydrocarbon-degrading consortia and pure bacterial strains using crude oil as carbon source. *Front. Microbiol.,* **2015**, *6*, 274. Available from: https://www.frontiersin.org/articles/10.3389/fmicb.2015.00274/full
[http://dx.doi.org/10.3389/fmicb.2015.00274] [PMID: 25904907]

[17]   Patel, J.; Borgohain, S.; Kumar, M.; Rangarajan, V.; Somasundaran, P.; Sen, R. Recent developments in microbial enhanced oil recovery. *Renew. Sustain. Energy Rev.,* **2015**, *52*, 1539-1558.
[http://dx.doi.org/10.1016/j.rser.2015.07.135]

[18]   Karlapudi, A.P.; Venkateswarulu, T.C.; Tammineedi, J.; Kanumuri, L.; Ravuru, B.K.; Dirisala, V.; Kodali, V.P. Role of biosurfactants in bioremediation of oil pollution : A review. *Petroleum,* **2018**, *4*(3), 241-249.
[http://dx.doi.org/10.1016/j.petlm.2018.03.007]

[19]   Itrich, N.R.; McDonough, K.M.; van Ginkel, C.G.; Bisinger, E.C.; LePage, J.N.; Schaefer, E.C.; Menzies, J.Z.; Casteel, K.D.; Federle, T.W. Widespread microbial adaptation to L -Glutamate- *N, N* - diacetate (L-GLDA) Following its market introduction in a consumer cleaning product. *Environ. Sci. Technol.,* **2015**, *49*(22), 13314-13321.
[http://dx.doi.org/10.1021/acs.est.5b03649] [PMID: 26465169]

[20]   Zhuang, W.Q.; Zhuang, W.Q.; Maszenan, A.M.; Tay, S.T. *Bacillus naphthovorans* sp. nov. from oil-contaminated tropical marine sediments and its role in naphthalene biodegradation. *Appl. Microbiol. Biotechnol.,* **2002**, *58*(4), 547-553.
[http://dx.doi.org/10.1007/s00253-001-0909-0] [PMID: 11954805]

[21]   Asia, I.O.; Enweani, I.B.; Eguavoen, I.O. Characterization and treatment of sludge from the petroleum industry. *Afr. J. Biotechnol.,* **2006**, *5*(5), 461-466.

[22]   Johnson, O.A.; Affam, A.C. Petroleum sludge treatment and disposal: A review. *Environ. Eng. Res.,* **2019**, *24*(2), 191-201.
[http://dx.doi.org/10.4491/eer.2018.134]

[23]   Karlapudi, A.P.; Venkateswarulu, T.C.; Tammineedi, J.; Kanumuri, L.; Ravuru, B.K.; Dirisala, V.; Kodali, V.P. Role of biosurfactants in bioremediation of oil pollution : A review. *Petroleum,* **2018**, *4*(3), 241-249.
[http://dx.doi.org/10.1016/j.petlm.2018.03.007]

[24]   De Almeida, D.G. Biosurfactants: Promising molecules for petroleum biotechnology advances. *Front Microbiol,* **2016**, 7. Available from: https://www.frontiersin.org/articles/10.3389/fmicb.2016.01718/full [cited 2021 May 14].

[25]   Al-Wahaibi, Y.; Joshi, S.; Al-Bahry, S.; Elshafie, A.; Al-Bemani, A.; Shibulal, B. Biosurfactant production by *Bacillus subtilis* B30 and its application in enhancing oil recovery. *Colloids Surf. B Biointerfaces,* **2014**, *114*, 324-333.
[http://dx.doi.org/10.1016/j.colsurfb.2013.09.022] [PMID: 24240116]

[26]   Anuradha, S.N. Structural and Molecular Characteristics of Lichenysin and Its Relationship with Surface Activity. In R. Sen (Ed.) Springer New York., *Biosurfactants,* **2010**; *672*, pp. 304–315.
[http://dx.doi.org/10.1007/978-1-4419-5979-9_23]

[27]   Long, X.; He, N.; He, Y.; Jiang, J.; Wu, T. Biosurfactant surfactin with pH-regulated emulsification activity for efficient oil separation when used as emulsifier. *Bioresour. Technol.,* **2017**, *241*, 200-206.
[http://dx.doi.org/10.1016/j.biortech.2017.05.120] [PMID: 28570884]

[28]   Feng, X.; Mussone, P.; Gao, S.; Wang, S.; Wu, S.Y.; Masliyah, J.H.; Xu, Z. Mechanistic study on

demulsification of water in diluted bitumen emulsions by ethylcellulose. *Langmuir,* **2010**, *26*(5), 3050-3057.
[http://dx.doi.org/10.1021/la9029563] [PMID: 20175568]

[29]   Anayo, O.F.; Scholastica, E.C.; Peter, O.C.; Nneji, U.G.; Obinna, A.; Mistura, L.O. The Beneficial Roles of *Pseudomonas* in Medicine, Industries, and Environment: A Review [Internet]. Pseudomonas Aeruginosa : An Armory An Armory In: *Intechopen*; **2019**.

Available from: https://www.intechopen. com/books/pseudomonas-aeruginosa-an-armory-within-the-beneficial-roles-of-pseudomonas-in-medi- cine-industries-and-environment-a-review [cited 2021 May 14].

[30]   Banat, I.M.; Makkar, R.S.; Cameotra, S.S. Potential commercial applications of microbial surfactants. *Appl. Microbiol. Biotechnol.,* **2000**, *53*(5), 495-508.
[http://dx.doi.org/10.1007/s002530051648] [PMID: 10855707]

[31]   Kim, J.; Yoo, G.; Lee, H.; Lim, J.; Kim, K.; Kim, C.W.; Park, M.S.; Yang, J.W. Methods of downstream processing for the production of biodiesel from microalgae. *Biotechnol. Adv.,* **2013**, *31*(6), 862-876.
[http://dx.doi.org/10.1016/j.biotechadv.2013.04.006] [PMID: 23632376]

[32]   Colin, V.L.; Bourguignon, N.; Gómez, J.S.; de Carvalho, K.G.; Ferrero, M.A.; Amoroso, M.J. Production of surface active compounds by a hydrocarbon degrading actinobacterium: Presumptive relationship with lipase activity. *Water Air Soil Pollut.,* **2017**, *228*(12), 454.
[http://dx.doi.org/10.1007/s11270-017-3623-y]

[33]   Zilber, I.K.; Rosenberg, E.; Gutnick, D. Incorporation of ^{32} P and Growth of Pseudomonad UP-2 on *n* - Tetracosane. *Appl. Environ. Microbiol.,* **1980**, *40*(6), 1086-1093.
[http://dx.doi.org/10.1128/aem.40.6.1086-1093.1980] [PMID: 16345671]

[34]   Sarafzadeh, P.; Hezave, A.Z.; Ravanbakhsh, M.; Niazi, A.; Ayatollahi, S. *Enterobacter cloacae* as biosurfactant producing bacterium: Differentiating its effects on interfacial tension and wettability alteration Mechanisms for oil recovery during MEOR process. *Colloids Surf. B Biointerfaces,* **2013**, *105*, 223-229.
[http://dx.doi.org/10.1016/j.colsurfb.2012.12.042] [PMID: 23376749]

[35]   Chaprão, M.J.; Ferreira, I.N.S.; Correa, P.F.; Rufino, R.D.; Luna, J.M.; Silva, E.J.; Sarubbo, L.A. Application of bacterial and yeast biosurfactants for enhanced removal and biodegradation of motor oil from contaminated sand. *Electron. J. Biotechnol.,* **2015**, *18*(6), 471-479.
[http://dx.doi.org/10.1016/j.ejbt.2015.09.005]

[36]   Dhanarajan, G.; Rangarajan, V.; Bandi, C.; Dixit, A.; Das, S.; Ale, K.; Sen, R. Biosurfactant-biopolymer driven microbial enhanced oil recovery (MEOR) and its optimization by an ANN-GA hybrid technique. *J. Biotechnol.,* **2017**, *256*, 46-56.
[http://dx.doi.org/10.1016/j.jbiotec.2017.05.007] [PMID: 28499818]

[37]   Sen, R. Biotechnology in petroleum recovery: The microbial EOR. *Pror. Energy Combust. Sci.,* **2008**, *34*(6), 714-724.
[http://dx.doi.org/10.1016/j.pecs.2008.05.001]

[38]   Pavia, MR; Ishoey, T Biotechnology in petroleum recovery: The microbial EOR. *Prog Energy Combust Sci.,* **2008**, *34*(6), 714-724.

[39]   Awan, AR (54) Systems and methods of microbial **2008**, 12.

[40]   Sekelsky, A.M.; Shreve, G.S. Kinetic model of biosurfactant enhanced hexadecane biodegradation by *Pseudomonas aeruginosa. Biotechnol. Bioeng.,* **1999**, *63*(4), 401-409.
[http://dx.doi.org/10.1002/(SICI)1097-0290(19990520)63:4<401::AID-BIT3>3.0.CO;2-S] [PMID: 10099620]

[41]   Grimberg, S.J.; Stringfellow, W.T.; Aitken, M.D. Quantifying the biodegradation of phenanthrene by *Pseudomonas stutzeri* P16 in the presence of a nonionic surfactant. *Appl. Environ. Microbiol.,* **1996**,

62(7), 2387-2392.
[http://dx.doi.org/10.1128/aem.62.7.2387-2392.1996] [PMID: 8779577]

[42]   Willumsen, P.A.; Nielsen, J.K.; Karlson, U. Degradation of phenanthrene-analogue azaarenes by Mycobacterium gilvum strain LB307T under aerobic conditions. *Appl. Microbiol. Biotechnol.,* **2001**, *56*(3-4), 539-544.
[http://dx.doi.org/10.1007/s002530100640] [PMID: 11549034]

[43]   Bruheim, P.; Bredholt, H.; Eimhjellen, K. Bacterial degradation of emulsified crude oil and the effect of various surfactants. *Can. J. Microbiol.,* **1997**, *43*(1), 17-22.
[http://dx.doi.org/10.1139/m97-003] [PMID: 9057292]

[44]   Bruheim, P; Eimhjellen, K Chemically emulsified crude oil as substrate for bacterial oxidation: Differences in species response. **1998**, *44*, 5.
[http://dx.doi.org/10.1139/w97-137]

[45]   Rosenberg, E.; Barkay, T.; Navon-Venezia, S.; Ron, E.Z. Role of *Acinetobacter bioemulsans* in petroleum degradation. In: *Novel Approaches for Bioremediation of Organic Pollution*; Fass, R.; Flashner, Y.; Reuveny, S., Eds.; Springer US: Boston, MA, **1999**; pp. 171-180.
[http://dx.doi.org/10.1007/978-1-4615-4749-5_17]

[46]   Marcoux, J.; Déziel, E.; Villemur, R.; Lépine, F.; Bisaillon, J.G.; Beaudet, R. Optimization of high molecular weight polycyclic aromatic hydrocarbons' degradation in a two-liquid phase bioreactor. *J. Appl. Microbiol.,* **2000**, *88*(4), 655-662.
[http://dx.doi.org/10.1046/j.1365-2672.2000.01011.x] [PMID: 10792524]

CHAPTER 9

A Review of Biosurfactant-Mediated Synthesis of Nanoparticles for Environmental Applications

Elina Khatua[1,2], Swastika Mallick[1,2] and Nilotpala Pradhan[1,2,*]

[1] *Environment and Sustainability Department, CSIR-Institute of Minerals and Materials Technology, Bhubaneswar, Odisha-751030, India*

[2] *Academy of Scientific and Innovative Research, Ghaziabad, Uttar Pradesh-201002, India*

Abstract: The potential of surfactants has been harnessed to fulfill human purposes for a long time. Biosurfactants are one of the promising bioactive molecules, produced by microorganisms, and subjected to intense research due to their chemical structure, diverse applications, and eco-friendly nature. Nanobiotechnology is an emerging scientific domain, encompassing various sectors like agriculture, medicine, bioremediation, food technology, *etc.* The discovery of biosurfactant coated nanoparticles has marked a breakthrough in the field of scientific research due to its cost-effectiveness and low toxicity nature. The present review emphasizes the role of discovered biosurfactants in nanoparticle synthesis and its application in the broad arena of nanotechnology and environment concerning issues.

Keywords: Biosurfactants, Glycolipids, Lipopeptides, Micro-emulsion, Nanoparticles, Surface active molecules.

INTRODUCTION

Anthropogenic activities are indiscriminately discharging industrial and household effluents into the environment. These effluents contain many chemicals which are toxic when they exceed a particular concentration. Some of these are not toxic but alter the relative concentration of species in a particular ecosystem causing havoc. One of such commodity is synthetic detergent, specifically phosphate detergent which has played a critical role in the eutrophication of inland lakes and ponds. These have caused the depletion of aquatic life and the destruction of the environment. Similarly, oil spillage is another such cause of environmental destruction. It is important to assess the biodegradable nature of

---

[*] **Corresponding author Nilotpala Pradhan:** Environment and Sustainability Department, CSIR-Institute of Minerals and Materials Technology, Bhubaneswar, Odisha-751030, India & Academy of Scientific and Innovative Research, Ghaziabad, Uttar Pradesh-201002, India; E-mail: npradhan@immt.res.in

Arun Kumar Pradhan and Manoranjan Arakha (Eds.)

commodity substances and search for future alternatives to save the environment. Microorganisms with high biodiversity are promising candidates for bioremediation. They are also known for the production of biosurfactants [1]. Biosurfactants are surface-active substances produced as secondary metabolites by a range of microbes, including bacteria, yeast, and fungi. The structural diversity of biosurfactants renders them useful in industries and the environment through biotechnological research and developments [2]. Bacteria manufacture surfactants for adsorbing, emulsifying, and solubilizing water-immiscible materials while taking such material as a source of food. For example, bacteria exposed to crude oil spillage area manages to disperse and emulsify the oil [1]. The biocompatibility, eco-friendly, and nontoxic nature of biosurfactants is due to their composition, which is made up of sugars, lipids, and proteins. The utility of biosurfactants as a greener, safer, and natural alternative to chemical surfactants, including synthesis of nanomaterials, has attracted the attention of scientists in the last decade [3].

Biosurfactants are amphiphilic surface-active compounds of microbial origin, carrying a hydrophilic head and hydrophobic tail [4]. Alcohol, carbohydrates, phosphates, carboxylic acid, and amino acids or peptides make up the hydrophilic parts of the head group, whereas branched, linear, hydroxylated, saturated, or unsaturated fatty acids make up the hydrophobic tail [5, 6]. They are produced primarily in aqueous media by aerophilic bacteria using either carbohydrates or hydrocarbons, fats, and oils as carbon sources [4]. They are better than synthetic surfactants as they are biodegradable, non-toxic, specific, and tolerant of higher temperatures, pH, and ionic strengths [7]. Production of biosurfactants is influenced by carbon and nitrogen supplies, as well as environmental variables such as pH, temperature, oxygen, and agitation rate during bacterial growth [6, 8].

CLASSIFICATION OF BIOSURFACTANTS

Biosurfactants are mostly produced from bacteria, fungi, and, yeast [9, 10]. Humans and plants also produce biosurfactants [11]. Dominating genera of bacteria reported for biosurfactant productions include *Pseudomonas, Bacillus,* and *Acinetobacter* [12]. Among yeasts' significant genera are *Torulopsis, Pseudozyma, Saccharomyces, Rhodotorula,* and *Kluyveromyces* [13 - 15]. Similarly, *Aspergillus, Ustilago, Fusarium, Trichoderma, and Penicillium* are well-reported biosurfactant-producing fungal genera in research publications [16]. Several species of fungi, yeast, and bacteria that produce biosurfactants are listed in Table **2**. The vast structural diversity allows different systems to classify microbial biosurfactants. According to their molecular weight, biosurfactants are divided into two categories: those with low molecular weights, like glycolipids, and those with large molecular weights, like lipopolysaccharides. According to the charge, low molecular weight biosurfactants are classified into non-ionic, anionic, cationic, and amphoteric biosurfactants. Another common method of classification is the categorization according to their chemical structure, such as glycolipids, lipopeptides, and proteins [17].

Classification Based on Molecular Weight

Biosurfactants are divided into two groups based on their molecular weights: low molecular weight biosurfactants and high molecular weight biosurfactants [18].

Low Molecular Weight Biosurfactants

These compounds generally include glycolipids, lipopeptides, rhamnolipids, trehalolipids, and sophorolipids; out of which glycolipids are the best studied to date [19]. These substances often exhibit lower surface and interfacial tension at air/water interfaces [4].

High Molecular Weight Biosurfactants

This group of compounds includes emulsan, alasan, polysaccharides, protein complexes, and liposan, collectively called bioemulsans [20], which stabilizes oil-in-water emulsion [21, 22]. At low concentrations, they act as highly efficient emulsifiers and exhibit substrate specificity [23]. These bio emulsifiers are of special interest in the cosmetic, food, or pharmaceutical sector, owing to their stable potential for emulsion [24].

Classification Based on Chemical Composition

Biosurfactants are classified into the following types according to their chemical composition [25].

Glycolipids

Glycolipids are made up of long-chain aliphatic or hydroxyl fatty acids and carbohydrates that are joined together by ester or ether groups [26]. The best-known glycolipids include rhamnolipids, trehalolipids, mannosylerithritol lipids, and sophorolipids [27, 28]. The carbohydrate part is constituted by mono/ di/tri or tetra saccharides, which include glucose, mannose, galactose, rhamnose, galactose sulfate, and glucuronic acid, depending upon the type of glycolipid [19].

Lipopeptides and Lipoproteins

These classes of biosurfactants include a large number of cyclic lipopeptides, which are composed of lipid groups attached to the polypeptide chain. They possess significant surface-active properties. Cyclic lipopeptides like decapeptide antibiotics (gramicidin) and lipopeptide antibiotics (polymyxin) produced from *Bacillus brevis* and *B. polymyxa* have surface active properties [27, 28]. They also include cyclic lipopeptide surfactin, which is one of the most active biosurfactants, produced by *Bacillus subtilis* [29]. Examples of lipopeptide biosurfactants include surfactin, lichenysin, iturin, fengycin, and serrawettin.

Fatty Acids, Phospholipids, and Neutral Lipids

There is a large number of bacteria and yeast that produce fatty acids and phospholipid biosurfactants when allowed to grow on an alkane substrate. Biosurfactants made from fatty acids are produced *via* oxidation. Complex fatty acids containing -OH groups and alkyl branches are also included in biosurfactants, *e.g.*, corynomuolic acids [30]. Fatty acids/ phospholipids/ neutral lipids (phosphatidylethanolamine, spiculisporic acid), polymeric biosurfactants (emulsan, alasan, biodispersan, liposan), and particulate biosurfactants (vesicles, whole-cell) can be considered under this group.

Polymeric Biosurfactants

This group of biosurfactants is polymeric heterosaccharide-containing proteins. Emulsan, alasan, liposan, mannoproteins, and polysaccharide complexes are best-studied polymeric surfactants [27].

Particulate Biosurfactants

Particulate biosurfactants include micro-emulsions produced by bacteria. A microemulsion is formed when certain bacterial species produce extracellular membrane vesicles that break hydrocarbons. These microemulsions aid in the microbial cell's assimilation of alkane; for example, vesicles of *Acinetobacter* sp. strain HO1-N [31].

Phospholipids

When certain microorganisms are grown on alkane substrates, the phospholipid increases significantly. Phospholipids are an important component in microbial membranes. For example, *Acinetobacter* sp. *HOI-N* produces phosphatidylethanolamine, a type of phospholipid when grown in hexadecane substrate [28].

TYPES OF MICROBES INVOLVED IN THE PRODUCTION OF BIOSURFACTANTS

There has been a continuous search for microorganisms with the potential to produce surfactants having high surface active or emulsifier properties. Bio emulsifier production is majorly influenced by the sources and the type of microorganisms. The distribution of microorganisms is universal, such as in air, water, and harsh environmental conditions.

Bacterial Biosurfactants

Bacteria are the major group of biosurfactant-producing organisms among all microorganisms. Different types of bacteria, including *Pseudomonas* sp., *Acinetobacter* sp., *Bacillus* sp., *etc.* are the most studied biosurfactant-producing microorganisms. They produce varieties of surface-active compounds like rhamnolipids, glycolipids, and lipopeptides. Some of the bacteria studied for biosurfactant production are listed in Table **1**.

Table 1. Biosurfactants produced from bacteria.

Biosurfactants	Name of Bacteria	References
Rhamnolipids	*Pseudomonas aeruginosa* ATCC 9027	[32]
	Bacillus thailandensis E264	[32]
	Serratia rubidaea SNAU02	[33]
Surfactin/Iturin	*Bacillus subtilis* SPB1	[34]
Subtilisin	*Bacillus subtilis*	[35]
	Bacillus subtilis JB1	[36]
Alasan	*Acinetobacter radioresistens* KA53	[37]
Lipopeptide	*Achromobacter* sp. HZ01	[38]
Pumilacidin	*Bacillus pumilus*	[39]
Serrawettin	*Serratia marcescens*	[40]
Carbohydrate-lipid complex	*Debaryomyces polymorphus*	[37]
Oligosaccharide-lipid complex	*Paenibacillus* sp. #510	[37]
Extracellular bioemulsans	*Gordonia* sp. BS29	[22]
Emulsan	*Acinetobacter calcoaceticus* PTCC 1316	[41]

Biosurfactants Produced by Yeast

Depending on the yeast strain reported, biosurfactants may be either glycolipids, protein-carbohydrate-lipid or protein-carbohydrate complexes, lipids, or fatty acids. Table **2** represents the biosurfactant-producing yeast and their chemical type.

Table 2. Biosurfactants produced from yeast.

Biosurfactant	Yeast Name	References
Glycolipids	*Absidia corymbifera* F-295	[42]
	Saccharomyces cerevisiae URM 6670	[43]
	Trichosporon montevideense CLOA72	[44]
	Wickerhamomyces anomalus CCMA 0358	[45]
	Cyberlindnera saturnus strain SBPN-27	[46]
	Candida batistae	[47]
Lipopeptide	*Candida lipolytica*	[8]
Lipoprotein	*Candida lipolytica* UCP 0988	[48]
Sophorolipids	*Rhodotorula babjevae* YS3	[49]
	Meyerozyma guilliermondii	[50]
	Candida bombicola	[51]
Mannosylerythritol lipid [MEL]	*Candida antarctica*	[52]
MEL A	*Pseudozyma antarctica* T-34	[52]
MEL B	*Pseudozyma* sp. KM-160	[53]
	Pseudozyma tsukubaensis JCM 10324^T	[54]
MEL C	*Pseudozyma graminicola* CBS 10092	[55]
Complex carbohydrate protein-lipid	*Candida tropicalis*	[56]
Complex carbohydrate protein-lipid	*Candida lipolytica* IA 1055	[56]

Biosurfactants by Fungi

Biosurfactants produced by fungi are not well-studied compared to bacteria and yeasts, due to their slower growth. They are good producers of biosurfactants, with emulsifiers forming a stable emulsion. They help in the dispersion of hydrophobic compounds, enabling them to be applied in different biotechnological applications. Biosurfactants derived from fungi can be of various types depending upon the nature of the molecule, such as lipopeptide, glycolipid, and complex carbohydrates. The list of biosurfactants produced from filamentous fungi is summarized in Table **3**.

Table 3. Biosurfactants produced by fungi.

Biosurfactants	Fungus Name	References
Lipopeptide	*Penicillium chrsogenum* SNP5	[57]
Glycolipid	*Aspergillus niger*	[58]
Complex carbohydrate/protein/lipid	*Cunninghmella echinulate*	[59]
Not identified	*Fusarium* sp.	[60]
	Fusarium, Penicillium and *Trichoderma*	[61]
	Aspergillus spp.	[62]
	Aspergillus fumigatus	[63]
	Rhizopus arrhizus	[64]

APPROACH OF NANOPARTICLE SYNTHESIS

Nanobiotechnology is an emerging area having potential applications in various fields. The typical size range for nanoparticles is 100 nm or less, at least in one dimension. There are numerous methods for the synthesis of nanoparticles, including physical, chemical, and biological. Biogenic synthesis of nanoparticles indicates the use of biological agents that may originate from plants, microbes, or their by-products [65]. In the physical or chemical methods of nanomaterial synthesis, reagents used are often toxic, hazardous, and have a long-lasting negative effect on the ecosystem. However, the green approach of nanoparticle synthesis has several advantages that include better stability compared to their chemical counterparts, environment-friendly, non-toxic nature, and affordability.

Role of Biosurfactants in Nanoparticle Synthesis

From the environmental sustainability point of view, microbial surfactants have recently emerged as a substitute for chemical surfactants. Other synthesis methods include various techniques like metal salt reduction, sol-gel chemistry, microemulsion, and photoreduction. Among all the techniques used in the bottom-up approach of nanoparticle synthesis, microemulsion has shown tremendous activity. Microemulsion generally consists of an amphipathic molecule having both hydrophilic and hydrophobic phases. Biomolecules, which are produced by a variety of biotic systems ranging from prokaryotes to eukaryotes, have been a fantastic topic of study in recent years [65]. The advantages of biosurfactants over chemical counterparts are due to their cost-effectiveness and easy scalability.

Microbial synthesis of nanoparticles may be intracellular, where nanoparticles are retained inside the cell as they are synthesized inside the cells, or extracellular, where nanoparticles are synthesized outside the cells by metabolites secreted by microorganisms. However, intracellular synthesis of the nanoparticle has a major drawback as it is tedious to extract nanoparticles from cells in pure forms along with retaining stability, and the process becomes cost-ineffective. Therefore, extracellular metabolites-based synthesis is the major focus and is studied extensively for nanoparticles' synthesis [66]. Fig. (**1**) shows the microemulsion method of nanoparticles' synthesis [67]. The ability of a biosurfactant mixture, including sophorolipid, rhamnolipid, and soy lecithin, for the generation of microemulsion and its prospective applications were described by Nguyen *et al.* [68]. Palanisamy and Raichur [69] reported the synthesis of spherical NiO nanoparticles with a particle size in the range of 47 ± 5 nm using rhamnolipids in water in oil microemulsion technique.

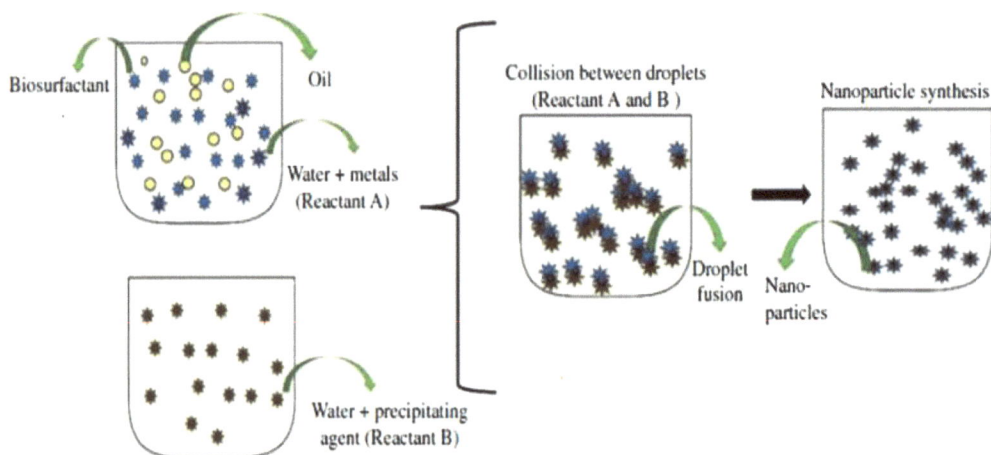

Fig. (1). Microemulsion method of nanoparticle synthesis [67].

Among biosurfactants, glycolipid and lipopeptide are commonly used for nanoparticle synthesis. Biosurfactants act as capping as well as reducing agents in the synthesis of metal nanoparticles such as Au, Ag, Cu, Ni, Zn, Fe, Co, *etc.* Rhamnolipid and sophorolipid, which are glycolipid biosurfactant, help in the uniform size distribution of synthesized nanoparticles by avoiding aggregation of nanoparticles. It also fixes the poor solubility of fatty acid capping agents. Nanoparticles are usually formed due to non-covalent interaction between hydrophobic moiety of the biosurfactants, leading to the formation of micelle or vesicles. The various self-arrangement patterns of these surface-active molecules lead to the formation of different morphological structures such as rod-shaped, hexagonal, cylindrical, lamellar, and spherical.

Use of Glycolipid Biosurfactant in Nanoparticle Synthesis

The broad class of glycolipid biosurfactants are used to synthesize various nanoparticles. Rhamnolipid is the most widely used glycolipid biosurfactant, followed by sophorolipids. Biosurfactants perform different tasks during the synthesis and fabrication of nanoparticles. They provide stability or functionalization by coating on the nanoparticles' surface. Biosurfactants can act as templates during the synthesis of nanoparticles, by providing stability or functionalizing them by coating on nanoparticles [70]. However, the most significant function of biosurfactant pertains to its use as a stabilizing agent. A glycolipid surfactant isolated from a marine actinobacteria *Brevibacterium casei* MSA19 associated with the sponge *Dendrilla nigra* was used for the silver nanoparticle synthesis by reverse micelle approach [71]. The surfactant acts as a stabilizer, and the synthesized nanoparticles were uniform in size and were stable for 2 months.

From a cost-effective point of view, biosurfactant is more suitable than synthetic surfactant. Rhamnolipids, a type of glycolipid produced from *Pseudomonas aeruginosa* LBI 2A1 using a low-cost carbon source (glycerol), were used for the silver nanoparticle synthesis under three different pH (from 5 to 9). The study showed that at higher pH, the nanoparticle shape and size were uniform and stable. Here, rhamnolipids act as a capping and stabilizing agent [72]. Similarly, another glycolipid biosurfactant isolated from *Pseudomonas aeruginosa* grown in a low-cost medium helps in silver nanoparticle stabilization [73]. The medium was formulated with 2.5% vegetable oil refinery residue and 2.5% corn steep liquor. Sowani *et al.* reported both silver and gold nanoparticles' synthesis by using cell-free supernatant of Actinomycete *Gordonia amicalis* HS-11 [74]. They used cell-free supernatant for the synthesis of gold and silver nanoparticles at pH 9 to obtain polydisperse and stable nanoparticles. They used thermostable glycolipid biosurfactants extracted from the bacteria and incubated with an aqueous solution of 1mM chloroauric acid [$HAuCl_4$] or silver nitrate [$AgNO_3$] solution synthesizing both gold and silver nanoparticles, which were monodispersed in nature. Glycolipid biosurfactant-mediated silver nanoparticles can also be synthesized under direct sunlight [75]. In this study, crude biosurfactant was mixed with $AgNO_3$ solution, and exposed to direct sunlight on a bright sunny day when the temperature ranged between 42°C to 45°C for 25 to 30 minutes. Spherical and spheroidal-shaped silver nanoparticles were synthesized, the size of which was in the range of 70-90nm. Sophorolipid and glycolipid act as stabilizing and capping agents in the synthesis of gold nanoparticles [76]. Here Basak *et al.* reported that acidic diacetate sophorolipid isolated from a yeast species *Cryptococcus* sp. VITGBN2 acts as a stabilizer in zinc oxide nanoparticle synthesis. The synthesis of nanoparticles was carried out by mixing $[ZnNO_3]_2.6H_2O$ and sophorolipids in a 1:1 molar ratio. Sodium hydroxide was added dropwise to form a white precipitate, followed by centrifugation and calcination at 300°C to obtain biosurfactant-functionalized zinc oxide nanoparticles. The size of the nanoparticle was in the range of 6 nm. It was the first reported paper that showed the dual nature of sophorolipid, both as a stabilizing and functionalizing agent. Shikha *et al.* reported sophorolipid production from the yeast *Stamerella bombicola* MTCC-1910 [77]. In this study, for the synthesis of sophorolipid capped gold nanoparticles [AuNp], 40μl of sophorolipid [100mg/ml] was added to 10 ml of $HAuCl_4$ [400μg/ml, pH 5.5±0.2]. The solution was mixed properly and incubated at 80°C till the appearance of a characteristic red color. To reduce the time for the synthesis of nanoparticles, a few drops of sodium borohydride ($NaBH_4$, 100mM) were added during the gold solution preparation. Synthesized nanoparticles were in the range of 30 to 50 nm. Ga'al *et al.* reported that Mannosylerythritol lipids (MEL) biosurfactant produced from the fungus *Ustilago maydis* act as both a stabilizing and reducing agent in silver nanoparticles' synthesis [78]. Bakur and co-workers synthesized gold nanoparticles (AuNp) in a similar way using MEL biosurfactant extracted from a strain of bacteria *Ustilago maydis* [79].

Use of Lipopeptide Biosurfactant in Nanoparticle Synthesis

Durval *et al.* used a fermentation medium for the bacteria *Bacillus cereus* UCP 1615 to extract lipopeptide biosurfactant [80]. The above experiment was carried out under UV irradiation for 8 hr., resulting in the aqueous dispersion of silver nanoparticles (AgNP) with a zeta potential of -23.4mV ± 1.4. Reddy *et al.* used surfactin biosurfactant isolated from *Bacillus subtilis* BBK006 to stabilize silver nanoparticle synthesis [81]. Olive oil and glucose were added to the fermentation medium as carbon sources. The cell-free extract was added to a mixture of sodium borohydride [$NaBH_4$] and $AgNO_3$ for the reaction. Stable nanosuspensions were produced at the end of the process, with a zeta potential of -38.7 mV. Eswari *et al.* reported lipopeptide biosurfactants for silver nanoparticle synthesis with an average particle size of 14 nm [82]. Yu *et al.* reported that a cell-free supernatant of *Bacillus subtilis* rich in lipopeptide surfactant, incubated with silver nitrate solution at pH 9, resulted in AgNP synthesis with a zeta potential of - 29.5 mV [83]. Plaza *et al.* used a similar method for lipopeptide-mediated silver nanoparticles using *Bacillus subtilis*T1 [84]. The media was formulated with agro-industrial waste, such as brewery effluents and molasses, as a low-cost substrate. Besides industrial waste, inexpensive lipid waste such as chicken tallow from slaughterhouses is used as a substrate for biosurfactant production. Radha *et al.* reported biosurfactant production from *Yarrowia lipolytica* MTCC 9520 yeast using chicken tallow as substrate [85]. Biosurfactant produced from yeast is a lipoprotein, which can synthesize AgNP with a zeta potential value of -22.2 mV. Nanoparticles were synthesized in water in an oil microemulsion phase process. Bezza *et al.* used lipopeptide biosurfactant extracted from *B. subtilis* to synthesize AgNP by reverse microemulsion method using butanol as a co-surfactant [86].

In another case, biosurfactants acted as stabilizing agents by controlling particle growth to an average size of 21 nm [87]. The microemulsion was formed by adding $AgNO_3$ in an iso-octane solution with lipopeptide biosurfactant and co-surfactant as butanol. In this case, Iso-octane was used as the oil phase in the preparation of two sequential emulsions. In the second microemulsion phase, $AgNO_3$ was substituted by borohydride in the same solvent solution system. The two solutions were mixed at 25°C, and the pH was maintained at 10 with sodium hydroxide. Nanoparticles were formed after 30 min and separated by centrifugation [87]. Lipopeptide biosurfactant extracted from *Acinetobacter junii* B6 is used to synthesize AuNP [88]. The lipopeptide biosurfactant was incubated in an aqueous solution of $HAuCl_4$ at a temperature of 60°C. The change in color of the solution to ruby red color confirmed the formation of gold nanoparticles with an average size ranging from 10 nm to 17 nm. Synthesized nanoparticles were mostly spherical and hexagonal and well dispersed in the aqueous solution. The biosurfactant serves a dual purpose as a stabilizing and capping agent during the synthesis of nanoparticles.

ENVIRONMENTAL APPLICATIONS OF BIOSURFACTANT BASED NANOPARTICLES

Environmental pollution encompassing the air, water, and soil has an adverse impact on the livelihood of animals. Nanotechnology, as an emerging field, has several applications in the environment in medicine, electronics, agriculture, *etc* [89]. It can be used in various environmental remediations like the removal of heavy metal contaminants, degradation of highly branched, long-chain polychlorinated biphenyl dye, polycyclic compounds, wastewater treatment (municipal and industrial effluents), in agriculture practices such as for the removal of herbicides, fertilizers, pesticides, *etc*. However, the effectiveness of nanoparticles depends upon their biodegradability, recyclability, nontoxicity, and cost-effectiveness, which limits their application. The re-aggregation nature of nanoparticles is reduced by the addition of biosurfactants, which ensures the monodispersity and stability of the nanoparticles. In addition to this, the amphiphilic nature of biosurfactant enhances its affinity for hydrophobic organic contaminants in the environment. Nehal and Singh [90] reported that lipopeptide biosurfactant-mediated silver nanoparticles could degrade low ethylene polythene cling films and high-density polythene up to 36.30 and 31.11%, respectively. A higher degree of degradation is due to the biocompatibility and stability of surfactant-coated silver nanoparticles which ensures its viability as plastic pollutant degradation on a large scale.

Besides polyethylene degradation, biosurfactant interacts with nanoparticles to decolorize toxic dyes like malachite and indeno [1,2,3-cd] pyrene [InP] dye. Iron oxide nanoparticles synthesized using extracellular polymeric substances (EPS) of *Alcaligens* sp. have shown potential biosurfactant activity [91]. These iron oxide nanoparticles showed $93.77 \pm 0.12\%$ decolorization of malachite green dye within 1.5 min. There were no adjuvants, catalysts, or variations in environmental conditions used for dye degradation. The dye concentration was 0.1mg/ml. Zinc sulfide nanoparticles synthesized using rhamnolipid biosurfactant as a capping agent were stable and eco-friendly in nature [92]. It had the potential to degrade 94% of direct brown methyl red (MR) dye in the textile industry wastewater and effluents. Benzo [a] pyrene [BaP] with five fused rings is one of the primary pollutants having carcinogenic, teratogenic, and mutagenic properties. The BaP degradation using rhamnolipid biosurfactant synthesized by the yeast consortia in the presence of zinc oxide nanoparticles has been demonstrated [93]. The maximal BaP degradation was reported to be about 82.67% at pH 7.0, temperature 30°C, and a shaking speed of 130 rpm in the presence of 2g/L zinc oxide nanoparticles. In a similar way, indeno [1,2,3-cd] pyrene [InP] dye, which is carcinogenic, can also be remediated using biosurfactant-coated iron nanoparticles. *Candida tropicalis* NN4 yeast had shown dye degradation with the biosurfactant identified as sophorolipids. Using a similar approach of statistical analysis, indeno [1, 2, 3-cd] pyrene [InP] dye degradation was about $90.68 \pm 0.7\%$ [94]. The predicted value of $90.68 \pm 0.4\%$ was in close approximation with the experimental value of $90.68 \pm 0.7\%$, which validates the model.

In addition to toxic dye degradation, biosurfactants can be used in the form of a nanocomposite with iron oxide nanoparticles and biochar, which acts as an excellent bioremediation agent due to its strong adsorption capacity [95]. Formulated biosurfactant/Iron oxide nanoparticle/biochar was used to clean up crude oil in polluted soil. It was reported that the biosurfactant was isolated from *Alcaligens faecalis* strain ADY25. Synthesized biosurfactant-iron oxide-biochar nano-composites were utilized to bioremediate soil contaminated with crude oil [10%w/w of soil] over a period of 35 days. There was 75% biodegradation of crude oil in polluted soil having BS/NP/BC in the ratio of 1:1:1.

CHALLENGES AND FUTURE PERSPECTIVE OF ENGINEERED NANOPARTICLES

In the history of human utilities, surfactants have played a major role owing to their surface active properties. Further, the surfactant-coated nanoparticle has tremendous applications, especially in the bioremediation of pollutants and sustainability of the environment. Although nanoparticles exhibit diverse applications in various sectors, many challenges need to be fixed, and many mechanisms need to be explored. Biosurfactant, as an agent for nanoparticle synthesis, is a great choice for its diverse functional group and cost-effective production. Biosurfactant as an extracellular metabolite reduces the cost tag of the downstream extraction process. Despite all the advantages of biosurfactants, in terms of environmental bioremediation, their transportation from one medium to another, like water, and soil, and their impact on the respective food chain and ecosystem remain to be investigated properly. In addition, most of the research related to biosurfactants is primarily based on a laboratory scale; its prospectus in environmental bioremediation and nanoparticle synthesis has to be explored on a large grade. Extensive studies need to be done for the production and evaluation of the compatible biomolecule from biosurfactants and optimal synthesis of nano-particles for better environmental remediation.

ETHICS APPROVAL

This article does not include any studies with human participants performed by any of the authors.

ACKNOWLEDGMENTS

The authors are thankful to the Director of the CSIR-Institute of Minerals and Materials Technology, Bhubaneswar, for permitting them to publish this review article. EK would like to thank the Council of Scientific and Industrial Research, Govt. of India, for the CSIR(NET)-JRF fellowship (Grant Number 16/06/2019[i] EU-V).

REFERENCES

[1]   Healy, M.G.; Devine, C.M.; Murphy, R. Microbial production of biosurfactants. *Resour. Conserv. Recycling,* **1996**, *18*(1-4), 41-57.
        [http://dx.doi.org/10.1016/S0921-3449(96)01167-6]

[2]   Varjani, S.J.; Upasani, V.N. Critical review on biosurfactant analysis, purification and characterization using rhamnolipid as a model biosurfactant. *Bioresour. Technol.,* **2017**, *232*, 389-397.
        [http://dx.doi.org/10.1016/j.biortech.2017.02.047] [PMID: 28238638]

[3]   Vecino, X.; Rodríguez-López, L.; Gudiña, E.J.; Cruz, J.M.; Moldes, A.B.; Rodrigues, L.R. Vineyard pruning waste as an alternative carbon source to produce novel biosurfactants by *Lactobacillus paracasei. J. Ind. Eng. Chem.,* **2017**, *55*(55), 40-49.
        [http://dx.doi.org/10.1016/j.jiec.2017.06.014]

[4]   Salihu, A; Abdulkadir, I An investigation for potential development on biosurfactants. *Academic. J.,* **2009**, *3*(5), 111-117.

[5]   Shekhar, S.; Sundaramanickam, A.; Balasubramanian, T. Biosurfactant Producing Microbes and their Potential Applications: A Review. *Crit. Rev. Environ. Sci. Technol.,* **2015**, *45*(14), 1522-1554.
        [http://dx.doi.org/10.1080/10643389.2014.955631]

[6]   Desai, J.D.; Banat, I.M. Microbial production of surfactants and their commercial potential. *Microbiol. Mol. Biol. Rev.,* **1997**, *61*(1), 47-64.
[PMID: 9106364]

[7]   Md, F. Biosurfactant: Production and Application. *J. Pet. Environ. Biotechnol.,* **2012**, *3*(4).
[http://dx.doi.org/10.4172/2157-7463.1000124]

[8]   Santos, D.; Rufino, R.; Luna, J.; Santos, V.; Sarubbo, L. Biosurfactants: Multifunctional Biomolecules of the 21st Century. *Int. J. Mol. Sci.,* **2016**, *17*(3), 401.
[http://dx.doi.org/10.3390/ijms17030401] [PMID: 26999123]

[9]   Mukherjee, S.; Das, P.; Sen, R. Towards commercial production of microbial surfactants. *Trends Biotechnol.,* **2006**, *24*(11), 509-515.
[http://dx.doi.org/10.1016/j.tibtech.2006.09.005] [PMID: 16997405]

[10]   Rocha e Silva, N.M.P.; Rufino, R.D.; Luna, J.M.; Santos, V.A.; Sarubbo, L.A. Screening of *Pseudomonas* species for biosurfactant production using low-cost substrates. *Biocatal. Agric. Biotechnol.,* **2014**, *3*(2), 132-139.
[http://dx.doi.org/10.1016/j.bcab.2013.09.005]

[11]   Geetha, S.J.; Banat, I.M.; Joshi, S.J. Biosurfactants: Production and potential applications in microbial enhanced oil recovery (MEOR). *Biocatal. Agric. Biotechnol.,* **2018**, *14*, 23-32.
[http://dx.doi.org/10.1016/j.bcab.2018.01.010]

[12]   Chrzanowski, Ł.; Ławniczak, Ł.; Czaczyk, K. Why do microorganisms produce rhamnolipids? *World J. Microbiol. Biotechnol.,* **2012**, *28*(2), 401-419.
[http://dx.doi.org/10.1007/s11274-011-0854-8] [PMID: 22347773]

[13]   Amaral, P.F.F.; da Silva, J.M.; Lehocky, M.; Barros-Timmons, A.M.V.; Coelho, M.A.Z.; Marrucho, I.M.; Coutinho, J.A.P. Production and characterization of a bioemulsifier from *Yarrowia lipolytica. Process Biochem.,* **2006**, *41*(8), 1894-1898.
[http://dx.doi.org/10.1016/j.procbio.2006.03.029]

[14]   Morita, T.; Fukuoka, T.; Imura, T.; Kitamoto, D. Mannosylerythritol lipids: production and applications. *J. Oleo Sci.,* **2015**, *64*(2), 133-141.
[http://dx.doi.org/10.5650/jos.ess14150] [PMID: 25748373]

[15]   Zinjarde, S.S.; Pant, A. Emulsifier from a tropical marine yeast, Yarrowia lipolytica NCIM 3589. *J. Basic Microbiol.,* **2002**, *42*(1), 67-73.
[http://dx.doi.org/10.1002/1521-4028(200203)42:1<67::AID-JOBM67>3.0.CO;2-M]   [PMID: 11925762]

[16]   Da Silva AF, Banat IM, Giachini AJ, Robl D. Fungal biosurfactants, from nature to biotechnological product: bioprospection, production and potential applications. *Bioprocess Biosyst Eng,* **2021**. *44*: 2003-34.
[http://dx.doi.org/10.9734/jamb/2019/v18i330170]

[17]   Uzoigwe, C.; Burgess, J.G.; Ennis, C.J.; Rahman, P.K.S.M. Bioemulsifiers are not biosurfactants and require different screening approaches. *Front. Microbiol.,* **2015**, *6*, 245.
[http://dx.doi.org/10.3389/fmicb.2015.00245] [PMID: 25904897]

[18]   Rosenberg, E.; Ron, E.Z. High- and low-molecular-mass microbial surfactants. *Appl. Microbiol. Biotechnol.,* **1999**, *52*(2), 154-162.
[http://dx.doi.org/10.1007/s002530051502] [PMID: 10499255]

[19]   Abbot V, Paliwal D, Sharma A, Sharma P. A review on the physicochemical and biological applications of biosurfactants in biotechnology and pharmaceuticals. *Heliyon,* **2022**, *8*, e10149.
[http://dx.doi.org/10.5958/2394-5478.2016.00002.9]

[20]   Franzetti, A.; Bestetti, G.; Caredda, P.; La Colla, P.; Tamburini, E. Surface-active compounds and their role in the access to hydrocarbons in *Gordonia* strains. *FEMS Microbiol. Ecol.,* **2008**, *63*(2), 238-248.

[http://dx.doi.org/10.1111/j.1574-6941.2007.00406.x] [PMID: 18070077]

[21]   Neu, T.R. Significance of bacterial surface-active compounds in interaction of bacteria with interfaces. *Microbiol. Rev.,* **1996**, *60*(1), 151-166.
[http://dx.doi.org/10.1128/mr.60.1.151-166.1996] [PMID: 8852899]

[22]   Franzetti, A.; Caredda, P.; Ruggeri, C.; Colla, P.L.; Tamburini, E.; Papacchini, M.; Bestetti, G. Potential applications of surface active compounds by *Gordonia* sp. strain BS29 in soil remediation technologies. *Chemosphere,* **2009**, *75*(6), 801-807.
[http://dx.doi.org/10.1016/j.chemosphere.2008.12.052] [PMID: 19181361]

[23]   Dastgheib, S.M.M.; Amoozegar, M.A.; Elahi, E.; Asad, S.; Banat, I.M. Bioemulsifier production by a halothermophilic Bacillus strain with potential applications in microbially enhanced oil recovery. *Biotechnol. Lett.,* **2008**, *30*(2), 263-270.
[http://dx.doi.org/10.1007/s10529-007-9530-3] [PMID: 17876532]

[24]   Henkel, M.; Hausmann, R. Diversity and classification of microbial surfactants. In: *Biobased Surfactants*; Synthesis, Properties, and Applications, **2019**; pp. 41-63.
[http://dx.doi.org/10.1016/B978-0-12-812705-6.00002-2]

[25]   Fenibo, E.O.; Douglas, S.I.; Stanley, H.O. A review on microbial surfactants: Production, classifications, properties and characterization. *J. Adv. Microbiol.,* **2019**, 1-22.
[http://dx.doi.org/10.9734/jamb/2019/v18i330170]

[26]   Shah, N.; Nikam, R.; Gaikwad, S.; Sapre, V.; Kaur, J. Biosurfactant: Types, detection methods, importance and applications. *Indian J. Microbiol. Res.,* **2016**, *3*(1), 5.
[http://dx.doi.org/10.5958/2394-5478.2016.00002.9]

[27]   Mnif I, Ellouz-Chaabouni S, Ghribi D. Glycolipid biosurfactants, main classes, functional properties and related potential applications in environmental biotechnology. *J. Polym .Environ.,* **2018**. 26: 2192-206.
[PMID: 9106364]

[28]   Muthusamy, K.; Gopalakrishnan, S.; Ravi, T.; Sivachidambaram, P. Biosurfactants: Properties, commercial production and application. *Curr. Sci.,* **2008**, *94*, 736-747.

[29]   Ron, E.Z.; Rosenberg, E. Natural roles of biosurfactants. Minireview. *Environ. Microbiol.,* **2001**, *3*(4), 229-236.
[http://dx.doi.org/10.1046/j.1462-2920.2001.00190.x] [PMID: 11359508]

[30]   Rahman, P.K.S.M.; Gakpe, E. Production, characterisation and applications of biosurfactants : Review. *Biotechnology,* **2008**, *7*(2), 360-370.
[http://dx.doi.org/10.3923/biotech.2008.360.370]

[31]   Käppeli, O.; Finnerty, W.R. Partition of alkane by an extracellular vesicle derived from hexadecane-grown *Acinetobacter. J. Bacteriol.,* **1979**, *140*(2), 707-712.
[http://dx.doi.org/10.1128/jb.140.2.707-712.1979] [PMID: 500568]

[32]   Díaz De Rienzo, M.A.; Kamalanathan, I.D.; Martin, P.J. Comparative study of the production of rhamnolipid biosurfactants by *B. thailandensis* E264 and *P. aeruginosa* ATCC 9027 using foam fractionation. *Process Biochem.,* **2016**, *51*(7), 820-827.
[http://dx.doi.org/10.1016/j.procbio.2016.04.007]

[33]   Nalini, S.; Parthasarathi, R. Optimization of rhamnolipid biosurfactant production from *Serratia rubidaea* SNAU02 under solid-state fermentation and its biocontrol efficacy against *Fusarium* wilt of eggplant. *Ann. Agrar. Sci.,* **2018**, *16*(2), 108-115.
[http://dx.doi.org/10.1016/j.aasci.2017.11.002]

[34]   Mnif, I.; Grau-Campistany, A.; Coronel-León, J.; Hammami, I.; Triki, M.A.; Manresa, A.; Ghribi, D. Purification and identification of *Bacillus subtilis* SPB1 lipopeptide biosurfactant exhibiting antifungal activity against *Rhizoctonia bataticola* and *Rhizoctonia solani. Environ. Sci. Pollut. Res. Int.,* **2016**, *23*(7), 6690-6699.

[http://dx.doi.org/10.1007/s11356-015-5826-3] [PMID: 26645234]

[35] Sutyak, K.E.; Wirawan, R.E.; Aroutcheva, A.A.; Chikindas, M.L. Isolation of the *Bacillus subtilis* antimicrobial peptide subtilosin from the dairy product-derived *Bacillus amyloliquefaciens*. *J. Appl. Microbiol.*, **2008**, *104*(4), 1067-1074.
[http://dx.doi.org/10.1111/j.1365-2672.2007.03626.x] [PMID: 17976171]

[36] Sung, J.H.; Ahn, S.J.; Kim, N.Y.; Jeong, S.K.; Kim, J.K.; Chung, J.K.; Lee, H.H. Purification, molecular cloning, and biochemical characterization of subtilisin JB1 from a newly isolated *Bacillus subtilis* JB1. *Appl. Biochem. Biotechnol.*, **2010**, *162*(3), 900-911.
[http://dx.doi.org/10.1007/s12010-009-8830-6] [PMID: 19902383]

[37] Adetunji, A.I.; Olaniran, A.O. *Production and potential biotechnological applications of microbial surfactants: An overview*; Saudi Journal of Biological Sciences. Elsevier B.V., **2021**, 28, pp. 669-679.

[38] Deng, M.C.; Li, J.; Hong, Y.H.; Xu, X.M.; Chen, W.X.; Yuan, J.P.; Peng, J.; Yi, M.; Wang, J.H. Characterization of a novel biosurfactant produced by marine hydrocarbon-degrading bacterium *Achromobacter* sp. HZ01. *J. Appl. Microbiol.*, **2016**, *120*(4), 889-899.
[http://dx.doi.org/10.1111/jam.13065] [PMID: 26788863]

[39] Banat, I.M.; Franzetti, A.; Gandolfi, I.; Bestetti, G.; Martinotti, M.G.; Fracchia, L.; Smyth, T.J.; Marchant, R. Microbial biosurfactants production, applications and future potential. *Appl. Microbiol. Biotechnol.*, **2010**, *87*(2), 427-444.
[http://dx.doi.org/10.1007/s00253-010-2589-0] [PMID: 20424836]

[40] Matsuyama, T.; Tanikawa, T.; Nakagawa, Y. *Serrawettins and Other Surfactants Produced by*; Serratia, **2011**, pp. 93-120.

[41] Amani, H.; Kariminezhad, H. Study on emulsification of crude oil in water using emulsan biosurfactant for pipeline transportation. *Petrol. Sci. Technol.*, **2016**, *34*(3), 216-222.
[http://dx.doi.org/10.1080/10916466.2015.1118500]

[42] Batrakov, S.G.; Konova, I.V.; Sheichenko, V.I.; Galanina, L.A. Glycolipids of the filamentous fungus Absidia corymbifera F-295. *Chem. Phys. Lipids*, **2003**, *123*(2), 157-164.
[http://dx.doi.org/10.1016/S0009-3084(02)00166-4] [PMID: 12691848]

[43] Ribeiro, B.G.; Guerra, J.M.C.; Sarubbo, L.A. Potential food application of a biosurfactant produced by *Saccharomyces cerevisiae* URM 6670. *Front. Bioeng. Biotechnol.*, **2020**, *8*, 434.
[http://dx.doi.org/10.3389/fbioe.2020.00434] [PMID: 32457894]

[44] Monteiro, A.S.; Coutinho, J.O.P.A.; Júnior, A.C.; Rosa, C.A.; Siqueira, E.P.; Santos, V.L. Characterization of new biosurfactant produced by *Trichosporon montevideense* CLOA 72 isolated from dairy industry effluents. *J. Basic Microbiol.*, **2009**, *49*(6), 553-563.
[http://dx.doi.org/10.1002/jobm.200900089] [PMID: 19810042]

[45] Souza, K.S.T.; Gudiña, E.J.; Azevedo, Z.; de Freitas, V.; Schwan, R.F.; Rodrigues, L.R.; Dias, D.R.; Teixeira, J.A. New glycolipid biosurfactants produced by the yeast strain *Wickerhamomyces anomalus* CCMA 0358. *Colloids Surf. B Biointerfaces*, **2017**, *154*, 373-382.
[http://dx.doi.org/10.1016/j.colsurfb.2017.03.041] [PMID: 28376391]

[46] Senthil Balan, S.; Ganesh Kumar, C.; Jayalakshmi, S. Physicochemical, structural and biological evaluation of Cybersan (trigalactomargarate), a new glycolipid biosurfactant produced by a marine yeast, *Cyberlindnera saturnus* strain SBPN-27. *Process Biochem.*, **2019**, *80*, 171-180.
[http://dx.doi.org/10.1016/j.procbio.2019.02.005]

[47] Konishi, M.; Maruoka, N.; Furuta, Y.; Morita, T.; Fukuoka, T.; Imura, T.; Kitamoto, D. Biosurfactant-producing yeasts widely inhabit various vegetables and fruits. *Biosci. Biotechnol. Biochem.*, **2014**, *78*(3), 516-523.
[http://dx.doi.org/10.1080/09168451.2014.882754] [PMID: 25036844]

[48] Santos, D.K.F.; Rufino, R.D.; Luna, J.M.; Santos, V.A.; Salgueiro, A.A.; Sarubbo, L.A. Synthesis and evaluation of biosurfactant produced by *Candida lipolytica* using animal fat and corn steep liquor. *J.*

Petrol. Sci. Eng., **2013**, *105*, 43-50.
[http://dx.doi.org/10.1016/j.petrol.2013.03.028]

[49]   Sen, S.; Borah, S.N.; Bora, A.; Deka, S. Production, characterization, and antifungal activity of a biosurfactant produced by *Rhodotorula babjevae* YS3. *Microb. Cell Fact.,* **2017**, *16*(1), 95.
[http://dx.doi.org/10.1186/s12934-017-0711-z] [PMID: 28558761]

[50]   Sharma, P.; Sangwan, S.; Kaur, H. Process parameters for biosurfactant production using yeast *Meyerozyma guilliermondii* YK32. *Environ. Monit. Assess.,* **2019**, *191*(9), 531.
[http://dx.doi.org/10.1007/s10661-019-7665-z] [PMID: 31375926]

[51]   Hirata, Y.; Ryu, M.; Oda, Y.; Igarashi, K.; Nagatsuka, A.; Furuta, T.; Sugiura, M. Novel characteristics of sophorolipids, yeast glycolipid biosurfactants, as biodegradable low-foaming surfactants. *J. Biosci. Bioeng.,* **2009**, *108*(2), 142-146.
[http://dx.doi.org/10.1016/j.jbiosc.2009.03.012] [PMID: 19619862]

[52]   Kitamoto, D.; Akiba, S.; Hiok, C.; Tabuchi, T. Extracellular accumulation of mannosylerythritol lipids by a strain of *Candida antarctica. Agric. Biol. Chem.,* **1990**, *54*(1), 31-36.
[http://dx.doi.org/10.1271/bbb1961.54.31]

[53]   Konishi, M.; Morita, T.; Fukuoka, T.; Imura, T.; Kakugawa, K.; Kitamoto, D. Production of different types of mannosylerythritol lipids as biosurfactants by the newly isolated yeast strains belonging to the genus *Pseudozyma. Appl. Microbiol. Biotechnol.,* **2007**, *75*(3), 521-531.
[http://dx.doi.org/10.1007/s00253-007-0853-8] [PMID: 17505770]

[54]   Fukuoka, T.; Morita, T.; Konishi, M.; Imura, T.; Kitamoto, D. A basidiomycetous yeast, *Pseudozyma tsukubaensis,* efficiently produces a novel glycolipid biosurfactant. The identification of a new diastereomer of mannosylerythritol lipid-B. *Carbohydr. Res.,* **2008**, *343*(3), 555-560.
[http://dx.doi.org/10.1016/j.carres.2007.11.023] [PMID: 18083152]

[55]   Morita, T.; Konishi, M.; Fukuoka, T.; Imura, T.; Yamamoto, S.; Kitagawa, M.; Sogabe, A.; Kitamoto, D. Identification of *Pseudozyma graminicola* CBS 10092 as a producer of glycolipid biosurfactants, mannosylerythritol lipids. *J. Oleo Sci.,* **2008**, *57*(2), 123-131.
[http://dx.doi.org/10.5650/jos.57.123] [PMID: 18198469]

[56]   Accorsini, F.R.; Mutton, M.J.R.; Lemos, E.G.M.; Benincasa, M.; Lemos, M.; Benincasa, M. Biosurfactants production by yeasts using soybean oil and glycerol as low cost substrate. *Braz. J. Microbiol.,* **2012**, *43*(1), 116-125.
[http://dx.doi.org/10.1590/S1517-83822012000100013] [PMID: 24031810]

[57]   Gautam, G. A Cost Effective Strategy for Production of Bio-surfactant from Locally Isolated *Penicillium chrysogenum* SNP5 and Its Applications. *J. Bioprocess. Biotech.,* **2014**, *4*(6).
[http://dx.doi.org/10.4172/2155-9821.1000177]

[58]   Silva, ACS da.; Santos PN, dos.; Silva TAL, e.; Andrade, RFS.; Campos-Takaki, GM. Biosurfactant production by fungi as a sustainable alternative. *Arq Inst Biol,* **2018**, *21*, 85.

[59]   Andrade Silva, N.; Luna, M.; Santiago, A.; Franco, L.; Silva, G.; de Souza, P.; Okada, K.; Albuquerque, C.; Silva, C.; Campos-Takaki, G. Biosurfactant-and-bioemulsifier produced by a promising Cunninghamella echinulata isolated from Caatinga soil in the northeast of Brazil. *Int. J. Mol. Sci.,* **2014**, *15*(9), 15377-15395.
[http://dx.doi.org/10.3390/ijms150915377] [PMID: 25257520]

[60]   Qazi, M.A.; Kanwal, T.; Jadoon, M.; Ahmed, S.; Fatima, N. Isolation and characterization of a biosurfactant-producing *Fusarium* sp. BS-8 from oil contaminated soil. *Biotechnol. Prog.,* **2014**, *30*(5), 1065-1075.
[http://dx.doi.org/10.1002/btpr.1933] [PMID: 24850435]

[61]   Méndez-Castillo, L.; Prieto-Correa, E.; Jiménez-Junca, C. Identification of fungi isolated from banana rachis and characterization of their surface activity. *Lett. Appl. Microbiol.,* **2017**, *64*(3), 246-251.
[http://dx.doi.org/10.1111/lam.12712] [PMID: 28060422]

[62] Morganna, R.; Cavalcanti, F.; Pereira, D.; Silva, D.; Calina De França Paz, M.; Farias De Queiroz, J.C. Screening, production and characterization of biosurfactants from caatingas filamentous fungi. *Int. J. Pharm. Sci. Invent.,* **2017**, *6*(5), 23-28.

[63] Castiglioni, G.L.; Bertolin, T.E.; Costa, J.A.V. Solid-state biosurfactant production by *Aspergillus fumigatus* using agricultural residues as substrate. *Quim. Nova,* **2009**, *32*(2), 292-295.
[http://dx.doi.org/10.1590/S0100-40422009000200005]

[64] Milagre, A.P.; Dayana, M.R.; Daylin, R.R.; Adriana, F.S.; Marcos, A.C.L.; Michele, F.S.; Rosileide, F.S.A.; Thayse, A.L.S.; André, L.C.M.A.S.; Galba, M.C-T. Development and improved selected markers to biosurfactant and bioemulsifier production by Rhizopus strains isolated from Caatinga soil. *Afr. J. Biotechnol.,* **2018**, *17*(6), 150-157.
[http://dx.doi.org/10.5897/AJB2017.16230]

[65] Pandit, C.; Roy, A.; Ghotekar, S.; Khusro, A.; Islam, M.N.; Emran, T.B.; Lam, S.E.; Khandaker, M.U.; Bradley, D.A. Biological agents for synthesis of nanoparticles and their applications. *J. King Saud Univ. Sci.,* **2022**, *34*(3), 101869.
[http://dx.doi.org/10.1016/j.jksus.2022.101869]

[66] Ovais, M.; Khalil, A.T.; Ayaz, M.; Ahmad, I.; Nethi, S.K.; Mukherjee, S. Biosynthesis of metal nanoparticles *via* microbial enzymes: A mechanistic approach. *International Journal of Molecular Sciences. MDPI AG,* **2018**, pp. 19.

[67] Sarma, H; Narasimha, M; Prasad, V. *Production and Role in Synthesis of Nanoparticles for Environmental Applications*; Wiley Online Library, **2021**.

[68] Nguyen, T.T.L.; Edelen, A.; Neighbors, B.; Sabatini, D.A. Biocompatible lecithin-based microemulsions with rhamnolipid and sophorolipid biosurfactants: Formulation and potential applications. *J. Colloid Interface Sci.,* **2010**, *348*(2), 498-504.
[http://dx.doi.org/10.1016/j.jcis.2010.04.053] [PMID: 20471022]

[69] Palanisamy, P.; Raichur, A.M. Synthesis of spherical NiO nanoparticles through a novel biosurfactant mediated emulsion technique. *Mater. Sci. Eng. C,* **2009**, *29*(1), 199-204.
[http://dx.doi.org/10.1016/j.msec.2008.06.008]

[70] Kiran GS, Selvin J, Manilal A, Sujith S. Biosurfactants as green stabilizers for the biological synthesis of nanoparticles. *Critical Reviews in Biotechnology,* **2011**, *31*, 354-364.

[71] Kiran, G.S.; Sabu, A.; Selvin, J. Synthesis of silver nanoparticles by glycolipid biosurfactant produced from marine Brevibacterium casei MSA19. *J. Biotechnol.,* **2010**, *148*(4), 221-225.
[http://dx.doi.org/10.1016/j.jbiotec.2010.06.012] [PMID: 20600381]

[72] Salazar-Bryam, A.M.; Yoshimura, I.; Santos, L.P.; Moura, C.C.; Santos, C.C.; Silva, V.L.; Lovaglio, R.B.; Costa Marques, R.F.; Jafelicci Junior, M.; Contiero, J. Silver nanoparticles stabilized by ramnolipids: Effect of pH. *Colloids Surf. B Biointerfaces,* **2021**, *205*, 111883.
[http://dx.doi.org/10.1016/j.colsurfb.2021.111883] [PMID: 34102528]

[73] Farias, C.B.B.; Silva, A.F.; Rufino, R.D.; Luna, J.M.; Souza, J.E.G.; Sarubbo, L.A. Synthesis of silver nanoparticles using a biosurfactant produced in low-cost medium as stabilizing agent. *Electron. J. Biotechnol.,* **2014**, *17*(3), 122-125.
[http://dx.doi.org/10.1016/j.ejbt.2014.04.003]

[74] Sowani, H.; Mohite, P.; Munot, H.; Shouche, Y.; Bapat, T.; Kumar, A.R.; Kulkarni, M.; Zinjarde, S. Green synthesis of gold and silver nanoparticles by an actinomycete *Gordonia amicalis* HS-11: Mechanistic aspects and biological application. *Process Biochem.,* **2016**, *51*(3), 374-383.
[http://dx.doi.org/10.1016/j.procbio.2015.12.013]

[75] Das, A.J.; Kumar, R.; Goutam, S.P. Sunlight irradiation induced synthesis of silver nanoparticles using glycolipid bio-surfactant and exploring the antibacterial activity. *J. Bioeng. Biomed. Sci.,* **2016**, *6*(5).
[http://dx.doi.org/10.4172/2155-9538.1000208]

[76] Basak, G.; Das, D.; Das, N. Dual role of acidic diacetate sophorolipid as biostabilizer for ZnO

nanoparticle synthesis and biofunctionalizing agent against *Salmonella enterica* and *Candida albicans.* *J. Microbiol. Biotechnol.*, **2014**, *24*(1), 87-96.
[http://dx.doi.org/10.4014/jmb.1307.07081] [PMID: 24150496]

[77]   Shikha, S.; Chaudhuri, S.R.; Bhattacharyya, M.S. Facile one pot greener synthesis of sophorolipid capped gold nanoparticles and its antimicrobial activity having special efficacy against gram negative *Vibrio cholerae*. *Sci. Rep.*, **2020**, *10*(1), 1463.
[http://dx.doi.org/10.1038/s41598-019-57399-3] [PMID: 31996706]

[78]   Ga'al, H.; Yang, G.; Fouad, H.; Guo, M.; Mo, J. Mannosylerythritol lipids mediated biosynthesis of silver nanoparticles: an eco-friendly and operative approach against chikungunya vector *Aedes albopictus*. *J. Cluster Sci.*, **2021**, *32*(1), 17-25.
[http://dx.doi.org/10.1007/s10876-019-01751-0]

[79]   Bakur, A.; Niu, Y.; Kuang, H.; Chen, Q. Synthesis of gold nanoparticles derived from mannosylerythritol lipid and evaluation of their bioactivities. *AMB Express*, **2019**, *9*(1), 62.
[http://dx.doi.org/10.1186/s13568-019-0785-6] [PMID: 31065818]

[80]   Durval, IJB; Meira, HM; de Veras, BO; Rufino, RD; Converti, A; Sarubbo, LA Green synthesis of silver nanoparticles using a biosurfactant from bacillus cereus ucp 1615 as stabilizing agent and its application as an antifungal agent. *Fermentation*, **2021**, *7*(4), 233.

[81]   Reddy AS, Chen C, Baker SC, Chen C, Jean J, Fan C, *et al.* Synthesis of silver nanoparticles using surfactin: A biosurfactant as stabilizing agent. *Materials Letters*, **2009**, *63*, 1227-30.

[82]   Eswari, JS; Dhagat, S; Mishra, P Biosurfactant assisted silver nanoparticle synthesis: A critical analysis of its drug design aspects. *Adv. Nat. Sci: Nanosci. Nanotechnol.*, **2018**, *9*, 045007.
[http://dx.doi.org/10.1088/2043-6254/aaec0e]

[83]   Yu, X; Li, J; Mu, D; Zhang, H; Liu, Q; Chen, G. Green synthesis and characterizations of silver nanoparticles with enhanced antibacterial properties by secondary metabolites of *Bacillus subtilis* [SDUM301120]. In: *Green Chemistry Letters and Reviews*; Taylor and Francis Ltd, **2021**; 14, pp. 189-202.

[84]   Płaza, G.A.; Chojniak, J.; Mendrek, B.; Trzebicka, B.; Kvitek, L.; Panacek, A.; Prucek, R.; Zboril, R.; Paraszkiewicz, K.; Bernat, P. Synthesis of silver nanoparticles by *Bacillus subtilis* T-1 growing on agro-industrial wastes and producing biosurfactant. *IET Nanobiotechnol.*, **2016**, *10*(2), 62-68.
[http://dx.doi.org/10.1049/iet-nbt.2015.0016] [PMID: 27074855]

[85]   Radha, P.; Suhazsini, P.; Prabhu, K.; Jayakumar, A.; Kandasamy, R. Chicken tallow, a renewable source for the production of biosurfactant by *Yarrowia lipolytica* MTCC9520, and its application in silver nanoparticle synthesis. *J. Surfactants Deterg.*, **2020**, *23*(1), 119-135.
[http://dx.doi.org/10.1002/jsde.12357]

[86]   Bezza, F.A.; Tichapondwa, S.M.; Chirwa, E.M.N. Fabrication of monodispersed copper oxide nanoparticles with potential application as antimicrobial agents. *Sci. Rep.*, **2020**, *10*(1), 16680.
[http://dx.doi.org/10.1038/s41598-020-73497-z] [PMID: 33028867]

[87]   Vecino, X.; Rodríguez-López, L.; Rincón-Fontán, M.; Cruz, J.M.; Moldes, A.B. Nanomaterials synthesized by biosurfactants. In: *Comprehensive Analytical Chemistry*; Elsevier B.V., **2021**; pp. 267-301.

[88]   Ohadi, M.; Forootanfar, H.; Dehghannoudeh, G.; Eslaminejad, T.; Ameri, A.; Shakibaie, M.; Najafi, A. Biosynthesis of gold nanoparticles assisted by lipopeptide biosurfactant derived from *Acinetobacter junii* B6 and evaluation of its antibacterial and cytotoxic activities. *Bionanoscience*, **2020**, *10*(4), 899-908.
[http://dx.doi.org/10.1007/s12668-020-00782-6]

[89]   Martínez, G.; Merinero, M.; Pérez-Aranda, M.; Pérez-Soriano, E.; Ortiz, T.; Villamor, E.; Begines, B.; Alcudia, A. Environmental impact of nanoparticles' application as an emerging technology: A review. *Materials*, **2020**, *14*(1), 166.
[http://dx.doi.org/10.3390/ma14010166] [PMID: 33396469]

[90] Nehal, N.; Singh, P. Role of nanotechnology for improving properties of biosurfactant from newly isolated bacterial strains from Rajasthan. In: *Materials Today*; Proceedings, **2021**; pp. 2555-2561.

[91] Sharma, B.; Menon, S.; Mathur, S.; Kumari, N.; Sharma, V. Decolorization of malachite green dye from aqueous solution using biosurfactant-stabilized iron oxide nanoparticles: process optimization and reaction kinetics. *Int. J. Environ. Sci. Technol.,* **2021**, *18*(7), 1739-1752.
[http://dx.doi.org/10.1007/s13762-020-02901-y]

[92] Hazra, C.; Kundu, D.; Chaudhari, A.; Jana, T. Biogenic synthesis, characterization, toxicity and photocatalysis of zinc sulfide nanoparticles using rhamnolipids from *Pseudomonas aeruginosa* BS01 as capping and stabilizing agent. *J. Chem. Technol. Biotechnol.,* **2013**, *88*(6), 1039-1048.
[http://dx.doi.org/10.1002/jctb.3934]

[93] Mandal, S.K.; Ojha, N.; Das, N. Optimization of process parameters for the yeast mediated degradation of benzo[a]pyrene in presence of ZnO nanoparticles and produced biosurfactant using 3-level Box-Behnken design. *Ecol. Eng.,* **2018**, *120*, 497-503.
[http://dx.doi.org/10.1016/j.ecoleng.2018.07.006]

[94] Ojha, N; Mandal, SK; Das, N Enhanced degradation of indeno[1,2,3-cd] pyrene using Candida tropicalis NN4 in presence of iron nanoparticles and produced biosurfactant: A statistical approach. *3 Biotech,* **2019**, *9*(3).

[95] Raji, Ro Development of biosurfactants-iron oxide nanoparticles-biochar formulation for remediation of crude oil contaminated soil. **2021**.

CHAPTER 10

Nanoliposome Mediated Heavy Metal Detection

Banishree Sahoo[1,*] and **Manoranjan Arakha**[1]

[1] *Centre for Biotechnology, Siksha 'O' Anusandhan (Deemed to be University), Bhubaneswar, Odisha-751003, India*

Abstract: The bulk of metal ions are carcinogens that significantly harm human health by producing free radicals. Therefore, the need for quick and accurate metal ion detection has been a matter of concern. However, the most dangerous metal ions are cadmium, arsenic, mercury chromium, and lead. Heavy metals are indestructible. Instead, they interact with living things directly or indirectly *via* the food chain. Metal ions have the potential to directly disrupt metabolic processes or to change into more hazardous forms inside the body. Nanotechnology is known as an emerging field that has been utilized for heavy metal analysis and removal from intricate matrices. Numerous methods based on nanomaterials have been developed for the analysis of heavy metals, including electrochemical, colorimetric, fluorescent, and biosensing technologies. As a result, in recent decades, there has been significant growth in the quest for more systematic nano-vesicular systems, like nanoliposomes, with elevated biocompatibility properties and directed releases. Additionally, nanoliposome have various applications such as drug delivery in the pharmaceutical industry, storage of food mainly cheeses, and dairy products for a long time in the food industry and cosmetics industry.

Keywords: Biocompatible, Drug delivery, Heavy metals, Nanotechnology, Nanoliposome.

INTRODUCTION

The world is in a disquieting condition, as industrialization increases the number of toxic metals in the environment. Metals are highly electrically conductive compounds that voluntarily give up one or more of their electrons to form cations. Heavy metal(loid)s are released into the surroundings through effluents from different industries such as paper, textile, tannery, mining, and pulp [1]. Heavy metals accumulate in birds, animals and human bodies over time. These heavy metals are characterized as "a group of toxic metals and metalloids associated with pollution and toxicity, having a density greater than 6 Mgm^{-3} and an atomic weight greater than that of iron" [2]. Out of 35 existing metals that are of concern

* **Corresponding author Banishree Sahoo:** Centre for Biotechnology, Siksha 'O' Anusandhan (Deemed to be University), Bhubaneswar, Odisha-751003, India; E-mail: sahoo.banishree1@gmail.com

Arun Kumar Pradhan and Manoranjan Arakha (Eds.)
All rights reserved-© 2024 Bentham Science Publishers

to us due to residential or occupational exposure, 23 are categorized as "heavy metals". The heavy metals are tellurium, antimony, bismuth, thallium, tin, arsenic, gold, cerium, lead, cadmium, cobalt, copper, iron, vanadium, chromium, gallium, mercury, nickel, platinum, silver, manganese, uranium, zinc, *etc*. Surprisingly, significant concentrations of any of these metals can cause subtle toxicity, despite the fact that little amounts of them are ubiquitous in our environment and diet, which has been demonstrated to be helpful to health. Surprisingly, significant concentrations of any of these metals can cause subtle toxicity, despite the fact that little amounts of them are ubiquitous in our environment and diet, which has been demonstrated to be helpful to health [3]. However, the deposition of higher concentrations of heavy metals accumulates in the soil, sediment, and water and from the contaminated area plants and animals absorbed the minerals, thus accumulating inside the body and entering into the food chains [4]. These heavy metals are assimilated and aggregated in different organs of the body and they can cross the blood-brain barrier and produce adverse effects on human health. Arsenic is transported *via* aquaglyceroporins, whereas cadmium enters *via* calcium channels [5]. Lead can enter the body *via* a divalent metal transporter or a calcium channel. Mercury utilizes a similar transporter through which it enters the cell. These metals can cause a wide range of disorders by damaging cellular enzymes, causing cardiovascular disease, DNA damage, obstructing protein synthesis routes and affecting the neurons and neurotransmitters [6]. Thus, identifying heavy metals that are major pollutants of the environment, such as mercury, cadmium, lead, arsenic, and chromium is very critical for protecting and preserving the environment and avoiding their adverse consequences on living organisms. Despite the direct toxicity, heavy metals are non-biodegradable in nature and are kept for a prolonged period of time [7].

Conventional methods for detecting heavy metals include neutron activation analysis [8], microprobes, atomic absorption spectroscopy [9], energy dispersive X-ray fluorescence [10], inductively coupled plasma atomic/optical emission spectrometry [11], X-ray fluorescence spectrometry [12] and flame atomic absorption spectrometry [13]. Although they can typically attain acceptable detection sensitivity, they are unavoidably constrained by high costs, heavy equipment, complex operation and lab use only. Ion exchange [11], precipitation, oxidation, membrane filtration and reverse osmosis [14] are all conventional strategies for heavy metal removal from wastewater. Hence these traditional approaches are limited to the laboratory, hence low-cost, simple, and quick analytical techniques that may be carried out *in situ* are required for broad and large-scale testing of heavy metals. Hence claimed that appropriate sensing techniques should be sensitive, inexpensive, user-friendly, specific, equipment-free, quick and resilient, and deliverable to end-users (the ASSURED criteria)

[14]. Their efficiency, however, is limited due to the drawbacks of high pH sensitivity, sludge contamination and corrosiveness [15].

Numerous electrolytic sensing techniques have been explored for the application in the detection of toxic contaminates throughout the few last decades.

Electrochemical sensing techniques employ the interaction between chemistry and electricity by monitoring output signals such as current, charge, frequency, potential, phase, *etc.* at the electrolyte-electrode interface. Electrochemical strategies are categorised into potentiostatic and potentiometric techniques. Potentiostatic electrochemical techniques regulate charge transfer at the electrolyte electrode interface by detecting species reduction/oxidation and analysing the observed current. Potentiometric approaches can be employed to identify analytes using zero current, with the voltage reported across the membrane being driven by an electromotive force and being directly proportional to the difference in sample concentration on either side of the membrane [16 - 20]. Both types of electrochemical sensors are used to detect quick, rapid and cost-effective heavy metals. Electrochemical detection methods are very sensitive, accurate, easy to handle and optimizable and can be used in the microlitre range. Since these sensor techniques are very costly and a lot of samples are required for the processing and high-end equipment. Hence recent focus on the simple material sensor which is rapid, sensitive, easy to handle and accurate [21].

Recent breakthroughs in nanotechnology, as well as new possibilities for nanomaterial production, have aided in the design and development of enhanced sensing techniques and portable gadgets. Nanotechnology is the use of multidisciplinary techniques that integrate chemistry, physics, biology and engineering principles to create nanoscale devices [22]. With the advancement of applications of nanotechnology, nanomaterials for heavy metal detection and removal have been delicately designed and fabricated offering numerous advantages over other traditional methods.

NANOPARTICLES MEDIATED HEAVY METAL DETECTION

Nanotechnology has made tremendous achievements and developments over the last few eras. Nanoparticles are classified as materials with at least one dimension smaller than 100 nm. They are classified into different classes based on types of material, dimension, *etc* [23]. Nanoparticles, particularly metal nanoparticles provide significant advantages in the realm of electrochemical sensing. Metallic nanoparticles can support the mass-transport rate and give quick electron transfer, the two of which increase the awareness of the cathodes used [24]. In this section, we will discuss the utilization of various sorts of metallic nanoparticles for the detection of most heavy metals.

Silver nanoparticles (AgNPs) are among the most important nanoparticles because of their low cost and distinctive physical and chemical attributes that make them helpful in a number of optical, catalytic, and chemical applications. For the purpose of detecting Cr^6, Cd^2, Hg^2, Cu^2, and Sb^3, silver nanoparticles have been mixed with a variety of substances. Two techniques such as electrodeposition and reduction have been used to create spherical AgNPs. In order to directly detect Cr (VI) using linear sweep voltammetry, using Nafion, Xing *et al.* developed a glassy carbon electrode with silver nanoparticles deposition on its surface. In the absence of further ion interference, the experiment produced a linear range between 2 and 230 ppb, and the limit of detection was 0.67 ppb. The sensor's applicability was evaluated using wastewater from a textile plant, and the concentration of Cr (VI) was 6.58 0.04 mg/L with a 99.5% recovery for spiked samples [25]. Silver nanoparticles biosynthesized from allium sativum extract was used for the detection of Cd^{2+} ions in waste water body and fertilizer industry effluents. When silver nanoparticles were mixed with contaminated water samples with Cd^{2+} ions, the visual colour change was found from brown to yellow and SPR of silver nanoparticles diminished [26]. For the detection of Hg^{2+} ions, silver nanoparticles were functionalised by 3-mercapto 1–2 propanediol that resulted in the change of colour from yellow to black [27]. Similarly, AgNPs biosynthesized from fried and dried mango leaves and displayed selective sensitivity for lead and mercury ions (Pb^{2+} and Hg^{2+}). Furthermore, AgNPs produced from green tea extract and pepper seed extract illustrated selective sensing characteristics for Zn^{2+}, Pb^{2+} and Hg^{2+} ions [28].

Gold nanoparticles (AuNPs) are the most widely used nanoparticles that are used for the detection of heavy metals through the electrochemical method. Their properties change according to the size, but AuNPs are known to be biocompatible at low toxicity [29]. A few reports reported that not only gold nanoparticles; but also, different materials associated with gold nanoparticles have been used for the detection of heavy metals like lead and mercury. The most prevalent methods for fabricating AuNPs are electrodeposition or reduction. However, the sizes and shapes of the gold nanoparticles vary depending on the synthesis conditions. Spherical shapes are the most commonly used for the detection of heavy metals. The spherical AuNPs have been synthesized having particle sizes ranging from 4 to 298 nm. Hassan *et al.* stated the amalgamation and application of various gold nanostructures for As detection (III) [30]. Nanoflowers were created by Ouyang *et al.* using a challenging procedure. They modified a glassy carbon electrode by first coating it with gold nanoparticles and then covering it with a layer of 3-mercaptopropyl-trimethoxysilane. The electrode was then submerged in an AuNPs solution to create a second layer of nanoparticles, and after being dipped in a solution of 4-pyridineethanethiol hydrochloride, pyridinium was bound to the NPs [22]. For comparison, Dutta *et al.* created spherical gold nanoparticles and gold nanostars and used them to detect Hg (II), As (III), and Pb (II) [31]. The ligands such as glutathione (GSH), cysteine and and Dichlorodiphenyltrichloroethane (DDT) were used in gold nano sensors for the detection of arsenic [32].

Over the last decade, Bismuth has been increasingly used in various areas of chemistry (catalysis, organic synthesis, clusters, *etc.*). Replacing mercury electrodes, bismuth is used as an electrode coating in electroanalytical chemistry because of its high peak resolution and low toxicity. Lee *et al.* used square wave anodic stripping voltammetry to detect Cd, Zn, and Pb ions using bismuth nano powder-modified electrodes. Spherical bismuth with varying particle size distributions was prepared in order to explore its effect on the sensitivity and detection limit of the studied metals. It was studied that decreasing the particle size from 406 to 166 nm improved both detection limit and sensitivity [33]. Using differential pulse anodic stripping voltammetry, Sahoo *et al.* modified a carbon paste electrode with graphene oxide and bismuth nanoparticles with diameters ranging from 40 to 100 nm for the detection of zinc, copper, cadmium, and lead ions. A linear concentration range of 20 to 120 mg/L was obtained, with limits of detection of 0.55, 2.8, 17, and 26 mg/L for Pb^2, Cd^2, Zn^2 and Cu^2 respectively. The electrode's performance was assessed in two separate water samples, lake and ground water, and divalent metal concentrations were found [34].

Different metal nanoparticles are modified with electrodes in electrochemical analysis. That is why these nanoparticles are used for the detection of copper, lead, mercury and cadmium. Veera Kumar *et al.* fabricated smaller crystals having an average particle size of 4-5 nm. They detected Pb^{2+}, Cd^{2+}, Hg^{2+} and Cu^{2+} using differential pulse voltammetry. The results showed that individual and simultaneous detections performed better. A linear response was found in the ion concentration ranges of 0.5-5.5, 0.5-8.9, 0.5-5.0, and 0.24-7.5 µM, with sensitivities of 66.7, 53.8, 41.1, and 500.3 mA mM1.cm2 and detection limits of 41, 50, 66, and 54 µA $µM^{-1}.cm^{-2}$, were observed for simultaneous detection of Pb^{2+}, Cd^{2+}, Hg^{2+} and Cu^{2+} [35]. Lee *et al.* modified tin particles (SnNP) with reduced graphene oxide on the glassy carbon electrode and used them for the detection of Pb^{2+}, Cd^{2+}, and Cu^{2+}. Sn NPs with a diameter of 50 nm were generated by electrochemically reducing Sn^{2+} with a graphene oxide solution. Individual metal ion analysis using square wave anodic stripping voltammetry yielded high stability and detection limits of 0.60nM, 0.63nM and 0.52nM, respectively. However, because intermetallic compound production is a possibility, simultaneous heavy metal analysis raised the detection limits to 6.77 nM, 7.56 nM and 5.62 nM, respectively [36]. Toghill *et al.* used linear sweep anodic stripping voltammetry to detect Cd^{2+} and Pb^{2+} using a BDD electrode modified with Sb nanoparticles. By using the electrochemical method, the nanoparticles were deposited on the electrode having an average particle size of 108-70 nm, but due to the toxicity of Sb, the team used the least amount of antimony during the experiment. According to the findings of this study, the addition of Sb nanoparticles had no effect on the separate detection of each analyte as compared to the bare BDD. However, simultaneous lead and cadmium detection were enhanced, and Pb did not hinder Cd nucleation on the electrode surface, as previously described [37].

IRON NANOPARTICLES MEDIATED BY PLANTS FOR HEAVY METAL DETECTION

Removal of Chromium

Recently phytonanotechnology is a trending and novel study for research in medical science that helps in the fabrication of metal nanoparticles by using plant extracts. The biosynthesized nanoparticles such as iron, silver, gold and copper are eco-friendly, less toxic and cost-effective.

The zero valent iron nanoparticles (ZVI NPs) were biosynthesized by using green tea extract mixed with a pre-prepared amount of iron sulfate solution and centrifuged at 150 rpm for 60 minutes. The biofabricated iron nanoparticles were used for the reduction of hexavalent chromium ions and their efficiency was found to be 99.9% under ideal condition. According to the findings of the research, the efficiency of the adsorbents decreased as the pH increased, and the absorbent doses should be approximately 0.12 g/L in order to achieve 100% efficacy [38].

The biosynthesis of iron nanoparticles from peanut skin is a cost-effective and eco-friendly process. For the amalgamation of zero valent iron nanoparticles (ZVI NPs), peanut skins were vacuum dried and powdered and some aliquots of extracts were mixed with iron source, ferric chloride. Then the solution was agitated with a magnetic stirrer in a nitrogen atmosphere. Then iron nanoparticles were biosynthesized and characterized with the help of UV-visible spectra, XRD, SEM and TEM. The biofabricated 0.25 g of Fe-NPs added to 10mg/L of hexavalent chromium solution achieved 100% effectiveness in a minute. In this way, heavy metal such as chromium was detected from waste water contaminants [39].

Removal of Lead

Agriculture waste biomass *i.e.* sineguelas waste (S-NaOH-NZVI) was used for the biosynthesis of nanoscale zero-valent iron (nZVI) to remove Pb (II) ions. Pure sineguelas wastes have limited Pb (II) absorbent capacity from the effluent. Hence to limit the agglomeration tendency, zero-valent iron particles are made into powder form. According to the TEM image, it was found that the biofabricated iron nanoparticles were found spherical in shape. The removal of Pb (II) efficacy gradually increased when the pH increased from 1 to 5 and remained unchanged at pH 5 to 9. Further, 0.15g dose of zero valent iron nanoparticle was found to be the best absorbent for the removal of pb (II) ions [40].

From the other study it was found that, iron NPs were synthesized from the *Synechocystis* sp. PCC6803 algae and ferric chloride solution. Then the mixture was centrifuged and the precipitation was washed in ethanol and dried. From the XRD data, the biosynthesized nanoparticles were crystallinity in nature. The absorption efficiency of iron nanoparticle is related to pH and it was the highest at pH 2.0 and absorption efficacy decreased at pH 6.0 for the detection of Pb (II). Pb (II) was completely absorbed and reached reaction equilibrium within one hour. In this way, the synthesized iron nanoparticles were used for waste water remediation as they absorb heavy metals such as chromium, cadmium and copper [41, 42].

Heavy Metal Detection Mechanism Using Green Nanoparticles

Recently colorimetric detection of heavy metals using green synthesized nanoparticles is gaining popularity owing to its user-friendliness, simplicity and sustainability [43]. The colorimetric detection of harmful heavy metals using green synthesised nanoparticles typically involves two methods. The first mechanism involves a redox interaction between the nanoparticles and the noxious heavy metal, which results in oxidation of the nanoparticles and a visible colour change. In this technique, heavy metals are reduced resulting in a blue shift in the UV-vis spectra and a decrease or disappearance of the peak of nanoparticles. In the second mechanism, nanoparticles and harmful heavy metals amalgamate during incubation as a result of complex formation occurred. This results in a redshift in the UV-vis spectra owing to an increase in nanoparticle size during aggregation, leading to colour shifts and so aids in visual identification. When silver nanoparticles were incubated with Hg^{2+} ions, they exhibited a hypsochromic shift (decrease in nanoparticles SPR), and the resulting colour of the solution became colourless, which was proportional to the quantity of Hg^{2+} ions present [44].

NANO-LIPOSOME AND NANOLIPOSOME MEDIATED HEAVY METAL DETECTION

In the 1960s, Bangham discovered that when some lipids are exposed to an aqueous medium, they form membranous structures and these structures were named as liposomes [45]. The produced vesicles are round in shape and they consist of an aqueous core that is surrounded by one or more phospholipid bilayers; this allows for the encapsulation of both lipophilic and hydrophilic active substances [46, 47]. Phospholipids with or without surfactants, cholesterol and other materials are used in their production. To modify some of the properties of biomembranes such as membrane permeability, surface charge and lipid stability in the bilayer, these are incorporated into bilayers of phospholipid [48, 49]. In this case, the presence of cholesterol improves mechanical strength of the membrane by inducing membrane elasticity and increasing lipid packing density. Furthermore, in the presence of biological fluids, liposomes prepared with cholesterol are anticipated to be more stable [50].

However, some limitations of liposomes are the inability to stay in systemic circulation for extended periods of time. When traditional liposomes are administered intravenously, the endothelial reticulum system detects them as foreign materials and eliminates them. Nanoliposomes, which have a similar composition to living cells, cause fewer problems than other nanoparticles (made of silicon, polymers and metals). However, attention has recently been paid to several factors affecting the safety of nanoliposomes, like surface charge, composition, size, stability, interaction with cells and incorporation into tissues as they represent critical factors in therapeutic use. Although ideal liposome dimensions for accessing cancer cells as compared to healthy tissues may be between 100 and 200 nm, however, ultra-small liposomes with diameters of 20-30 nm have shown significant advantages [51].

Broadly, bilayer lipid vesicles are known as nanoliposomes as shown in Fig. (1), that have nanometric size ranges [52]. These systems can encapsulate hydrophobic and hydrophilic compounds separately or simultaneously due to their bilayer structure which is composed of aqueous and lipidic sections. Even though these systems have the potential to be drug delivery vehicles, they face substantial obstacles when used for commercial purposes because of their limited high temperature sensitivity, physical stability and sensitivity to pH changes. Nonetheless, abundant studies have reported surface modification to improve storage and stability [53]. It is reported that the amount of entrapped material should be less than the amount required without the encapsulation for improved stability and targeting. This could be useful when working with expensive bioactive compounds. Furthermore, the use of inexpensive and natural ingredients such as egg yolk, soy, milk, and sunflower for nanoliposome preparation is possible, optimising the cost-effectiveness of the final product [54]. Several clinical preliminaries have reported that nanoliposomes are excellent agents for a variety of delivery systems, including anti-fungal, anti-cancer, gene medicine delivery, anti-biotic drugs, anti-inflammatory and anaesthetic drug delivery [55]. The benefits of nanoliposome-based formulations over non-nanoliposome-based formulations for topical, oral and intramuscular medication administration are discussed below:

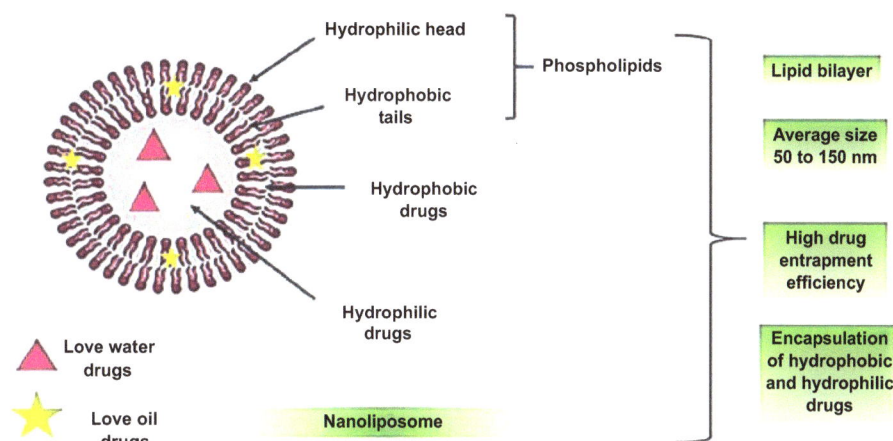

Fig. (1). Diagrammatic illustration of the amphiphilic structure of nanoliposomes' bilayers for the trapping of hydrophobic and hydrophilic pharmaceuticals. Some of these systems' outstanding features are listed on the right. Phospholipids like phosphatidylcholine, phosphatidylserine, or phosphatidylethanolamine make up the majority of the liposomal structure. However, it is frequently done to include cholesterol in the liposomal formulation in order to give the lipid membrane stability and rigidity [56].

Nanoliposomes have been coupled with various therapeutic methods to enhance their mechanism of action [57]. In this context, Gelfuso *et al.* investigated that when voriconazole-based nanoliposomes were coupled with iontophoresis, their effectiveness increases for the treatment of fungal keratitis. Furthermore, from the morphological studies using Transmission Electron Microscopy (TEM) it was revealed that they were oval in shape and the size was close to 100 nm. These findings established that, in comparison to commercial voriconazole medications, nanoliposomes had superior stability and potency for voriconazole passive administration. Both carriers have been effectively utilised in biomedical research because of their drug delivery mechanisms and release behaviours [52, 56, 58]. Nanoliposomes can be useful in the detection of heavy metals. In this context, Priyadarshini *et al.* synthesized silver nanoparticles using lipopeptide biosurfactants. The formulated silver nanoliposome vesicles can be used for the detection of Hg^{2+} ions selectively compared to other traditional methods [59].

OTHER APPLICATIONS OF NANOLIPOSOME

Biomedical Application

The application of nanostructured materials in the biomedical sciences is focused on the development of innovative strategies for drug design, disease detection, and drug delivery, with the goal of enhancing medication bioavailability by appropriate surface modification. In order to produce innovative vaccines, pharmaceuticals, cancer therapy advancements, photodynamic therapy, and even as a tool for disease detection, nanoliposomes have been thoroughly investigated in order to understand their interaction effects in diverse cultures, strains, and animal models. Nanoliposomes have been extensively explored in order to understand their interaction effects in various cultures, strains, and animal models for the development of novel vaccines, medications, cancer therapy improvements, photodynamic and even as a tool for disease detection. Among the ongoing biomedical therapies, cutaneous (20 nm), exemplification of calothrixin B as an anticancer agent (108nm) [60], gastrointestinal disorders (145 nm) [61, 62] chemotherapeutic sensitization of glioblastoma (75 nm) [63] and fungal infections (100 nm) [64] are listed as some of the effective instances of nanoliposomes as drug delivery systems [56]. Nanoliposomes, as drug carriers, can improve *in vivo* drug bioavailability and stability by limiting therapeutic interactions with undesired molecules and minimising hazardous side effects [65]. Nanoliposomes have the advantages of low toxicity, biocompatibility, the capacity to change the size and surface, and biodegradability [66]. To reduce the likelihood that macrophages may recognise liposomes, it is possible to cover them with biocompatible materials like polyethene glycol (PEG) (known as stealth liposomes). This technique significantly improved circulation half-life and liposomal stability [67]. Furthermore, liposomes have ligands on their surfaces that can recognise and bind to a specific group of cells.

Nanoliposomes are regarded as one of the most biocompatible nanocarriers utilised for designated pharmaceutical administration because of their potential to boost the biodistribution and bioavailability of the specified encapsulated drug site employed for targeted pharmacological administration. Nanoliposome-based treatment can accomplish active or triggered processes. The surface of the nanoliposome is made by ligands and antibodies in the first type, whereas drug release is achieved by stimuli sensitive in the second type [68]. Internal medication triggers include enzyme, pH or hormone levels, glucose, tiny biomolecules, or a redox gradient associated with the pathogenic elements of the illness. External stimuli such as ultrasound (US), light, magnetic fields and heat are also employed to initiate medication release at the diseases site. A smart DDS, according to Darvin *et al.* (2019), can reach a specific place where the medicine is meant to release. It can also release the medication in reaction to certain stimuli (*e.g.*, light, electric field, temperature, pH, magnetic, redox, ultrasound, and enzyme). As a result of this capability, they are intelligent systems capable of integrated sensing, self-regulation, activation and monitoring by the environment and stimuli [56, 69].

Skin-Curative Potential of Nanoliposomes

The largest and the most crucial organ for delivering systemic and tropical medications is the skin. It serves as a passive barrier to chemicals that can penetrate the body to safeguard it from the external harm. On the other hand, because of its exposure to the environment, it is more susceptible to damage and injury. Lesions are frequently caused by skin [70].The skin's main barrier, the Stratum corneum (SC), is made up of 15–20 layers of degenerated epidermal cells. Ceramides, cholesterol, and fatty acids abound in this barrier. Considering these ideas, nanoliposomes become a viable option for the delivery of topical medications. To improve active material penetration into the epidermal layers, nanoliposomes are widely used [71]. In order to administer TC topically for the treatment of acne, Hasanpouri *et al*. investigated the usage of nanoliposomes and nanotransferrosomes. The TC-loaded liposomal formulation's particle size and distribution were 74.8 ± 9.5 nm, with a polydispersity index (PDI) of 0.26 ± 0.03, and a mean zeta potential value of 17.2 ± 5.2 mV, which was below the critical threshold of 30 and suggested colloidal instability [72].

Application of Nanoliposome in Food Industry

Despite the fact that nanotechnology applications in the food industry are still in their early stages, significant progress has been made. So far, the principal applications have focused on modifying the texture of food components, producing unique sensations and taste, encapsulating food components or additives, managing flavour release, and improving nutritional component bioavailability [73].

Most of the epitome technologies utilized in the food business today depend on biopolymer frameworks made out of starches, sugars, proteins, gums, dextrins, synthetics and alginates [74]. On the other hand, liposomes have acquired significance due to the previously mentioned specific highlights [75]. Based on the findings of liposomal studies in medical and pharmaceutical research and applications (such as cancer treatments, drug delivery, gene therapy, and so on), food scientists have begun to use liposomes for the controlled delivery of functional ingredients such as peptides, enzymes, vitamins, and flavours in a variety of food applications [74]. Some of the parts highlight the application of lipid vesicles in the food industry.

Application in Dairy Products

Encapsulation technologies have various potential uses in the dairy sector, ranging from protecting delicate molecules and chemicals to enhancing the adequacy of food additives. A lot of research papers indicate that using nanocarrier systems can change the pharmacokinetic properties of medicines, plants, vitamins, and even enzymes [76, 77]. Furthermore, certain lipid vesicles have been used in dairy products for the selective delivery of encapsulated substances [78]. Up to this point, the significant purposes have led to alter the texture of food components, assist in cheese maturing, produce novel tastes and sensations, manage flavour release, and boost nutritional element absorption and bioavailability. The underlying utilization of lipid-based vesicles in the dairy area was in the cheese ripening process [79]. Besides, other lipidic transporters, for example, nanoliposomes, have been utilized to ensnare or epitomize enzymes, antimicrobials, and minerals (*e.g.*, iron, calcium) in different dairy items [80].

The dairy business benefits enormously from nanoliposomes' capacity to move epitomized material to the explicit region of the food framework. To give many different types of cheese the right smell, flavor, and texture, an intricate chemical process called cheese maturing utilizes chemical reactions like lipolysis, glycolysis, and proteolysis. The main phase of cheese maturing, proteolysis, changes the flavour and aroma of the dairy items. There are endeavours to abridge this period by adding enzymes, flavouring agents, preservatives and texture-enhancing ingredients by encapsulating cheese as the fact that cheese maturing process is extended and costly and there is more than adequate time for the activity and growth of spoiling organisms [81]. Proteinase is the most normally used enzyme in cheese maturing, in any case, lipase and flavorzyme are additionally used somewhat (epitomized or in their free structure). Encapsulated bacterial or fungal proteinase has been utilized to speed up the maturing of cheddar without creating off-tastes or causing textural deficiencies [82]. In media, enzymes are more stable. Moreover, for example, thermostabilized (*e.g.*, sugars) can be epitomized with the enzyme to safeguard it from high temperatures during food handling processes [83]. The timing, length, and rate at which the enzyme is delivered may be generally further improved by tweaking the arrangement of nanoliposomes. The ever-evolving arrival of the exemplified material in the cheese framework forestalls the making of unpleasant flavour created by free chemicals. The expansion of unencapsulated enzymes to enzyme drain adversely affects the substance qualities of cheese curd and decreases the proficiency of cheese made inferable from the quick hydrolysis of the casein. The effect of nanoliposomal protease on the qualities of cheese curd is dependent on the sort of enzyme, its entrapment efficiency, and the concentration of vesicles in the curd.

Flavorzyme, a peptide generated by *Aspergillus oryzae* that is widely used for the hydrolysis of various proteins, is another enzyme used in the dairy sector [84]. The heating approach was used to introduce flavourzyme into nanoliposomes in a study done by Jahadi and colleagues [85]. The enzyme- stacked vesicles were combined with cow milk, resulting in white-tenderized cheese. The proteolysis degree, chemical substance items in whey in addition to curd, and yield of the item were undeniably examined. The flavourzyme enzyme was entrapped with an effectiveness of around 25%, and the nano liposomal formulation had no discernible effect on the cheese's sensory properties. Besides, when contrasted with the control test, the synthetic cosmetics of both whey and curd stayed unaltered [66]. One more review proposes, flavourzyme-stacked nanoliposomes were mixed to white cheese, and the impact of a few boundaries (like bringing duration, enzyme concentration, and ripening period) on taste qualities and proteolytic action was surveyed.

Nanoliposomes can be used to fortify various products with health-promoting substances in addition to encapsulating enzymes. The dairy business, and the food business specifically, endeavour to strengthen its items with vitamins, minerals, and other gainful fixings because of the developing interest in food varieties with high health benefits. Several studies have discussed the use of lipid vesicles to add encapsulated vitamins to food products [86, 87]. The development and maintenance of bone and cartilage tissue depend heavily on vitamin D. The following strategies can be used to provide this vitamin to cheese; the translucent type of vitamin D being dissolved in milk cream and then being combined with milk; vitamin D being used in an appropriate carrier system; or vitamin D being encapsulated [88]. When cheese products were boosted utilising the three methods mentioned above, vitamin D retention during storage was 42.7%, 40.5%, and 61.5%, respectively. Vitamin D is clearly the most effectively delivered when it is encapsulated. The best vitamin D stability was achieved during the production and maturing of cheddar more than a 7-month time frame by involving lipid vesicles for the epitome and controlled arrival of nutrient D [89].The bilayer lipid vesicles have additionally been displayed to co-typify L-ascorbic acid (in the fluid stage) and vitamin E (in the lipid stage). This is an example of a possible technique for incorporating two vitamins into the food systems [90].

Application of Nanoliposome in Cosmetics

In the recent era, because of its remarkable features, such as biocompatibility, biodegradability, and nanosized structure, nanoliposomes have piqued the interest of many researchers [91]. As a result of these features, they may be widely and rapidly employed in a variety of disciplines, particularly as carriers of active substances in agriculture, nano-therapy, food technology, and cosmetics. There are several beneficial targets (hair, skin, and mucosal surfaces) for the delivery of active chemicals such as medicines and botanicals [92].

The nano products that are used in cosmetics are as follows:

Fullerenes

Nanotechnology may be used to create novel materials, such as carbon fullerenes. These small carbon spheres are said to possess anti-ageing properties.

Sunscreens

Titanium dioxide and zinc oxide nanoparticles are used for the manufacturing of sunscreen lotion.

Nanoemulsions and Nanosomes

Examples of nanoemulsions and nanosomes are antioxidants and vitamins which protect active ingredient. In this way, nanoliposomes can be used for the formulation of different cosmetics.

CONCLUSION

Nanotechnology has advanced significantly over the last few decades, resulting in the manufacture of naturally occurring nanoliposomes for culinary and therapeutic purposes. As an outcome, this work summed up that nanoliposomes are a promising technique for nano encapsulating either lipophilic (EOs) or hydrophilic (CE) bioactive concentrates. Such nanoliposomes may be utilized in food applications, for example, the commercialization of epitomized normal additives to further develop the timeframe of realistic usability of food items. Nanoliposomes can be utilized for the improvement of significant medications for the therapy of contaminations, malignant growth and irritation. Nanoliposomes can be used for the advancement of valuable medications for the treatment of infections, cancer and inflammation. In this way, nanoliposomes can be used for various applications that will be helpful for the mankind.

REFERENCES

[1]   Sharma, P.; Dutta, D.; Udayan, A.; Kumar, S. Industrial wastewater purification through metal pollution reduction employing microbes and magnetic nanocomposites. *J. Environ. Chem. Eng.*, **2021**, *9*(6), 106673.
        [http://dx.doi.org/10.1016/j.jece.2021.106673]

[2]   Ganeshamurthy, A.N.; Varalakshmi, L.R.; Sumangala, H.P. Environmental risks associated with heavy metal contamination in soil, water and plants in urban and periurban agriculture. *J. Hortic. Sci.*, **2008**, *3*(1), 1-29.
        [http://dx.doi.org/10.24154/jhs.v3i1.589]

[3]   Gumpu, M.B.; Sethuraman, S.; Krishnan, U.M.; Rayappan, J.B.B. A review on detection of heavy metal ions in water : An electrochemical approach. *Sens. Actuators B Chem.*, **2015**, *213*, 515-533.
        [http://dx.doi.org/10.1016/j.snb.2015.02.122]

[4]   Jaishankar, M.; Tseten, T.; Anbalagan, N.; Mathew, B.B.; Beeregowda, K.N. Toxicity, mechanism and health effects of some heavy metals. *Interdiscip. Toxicol.,* **2014**, *7*(2), 60-72.
        [http://dx.doi.org/10.2478/intox-2014-0009] [PMID: 26109881]

[5]   Nivetha, N.; Srivarshine, B.; Sowmya, B.; Rajendiran, M.; Saravanan, P.; Rajeshkannan, R.; Rajasimman, M.; Pham, T.H.T.; Shanmugam, V.; Dragoi, E.N. A comprehensive review on bio-stimulation and bio-enhancement towards remediation of heavy metals degeneration. *Chemosphere,* **2023**, *312*(Pt 1), 137099.
        [http://dx.doi.org/10.1016/j.chemosphere.2022.137099] [PMID: 36372332]

[6]   Balali-Mood, M.; Naseri, K.; Tahergorabi, Z.; Khazdair, M.R.; Sadeghi, M. Toxic mechanisms of five heavy metals: Mercury, lead, chromium, cadmium, and arsenic. *Front. Pharmacol.,* **2021**, *12*, 643972.
        [http://dx.doi.org/10.3389/fphar.2021.643972] [PMID: 33927623]

[7]   Chauhan, S.; Dahiya, D.; Sharma, V.; Khan, N.; Chaurasia, D.; Nadda, A.K.; Varjani, S.; Pandey, A.; Bhargava, P.C. Advances from conventional to real time detection of heavy metal(loid)s for water monitoring: An overview of biosensing applications. *Chemosphere,* **2022**, *307*(Pt 4), 136124.
        [http://dx.doi.org/10.1016/j.chemosphere.2022.136124] [PMID: 35995194]

[8]   El-Araby, E.H.; Abd El-Wahab, M.; Diab, H.M.; El-Desouky, T.M.; Mohsen, M. Assessment of Atmospheric heavy metal deposition in North Egypt aerosols using neutron activation analysis and optical emission inductively coupled plasma. *Appl. Radiat. Isot.,* **2011**, *69*(10), 1506-1511.
        [http://dx.doi.org/10.1016/j.apradiso.2011.06.005] [PMID: 21723139]

[9]   Tuzen, M.; Soylak, M. Multi-element coprecipitation for separation and enrichment of heavy metal ions for their flame atomic absorption spectrometric determinations. *J. Hazard. Mater.,* **2009**, *162*(2-3), 724-729.
        [http://dx.doi.org/10.1016/j.jhazmat.2008.05.087] [PMID: 18584957]

[10]   Obiajunwa, E.; Pelemo, D.A.; Owolabi, S.A. Characterisation of heavy metal pollutants of soils and sediments around a crude-oil production terminal using EDXRF. *Nucl. Instrum. Methods Phys.Res. B.Beam Interactions with Materials and Atoms,* **2002**, *194*(1), 61-64.

[11]   Moor, C.; Lymberopoulou, T.; Dietrich, V.J. Determination of heavy metals in soils, sediments and geological materials by ICP-AES and ICP-MS. *Mikrochim. Acta,* **2001**, *136*(3-4), 123-128.
        [http://dx.doi.org/10.1007/s006040170041]

[12]   O'Neil, G.D.; Newton, M.E.; Macpherson, J.V. Direct identification and analysis of heavy metals in solution (Hg, Cu, Pb, Zn, Ni) by use of *in situ* electrochemical X-ray fluorescence. *Anal. Chem.,* **2015**, *87*(9), 4933-4940.
        [http://dx.doi.org/10.1021/acs.analchem.5b00597] [PMID: 25860820]

[13]   Sohrabi, M.R.; Matbouie, Z.; Asgharinezhad, A.A.; Dehghani, A. Solid phase extraction of Cd(II) and Pb(II) using a magnetic metal-organic framework, and their determination by FAAS. *Mikrochim. Acta,* **2013**, *180*(7-8), 589-597.
        [http://dx.doi.org/10.1007/s00604-013-0952-4]

[14]   Venkateswara Raju, C.; Hwan Cho, C.; Mohana Rani, G.; Manju, V.; Umapathi, R.; Suk Huh, Y.; Pil Park, J. Emerging insights into the use of carbon-based nanomaterials for the electrochemical detection of heavy metal ions. *Coord. Chem. Rev.,* **2023**, *476*, 214920.
        [http://dx.doi.org/10.1016/j.ccr.2022.214920]

[15]   Gong, Z.; Chan, H.T.; Chen, Q.; Chen, H. Application of nanotechnology in analysis and removal of heavy metals in food and water resources. *Nanomaterials,* **2021**, *11*(7), 1792.
        [http://dx.doi.org/10.3390/nano11071792] [PMID: 34361182]

[16]   Singh, S.; Wang, J.; Cinti, S. Review—an overview on recent progress in screen-printed electroanalytical (bio) sensors. *ECS Sensors Plus,* **2022**, *1*(2), 023401.
        [http://dx.doi.org/10.1149/2754-2726/ac70e2]

[17]   Lu, Y.; Liang, X.; Niyungeko, C.; Zhou, J.; Xu, J.; Tian, G. A review of the identification and

detection of heavy metal ions in the environment by voltammetry. *Talanta,* **2018**, *178*, 324-338.
[http://dx.doi.org/10.1016/j.talanta.2017.08.033] [PMID: 29136830]

[18] Umapathi, R.; Ghoreishian, S.M.; Sonwal, S.; Rani, G.M.; Huh, Y.S. Portable electrochemical sensing methodologies for on-site detection of pesticide residues in fruits and vegetables. *Coord. Chem. Rev.,* **2022**, *453*, 214305.
[http://dx.doi.org/10.1016/j.ccr.2021.214305]

[19] Umapathi, R.; Park, B.; Sonwal, S.; Rani, G.M.; Cho, Y.; Huh, Y.S. Advances in optical-sensing strategies for the on-site detection of pesticides in agricultural foods. *Trends Food Sci. Technol.,* **2022**, *119*, 69-89.
[http://dx.doi.org/10.1016/j.tifs.2021.11.018]

[20] Umapathi, R.; Rani, G.M.; Kim, E.; Park, S-Y.; Cho, Y.; Huh, Y.S. Sowing kernels for food safety: Importance of rapid on-site detction of pesticide residues in agricultural foods. *Food Front.,* **2022**, *3*(4), 666-676.
[http://dx.doi.org/10.1002/fft2.166]

[21] Nayak, S.; Goveas, L.C.; Kumar, P.S.; Selvaraj, R.; Vinayagam, R. Plant-mediated gold and silver nanoparticles as detectors of heavy metal contamination. *Food Chem. Toxicol.,* **2022**, *167*, 113271.
[http://dx.doi.org/10.1016/j.fct.2022.113271] [PMID: 35792219]

[22] He, X.; Deng, H.; Hwang, H.M. The current application of nanotechnology in food and agriculture. *Yao Wu Shi Pin Fen Xi,* **2019**, *27*(1), 1-21.
[PMID: 30648562]

[23] Hyder, A.; Buledi, J.A.; Nawaz, M.; Rajpar, D.B.; Shah, Z.H.; Orooji, Y.; Yola, M.L.; Karimi-Maleh, H.; Lin, H.; Solangi, A.R. Identification of heavy metal ions from aqueous environment through gold, Silver and Copper Nanoparticles: An excellent colorimetric approach. *Environ. Res.,* **2022**, *205*, 112475.
[http://dx.doi.org/10.1016/j.envres.2021.112475] [PMID: 34863692]

[24] Sawan, S.; Maalouf, R.; Errachid, A.; Jaffrezic-Renault, N. Metal and metal oxide nanoparticles in the voltammetric detection of heavy metals: A review. *Trends Analyt. Chem.,* **2020**, *131*, 116014.
[http://dx.doi.org/10.1016/j.trac.2020.116014]

[25] Xing, S.; Xu, H.; Chen, J.; Shi, G.; Jin, L. Nafion stabilized silver nanoparticles modified electrode and its application to Cr(VI) detection. *J. Electroanal. Chem.,* **2011**, *652*(1-2), 60-65.
[http://dx.doi.org/10.1016/j.jelechem.2010.03.035]

[26] Aravind, A.; Sebastian, M.; Mathew, B. Green silver nanoparticles as a multifunctional sensor for toxic Cd(ii) ions. *New J. Chem.,* **2018**, *42*(18), 15022-15031.
[http://dx.doi.org/10.1039/C8NJ03696A]

[27] Maiti, S.; Barman, G.; Konar Laha, J. Detection of heavy metals (Cu^{2+}, Hg^{2+}) by biosynthesized silver nanoparticles. *Appl. Nanosci.,* **2016**, *6*(4), 529-538.
[http://dx.doi.org/10.1007/s13204-015-0452-4]

[28] Karthiga, D.; Anthony, S.P. Selective colorimetric sensing of toxic metal cations by green synthesized silver nanoparticles over a wide pH range. *RSC Advances,* **2013**, *3*(37), 16765-16774.
[http://dx.doi.org/10.1039/c3ra42308e]

[29] Zeng, S.; Yong, K.T.; Roy, I.; Dinh, X-Q.; Yu, X.; Luan, F. A review on functionalized gold nanoparticles for biosensing applications. *Plasmonics,* **2011**, *6*(3), 491-506.
[http://dx.doi.org/10.1007/s11468-011-9228-1]

[30] Hassan, S.S.; Sirajuddin, ; Solangi, A.R.; Kazi, T.G.; Kalhoro, M.S.; Junejo, Y.; Tagar, Z.A.; Kalwar, N.H. Nafion stabilized ibuprofen–gold nanostructures modified screen printed electrode as arsenic(III) sensor. *J. Electroanal. Chem.,* **2012**, *682*, 77-82.
[http://dx.doi.org/10.1016/j.jelechem.2012.07.006]

[31] Dutta, S.; Strack, G.; Kurup, P. Gold nanostar electrodes for heavy metal detection. *Sens. Actuators B*

Chem., **2019**, *281*, 383-391.
[http://dx.doi.org/10.1016/j.snb.2018.10.111]

[32] Chowdury, M.; Walji, N.; Mahmud, M.; MacDonald, B. based microfluidic device with a gold nanosensor to detect arsenic contamination of groundwater in Bangladesh. *Micromachines,* **2017**, *8*(3), 71.
[http://dx.doi.org/10.3390/mi8030071]

[33] Lee, G.J.; Kim, C.K.; Lee, M.K. Simultaneous voltammetric determination of Zn, Cd and Pb at bismuth nanopowder electrodes with various particle size distributions, An Internat. *J. Devoted Fundamen. Prac. Asp. Electroanal.,* **2010**, *22*(5), 530-535.

[34] Sahoo, P.K.; Panigrahy, B.; Sahoo, S.; Satpati, A.K.; Li, D.; Bahadur, D. *In situ* synthesis and properties of reduced graphene oxide/Bi nanocomposites: As an electroactive material for analysis of heavy metals. *Biosens. Bioelectron.,* **2013**, *43*, 293-296.
[http://dx.doi.org/10.1016/j.bios.2012.12.031] [PMID: 23334218]

[35] Veerakumar, P.; Veeramani, V.; Chen, S.M.; Madhu, R.; Liu, S.B. Palladium nanoparticle incorporated porous activated carbon: electrochemical detection of toxic metal ions. *ACS Appl. Mater. Interfaces,* **2016**, *8*(2), 1319-1326.
[http://dx.doi.org/10.1021/acsami.5b10050] [PMID: 26700093]

[36] Lee, P.M.; Chen, Z.; Li, L.; Liu, E. Reduced graphene oxide decorated with tin nanoparticles through electrodeposition for simultaneous determination of trace heavy metals. *Electrochim. Acta,* **2015**, *174*, 207-214.
[http://dx.doi.org/10.1016/j.electacta.2015.05.092]

[37] Toghill, K.E.; Xiao, L.; Wildgoose, G.G.; Compton, R.G. Electroanalytical determination of cadmium (II) and lead (II) using an antimony nanoparticle modified boron-doped diamond electrode. *Electroanalysis,* **2009**, *21*(10), 1113-1118.
[http://dx.doi.org/10.1002/elan.200904547]

[38] Yang, J.; Wang, S.; Xu, N.; Ye, Z.; Yang, H.; Huangfu, X. Synthesis of montmorillonite-supported nano-zero-valent iron *via* green tea extract: Enhanced transport and application for hexavalent chromium removal from water and soil. *J. Hazard. Mater.,* **2021**, *419*, 126461.
[http://dx.doi.org/10.1016/j.jhazmat.2021.126461] [PMID: 34186421]

[39] Pan, Z.; Lin, Y.; Sarkar, B.; Owens, G.; Chen, Z. Green synthesis of iron nanoparticles using red peanut skin extract: Synthesis mechanism, characterization and effect of conditions on chromium removal. *J. Colloid Interface Sci.,* **2020**, *558*, 106-114.
[http://dx.doi.org/10.1016/j.jcis.2019.09.106] [PMID: 31585219]

[40] Arshadi, M.; Soleymanzadeh, M.; Salvacion, J.W.L.; SalimiVahid, F. Nanoscale Zero-Valent Iron (NZVI) supported on sineguelas waste for Pb(II) removal from aqueous solution: Kinetics, thermodynamic and mechanism. *J. Colloid Interface Sci.,* **2014**, *426*, 241-251.
[http://dx.doi.org/10.1016/j.jcis.2014.04.014] [PMID: 24863789]

[41] Shen, L.; Wang, J.; Li, Z.; Fan, L.; Chen, R.; Wu, X.; Li, J.; Zeng, W. A high-efficiency Fe2O3@Microalgae composite for heavy metal removal from aqueous solution. *J. Water Process Eng.,* **2020**, *33*, 101026.
[http://dx.doi.org/10.1016/j.jwpe.2019.101026]

[42] Thilakan, D.; Patankar, J.; Khadtare, S. Plant-derived iron nanoparticles for removal of heavy metals. *Inter. J. Chem. Eng.,* **2022**.

[43] Annadhasan, M.; Rajendiran, N. Highly selective and sensitive colorimetric detection of Hg(ii) ions using green synthesized silver nanoparticles. *RSC Advances,* **2015**, *5*(115), 94513-94518.
[http://dx.doi.org/10.1039/C5RA18106B]

[44] Annadhasan, M.; Muthukumarasamyvel, T.; Sankar Babu, V.R.; Rajendiran, N. Green synthesized silver and gold nanoparticles for colorimetric detection of Hg^{2+}, Pb^{2+}, and Mn^{2-} in aqueous medium. *ACS Sustain. Chem.& Eng.,* **2014**, *2*(4), 887-896.

[http://dx.doi.org/10.1021/sc400500z]

[45] Bangham, A.D.; Standish, M.M.; Watkins, J.C. Diffusion of univalent ions across the lamellae of swollen phospholipids. *J. Mol. Biol.,* **1965**, *13*(1), 238-IN27.
[http://dx.doi.org/10.1016/S0022-2836(65)80093-6] [PMID: 5859039]

[46] Mozafari, M.R.; Khosravi-Darani, K.; Borazan, G.G.; Cui, J.; Pardakhty, A.; Yurdugul, S. Encapsulation of food ingredients using nanoliposome technology. *Int. J. Food Prop.,* **2008**, *11*(4), 833-844.
[http://dx.doi.org/10.1080/10942910701648115]

[47] Çağdaş, M.; Sezer, A.D.; Bucak, S. Liposomes as potential drug carrier systems for drug delivery. *Applic. Nanotechnol. Drug Delivery,* **2017**, *1*, 1-50.

[48] Demetzos, C. Differential Scanning Calorimetry (DSC): a tool to study the thermal behavior of lipid bilayers and liposomal stability. *J. Liposome Res.,* **2008**, *18*(3), 159-173.
[http://dx.doi.org/10.1080/08982100802310261] [PMID: 18770070]

[49] Singh, J.; Garg, T.; Rath, G.; Goyal, A.K. Advances in nanotechnology-based carrier systems for targeted delivery of bioactive drug molecules with special emphasis on immunotherapy in drug resistant tuberculosis – a critical review. *Drug Deliv.,* **2016**, *23*(5), 1676-1698.
[http://dx.doi.org/10.3109/10717544.2015.1074765] [PMID: 26289212]

[50] Bangham, A.D. Liposomes: The Babraham connection. *Chem. Phys. Lipids,* **1993**, *64*(1-3), 275-285.
[http://dx.doi.org/10.1016/0009-3084(93)90071-A] [PMID: 8242839]

[51] Taléns-Visconti, R.; Díez-Sales, O.; de Julián-Ortiz, J.V.; Nácher, A. Nanoliposomes in Cancer Therapy: Marketed products and current clinical trials. *Int. J. Mol. Sci.,* **2022**, *23*(8), 4249.
[http://dx.doi.org/10.3390/ijms23084249] [PMID: 35457065]

[52] Khorasani, S.; Danaei, M.; Mozafari, M.R. Nanoliposome technology for the food and nutraceutical industries. *Trends Food Sci. Technol.,* **2018**, *79*, 106-115.
[http://dx.doi.org/10.1016/j.tifs.2018.07.009]

[53] Pathak, Y.V.; Hanini, A.; Shah, A. *Surface modification of nanoparticles for targeted drug delivery, Surface Modification nanoparticles for targeted drug delivery*; Springer, **2019**.
[http://dx.doi.org/10.1007/978-3-030-06115-9]

[54] Khosravi-Darani, K. *Inter. J. Nanosci. Nanotechnol.,* **2010**, *6*(1), 3-13.

[55] Allen, T.M.; Cullis, P.R. Liposomal drug delivery systems: From concept to clinical applications. *Adv. Drug Deliv. Rev.,* **2013**, *65*(1), 36-48.
[http://dx.doi.org/10.1016/j.addr.2012.09.037] [PMID: 23036225]

[56] Aguilar-Pérez, K.M.; Avilés-Castrillo, J.I.; Medina, D.I.; Parra-Saldivar, R.; Iqbal, H.M.N. Insight into nanoliposomes as smart nanocarriers for greening the twenty-first century biomedical settings. *Front. Bioeng. Biotechnol.,* **2020**, *8*, 579536.
[http://dx.doi.org/10.3389/fbioe.2020.579536] [PMID: 33384988]

[57] Gelfuso, G.M.; Ferreira-Nunes, R.; Dalmolin, L.F.; Dos S Ré, A.C.; Dos Santos, G.A.; de Sá, F.A.P.; Cunha-Filho, M.; Alonso, A.; Neto, S.A.M.; Anjos, J.L.V.; Aires, C.P.; Lopez, R.F.V.; Gratieri, T. Iontophoresis enhances voriconazole antifungal potency and corneal penetration. *Int. J. Pharm.,* **2020**, *576*, 118991.
[http://dx.doi.org/10.1016/j.ijpharm.2019.118991] [PMID: 31884059]

[58] Subramani, T.; Ganapathyswamy, H. An overview of liposomal nano-encapsulation techniques and its applications in food and nutraceutical. *J. Food Sci. Technol.,* **2020**, *57*(10), 3545-3555.
[http://dx.doi.org/10.1007/s13197-020-04360-2] [PMID: 32903987]

[59] Priyadarshini, E.; Pradhan, N.; Pradhan, A.K.; Pradhan, P. Label free and high specific detection of mercury ions based on silver nano-liposome. *Spectrochim. Acta A Mol. Biomol. Spectrosc.,* **2016**, *163*, 127-133.
[http://dx.doi.org/10.1016/j.saa.2016.03.040] [PMID: 27045785]

[60]   Yingyuad, P.; Sinthuvanich, C.; Leepasert, T.; Thongyoo, P.; Boonrungsiman, S. Preparation, characterization and *in vitro* evaluation of calothrixin B liposomes. *J. Drug Deliv. Sci. Technol.,* **2018**, *44*, 491-497.
[http://dx.doi.org/10.1016/j.jddst.2018.02.010]

[61]   Chen, Y.; Xia, G.; Zhao, Z.; Xue, F.; Gu, Y.; Chen, C.; Zhang, Y. 7,8-Dihydroxyflavone nano-liposomes decorated by crosslinked and glycosylated lactoferrin: storage stability, antioxidant activity, *in vitro* release, gastrointestinal digestion and transport in Caco-2 cell monolayers. *J. Funct. Foods,* **2020**, *65*, 103742.
[http://dx.doi.org/10.1016/j.jff.2019.103742]

[62]   Yi, J.; Lam, T.I.; Yokoyama, W.; Cheng, L.W.; Zhong, F. Controlled release of β-carotene in β-lactoglobulin-dextran-conjugated nanoparticles' *in vitro* digestion and transport with Caco-2 monolayers. *J. Agric. Food Chem.,* **2014**, *62*(35), 8900-8907.
[http://dx.doi.org/10.1021/jf502639k] [PMID: 25131216]

[63]   Papachristodoulou, A.; Signorell, R.D.; Werner, B.; Brambilla, D.; Luciani, P.; Cavusoglu, M.; Grandjean, J.; Silginer, M.; Rudin, M.; Martin, E.; Weller, M.; Roth, P.; Leroux, J.C. Chemotherapy sensitization of glioblastoma by focused ultrasound-mediated delivery of therapeutic liposomes. *J. Control. Release,* **2019**, *295*, 130-139.
[http://dx.doi.org/10.1016/j.jconrel.2018.12.009] [PMID: 30537486]

[64]   Bhagat, S.; Parikh, Y.; Singh, S.; Sengupta, S. A novel nanoliposomal formulation of the FDA approved drug Halofantrine causes cell death of Leishmania donovani promastigotes *in vitro*. *Colloids Surf. A Physicochem. Eng. Asp.,* **2019**, *582*, 123852.
[http://dx.doi.org/10.1016/j.colsurfa.2019.123852]

[65]   Díaz, M.; Vivas-Mejia, P. Nanoparticles as drug delivery systems in cancer medicine: emphasis on RNAi-containing nanoliposomes. *Pharmaceuticals,* **2013**, *6*(11), 1361-1380.
[http://dx.doi.org/10.3390/ph6111361] [PMID: 24287462]

[66]   Doll, T.A.P.F.; Raman, S.; Dey, R.; Burkhard, P. Nanoscale assemblies and their biomedical applications. *J. R. Soc. Interface,* **2013**, *10*(80), 20120740.
[http://dx.doi.org/10.1098/rsif.2012.0740] [PMID: 23303217]

[67]   Torchilin, V.P. Recent advances with liposomes as pharmaceutical carriers. *Nat. Rev. Drug Discov.,* **2005**, *4*(2), 145-160.
[http://dx.doi.org/10.1038/nrd1632] [PMID: 15688077]

[68]   Mirza, Z.; Karim, S. Nanoparticles-based drug delivery and gene therapy for breast cancer: Recent advancements and future challenges. In: *Seminars in cancer biology, Academic Press*; Elsevier, **2021**.
[http://dx.doi.org/10.1016/j.semcancer.2019.10.020]

[69]   Wang, Y.; Kohane, D.S. External triggering and triggered targeting strategies for drug delivery. *Nat. Rev. Mater.,* **2017**, *2*(6), 17020.
[http://dx.doi.org/10.1038/natrevmats.2017.20]

[70]   Wang, W.; Lu, K.; Yu, C.; Huang, Q.; Du, Y.Z. Nano-drug delivery systems in wound treatment and skin regeneration. *J. Nanobiotechnology,* **2019**, *17*(1), 82.
[http://dx.doi.org/10.1186/s12951-019-0514-y] [PMID: 31291960]

[71]   González-Rodríguez, M.L.; Rabasco, A.M. Charged liposomes as carriers to enhance the permeation through the skin. *Expert Opin. Drug Deliv.,* **2011**, *8*(7), 857-871.
[http://dx.doi.org/10.1517/17425247.2011.574610] [PMID: 21557706]

[72]   Hamishehkar, H.; Hasanpouri, A.; Lotfipour, F.; Ghanbarzadeh, S. Improvement of dermal delivery of tetracycline using vesicular nanostructures. *Res. Pharm. Sci.,* **2018**, *13*(5), 385-393.
[http://dx.doi.org/10.4103/1735-5362.236831] [PMID: 30271440]

[73]   Reza Mozafari, M.; Johnson, C.; Hatziantoniou, S.; Demetzos, C. Nanoliposomes and their applications in food nanotechnology. *J. Liposome Res.,* **2008**, *18*(4), 309-327.

[http://dx.doi.org/10.1080/08982100802465941] [PMID: 18951288]

[74]   Taylor, T.M.; Weiss, J.; Davidson, P.M.; Bruce, B.D. Liposomal nanocapsules in food science and agriculture. *Crit. Rev. Food Sci. Nutr.,* **2005**, *45*(7-8), 587-605.
[http://dx.doi.org/10.1080/10408390591001135] [PMID: 16371329]

[75]   Kirby, C.J.; Whittle, C.J.; Rigby, N.; Coxon, D.T.; Law, B.A. Stabilization of ascorbic acid by microencapsulation in liposomes. *Int. J. Food Sci. Technol.,* **1991**, *26*(5), 437-449.
[http://dx.doi.org/10.1111/j.1365-2621.1991.tb01988.x]

[76]   Yang, Y.J.; Liu, X.W.; Kong, X.J.; Qin, Z.; Li, S.H.; Jiao, Z.H.; Li, J.Y. An LC–MS/MS method for the quantification of diclofenac sodium in dairy cow plasma and its application in pharmacokinetics studies. *Biomed. Chromatogr.,* **2019**, *33*(7), e4520.
[http://dx.doi.org/10.1002/bmc.4520] [PMID: 30817844]

[77]   Koziolek, M.; Alcaro, S.; Augustijns, P.; Basit, A.W.; Grimm, M.; Hens, B.; Hoad, C.L.; Jedamzik, P.; Madla, C.M.; Maliepaard, M.; Marciani, L.; Maruca, A.; Parrott, N.; Pávek, P.; Porter, C.J.H.; Reppas, C.; van Riet-Nales, D.; Rubbens, J.; Statelova, M.; Trevaskis, N.L.; Valentová, K.; Vertzoni, M.; Čepo, D.V.; Corsetti, M. The mechanisms of pharmacokinetic food-drug interactions : A perspective from the UNGAP group. *Eur. J. Pharm. Sci.,* **2019**, *134*, 31-59.
[http://dx.doi.org/10.1016/j.ejps.2019.04.003] [PMID: 30974173]

[78]   Srinivasan, V.; Chavan, S.; Jain, U. Liposomes for nanodelivery systems in food products. In: *Nanoscience for Sustainable Agriculture, Nanosci. Sustain. Agricul*; Springer, **2019**; pp. 627-638.
[http://dx.doi.org/10.1007/978-3-319-97852-9_24]

[79]   Law, B.A.; King, J.S. Use of liposomes for proteinase addition to Cheddar cheese. *J. Dairy Res.,* **1985**, *52*(1), 183-188.
[http://dx.doi.org/10.1017/S0022029900024006]

[80]   Zarrabi, A.; Alipoor Amro Abadi, M.; Khorasani, S.; Mohammadabadi, M.R.; Jamshidi, A.; Torkaman, S.; Taghavi, E.; Mozafari, M.R.; Rasti, B. Nanoliposomes and tocosomes as multifunctional nanocarriers for the encapsulation of nutraceutical and dietary molecules. *Molecules,* **2020**, *25*(3), 638.
[http://dx.doi.org/10.3390/molecules25030638] [PMID: 32024189]

[81]   Khanniri, E.; Bagheripoor-Fallah, N.; Sohrabvandi, S.; Mortazavian, A.M.; Khosravi-Darani, K.; Mohammad, R. Application of liposomes in some dairy products. *Crit. Rev. Food Sci. Nutr.,* **2016**, *56*(3), 484-493.
[http://dx.doi.org/10.1080/10408398.2013.779571] [PMID: 25574577]

[82]   Kheadr, E.E.; Vuillemard, J.C.; El Deeb, S.A. Accelerated Cheddar cheese ripening with encapsulated proteinases. *Int. J. Food Sci. Technol.,* **2000**, *35*(5), 483-495.
[http://dx.doi.org/10.1046/j.1365-2621.2000.00398.x]

[83]   Sobel, R.; Versic, R.; Gaonkar, A.G. Introduction to microencapsulation and controlled delivery in foods. In: *Microencapsulation in the food industry, InMicroencapsulation in the food industry*; Elsevier, **2014**; pp. 3-12.
[http://dx.doi.org/10.1016/B978-0-12-404568-2.00001-7]

[84]   Merz, M.; Eisele, T.; Berends, P.; Appel, D.; Rabe, S.; Blank, I.; Stressler, T.; Fischer, L. Flavourzyme, an enzyme preparation with industrial relevance: automated nine-step purification and partial characterization of eight enzymes. *J. Agric. Food Chem.,* **2015**, *63*(23), 5682-5693.
[http://dx.doi.org/10.1021/acs.jafc.5b01665] [PMID: 25996918]

[85]   Jahadi, M.; Khosravi-Darani, K.; Ehsani, M.R.; Mozafari, M.R.; Saboury, A.A.; Pourhosseini, P.S. The encapsulation of flavourzyme in nanoliposome by heating method. *J. Food Sci. Technol.,* **2015**, *52*(4), 2063-2072.
[http://dx.doi.org/10.1007/s13197-013-1243-0] [PMID: 25829586]

[86]   Maurya, V.K.; Bashir, K.; Aggarwal, M. Vitamin D microencapsulation and fortification: Trends and technologies. *J. Steroid Biochem. Mol. Biol.,* **2020**, *196*, 105489.

[http://dx.doi.org/10.1016/j.jsbmb.2019.105489] [PMID: 31586474]

[87]   Rovoli, M.; Pappas, I.; Lalas, S.; Gortzi, O.; Kontopidis, G. *In vitro* and *in vivo* assessment of vitamin A encapsulation in a liposome–protein delivery system. *J. Liposome Res.,* **2019**, *29*(2), 142-152.
[http://dx.doi.org/10.1080/08982104.2018.1502314] [PMID: 30187807]

[88]   Mohammadi, R.; Mahmoudzadeh, M.; Atefi, M.; Khosravi-Darani, K.; Mozafari, M.R. Applications of nanoliposomes in cheese technology. *Int. J. Dairy Technol.,* **2015**, *68*(1), 11-23.
[http://dx.doi.org/10.1111/1471-0307.12174]

[89]   Banville, C.; Vuillemard, J.C.; Lacroix, C. Comparison of different methods for fortifying Cheddar cheese with vitamin D. *Int. Dairy J.,* **2000**, *10*(5-6), 375-382.
[http://dx.doi.org/10.1016/S0958-6946(00)00054-6]

[90]   Mozafari, M.R.; Flanagan, J.; Matia-Merino, L.; Awati, A.; Omri, A.; Suntres, Z.E.; Singh, H. Recent trends in the lipid-based nanoencapsulation of antioxidants and their role in foods. *J. Sci. Food Agric.,* **2006**, *86*(13), 2038-2045.
[http://dx.doi.org/10.1002/jsfa.2576]

[91]   Fakhravar, Z.; Ebrahimnejad, P.; Daraee, H.; Akbarzadeh, A. Nanoliposomes: Synthesis methods and applications in cosmetics. *J. Cosmet. Laser Ther.,* **2016**, *18*(3), 174-181.
[http://dx.doi.org/10.3109/14764172.2015.1039040] [PMID: 25968161]

[92]   Golmohammadzadeh, S.; Jaafari, M.R.; Hosseinzadeh, H. Does saffron have antisolar and moisturizing effects? *Iran. J. Pharm. Res.,* **2010**, *9*(2), 133-140.
[PMID: 24363719]

CHAPTER 11

Omics Perspectives Regarding Biosurfactant Biosynthesis and the Suitability of Site Bioremediation and Developments

Arabinda Jena[1,*] and **Sameer Ranjan Sahoo[2]**

[1] *Fisheries, UNDP (Collaboration with Directorate Fisheries), Cuttack, Odisha-753001, India*

[2] *Centre for Biotechnology, Siksha 'O' Anusandhan (Deemed to be University), Bhubaneswar, Odisha-751003, India*

Abstract: Modern compounds are called biosurfactants. Their application(s) interfere with day-to-day activities of human beings. The economics of production place a significant restriction on the broad applicability of biosurfactant(s). There can be many ways to overcome this. This study primarily focuses on current technical advancements in biosurfactant research. One of the innovations is the application of metabolomic and sequence-based omics approaches, which have evolved into a high-throughput essential tool for the detection of biosurfactant producers. Many bacteria produce ethanol, microbial lipids, polyhydroxyalkanoates, and other value-added compounds in addition to biosurfactants. The use of metabolic engineering techniques can further address restrictions while also improving the overall process's economics. The yield of biosurfactants is increased by the tailoring technique, which enables variation in the composition of the congeners produced. By enhancing their bioavailability or water solubility, bio-based surfactants have demonstrated promising effects in reducing pollution in terrestrial and aquatic habitats. Due to the expanding market for biosurfactants, this investigation identified technologically feasible developments in biosurfactant research that might help researchers create more trustworthy and secure technologies.

Keywords: Biosurfactants, Bioremediation, Metabolomics, Metagenomics, Metabolic engineering.

INTRODUCTION

Nearly all everyday chores include surfactants, which are chemicals or petroleum-based substances having tension-active characteristics [1, 2]. Microbial

---

[*] **Corresponding author Arabinda Jena:** Fisheries, UNDP (Collaboration with Directorate Fisheries), Cuttack, Odisha-753001, India; E-mail: arabindajena1994@gmail.com

Arun Kumar Pradhan and Manoranjan Arakha (Eds.)

surfactants have drawn the attention of researchers due to environmental concerns and the advantages microbial surfactants offer over synthetic ones, such as their biodegradability, low toxicity, stability in a variety of temperatures and pH levels, and salt stability [3 - 5]. As a result, biosurfactants are now widely used in a range of disciplines, including the environment, food, and biomedicine [6, 7].

The global market for commercial biosurfactants is expected to develop at a 5.6% CAGR from 2017 to 2022, reaching $5.52 billion [8]. In 2016, the market for biosurfactants was valued at $30.64 billion, and by 2021, it is anticipated to reach $39.86 billion. In 2023, the biosurfactant market earned US$ 1.8 billion in sales, with an 8% growth to US$ 2.6 billion projected in 2023. This yielded 540 kilotonnes of biosurfactants [9].

The market for these compounds is still undeveloped despite growing interest in biosurfactants due to the economics of manufacture and the availability of less-priced raw ingredients [10]. Other established methods for identifying effective biosurfactant makers from certain environmental niches require work and time, including the blue agar plate test, drop collapse assay, hemolytic assay, oil spreading assay, surface tension measurement, and emulsification assay [11 - 13]. Therefore, it is crucial to employ high throughput omics techniques in the study of biosurfactants. Depending on the sample under examination, the omics approach might be classed as metagenomics, metatranscriptomics, metaproteomics, or metabolomics. Total macro- and micro-molecules, as well as DNA, RNA, and proteins, are all present (Fig. **1**). These methods yield astonishing amounts of information on an organism's genetic makeup, metabolic profile, and functionally active portion [14, 15].

Metabolomic studies have been carried out to examine the effects of biosurfactants on the bioremediation of petroleum-contaminated wastelands [16]. This demonstrated the design strategy's effectiveness in reducing the abiotic stress on the plants. The *Rhodococcus* sp. I2R strain was grown in 22 different settings, and the metabolomic approach discovered more than 30 different functional groups, including glycolipids. Herpes simplex virus and human coronavirus were both resistant to the antiviral and anticancer effects of the active component [17]. Moreover, new biosurfactants have been discovered using metagenomic methods [18]. With advanced genomic analysis it became economical along with improvements in next-generation [19]. This approach was used to identify the relevant biomolecules by comparing the protein-coding sequences' sequence homology to the reference database. A variety of technologies have been developed to aid in the identification of secondary metabolites of interest. Anti-SMASH (the antibiotic and secondary metabolite analysis shell) is one such tool that uses sequencing data to annotate and detect gene clusters. The *Serratia marcescens* Db11 genes associated with biosurfactants have been found using the anti-SMASH tool [20].

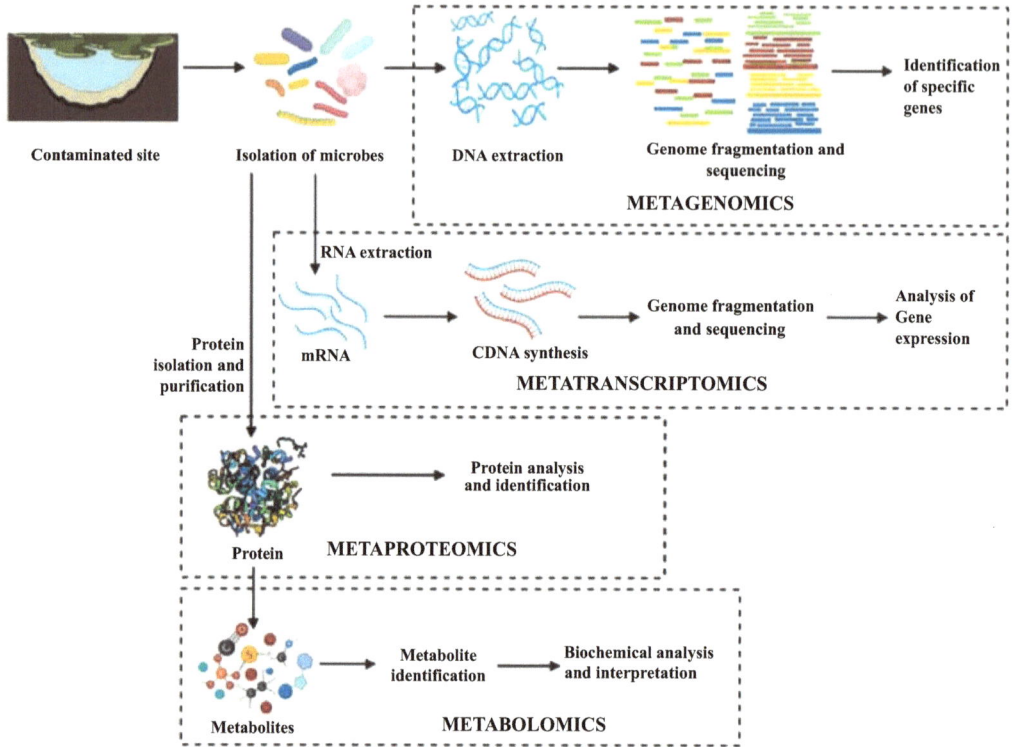

Fig. (1). Biosurfactant study using omics techniques.

Aside from that, there are limitations on the type and the amount of biosurfactants that can be produced for various uses that have been greatly enhanced by tailoring and engineering approaches. In some investigations, biosurfactant producers that weren't local to the area were created by modifying their genomes [21, 22]. Low Rhamnolipid levels were seen in *Pseudomonas aeruginosa* strain ATCC 9027, but with the addition of the rhlAB-R operon-containing plasmid, the Rhamnolipid titer increased to levels comparable to strain *Pseudomonas aeruginosa* PAO1. The ATCC 9027 strain also failed to make di Rhamnolipid because genetic changes were made to produce mono rhamnolipids [21].

In terms of market acceptance and economic viability, the use of these methodologies has profoundly impacted and enhanced biosurfactant research. This article went into detail about how biosurfactants are produced alongside other key industrial goods like bioplastics. It has been discussed how to find the sources of biosurfactants using metabolomics and next-generation sequencing techniques. The plan for boosting biosurfactant production was also covered in length, as was how biosurfactants affect ecosystems on land and at sea. A compilation analysis of the advancements in methodologies for identifying biosurfactant makers has not before been published, to the best of our understanding. We have described the obstacles and their potential solutions.

CO-PRODUCTION OF VALUE-ADDED PRODUCTS AND BIOSURFACTANTS

Biosurfactants, according to experts, are amphiphilic molecules with a wide range of structural variations that have multifunctional qualities in several sectors, including the cosmetic, food processing, pharmaceutical, textile, and paint industries [23, 24]. In addition to their exclusive synthesis by selected microorganisms, biosurfactants are effectively co-produced by a broad range of bacteria such as ethanol, microbial lipids, polyhydroxyalkanoates, and other particular added-value compounds. To turn inexpensive substrates into expensive products, many microorganisms were used as microbial biorefineries [25]. In the same growth conditions as pectinases such as pectin lyase (PNL), pectate lyase (PEL), and polygalacturonase, *Bacillus subtilis* BKDS1 produces biosurfactant (PG) [26].

Bacillus methylotrophic DCS1 produced a unique amylase and lipopeptide biosurfactant during 48 hours of incubation at 25 °C and 150 rpm using glutamic acid (5 g/L) and potato starch (10 g/L) as nitrogen and carbon sources. The amylase enzyme that was produced was naturally alkaline, showed great stability, and was compatible with various solid/liquid surfactants.

The biosurfactant created concurrently also demonstrated outstanding stability over a range of salinity, pH, and temperature conditions [27]. *Aspergillus niger* synthesized a basic enzyme complex using the solid-state fermentation of rice bran and maize cob as substrates.

Additionally, the simultaneous generation of rhamnolipids and ethanol by *Saccharomyces cerevisiae* and *P. aeruginosa* utilizing sugarcane bagasse was observed. After 86 hours of fermentation, 9.1 g/L Rhamnolipids with a surface tension of 35 mN/m, an emulsification index of 84%, and 8.4 g/L ethanol were produced in this fermentation system [28].

A facultative anaerobe is known to simultaneously synthesize polyhydroxyalkanoate (PHA) and Rhamnolipids. During the stationary phase of development, it generated -hydroxyalkanoic acid, a precursor to both Rhamnolipid and PHA that allowed Rhamnolipids and PHA to be produced simultaneously in a single fermentation process. This procedure produced 5.81

g/L of Rhamnolipids and 4.2 g/L or 5% (v/v) of PHA under ideal conditions of carbon: nitrogen ratio (5:1) at 6.5 pH.

A complete dark phase was maintained because PHA and Rhamnolipids are generated by the dark fermentative bacteria *E. aerogenes*. There have also been instances of biogas coexisting with a variety of liquid metabolites during dark fermentation. It was discovered that PHA was released into the broth during fermentation whereas Rhamnolipid was accumulated inside the microbial cell.

Because of this, there were no issues with the coproduction or the following product retrieval [29]. For the simultaneous generation of PHA and rhamnolipids, Thermus thermophilus HB8 employed glucose or sodium gluconate as a carbon source [30].

The microbial cell produces 34.8 weight percent of 3-hydroxyoctanoate, a monomer of PHA co-polymers, while releasing 0.2 g/L of rhamnolipids into the liquid fermentation environment. PHA and rhamnolipid formation are both impacted by the initial phosphate level (as PO3-4) in the medium.

After 72 hours of culture at a starting phosphate (PO3-4) concentration of 25 mM, PHAs and rhamnolipid accumulation of > 300 mg/L and > 200 mg/L, respectively, were attained. A GC analysis showed that the microbial cell had accumulated amounts of the PHA copolymer 3-hydroxy decanoate. Several congeners with varied lipid chain lengths, such as mono-, di-, and di-rhamno-monolipid, as well as particular saturated or unsaturated fatty acids, were investigated in the LCMS data [31, 32].

When grown in a single-stage batch culture with 7 g/L hydrolyzed palm oil (glycerol and fatty acids) at 30 °C and 28 °C, respectively, *P. aeruginosa* IFO3924 simultaneously generated 36% PHA and 0.4 g/L rhamnolipid. Mcl-PHA was created through oxidation and fatty acid *de novo* synthesis. In the proximity of the enzyme HAA synthetase (rhlA), 3-(3- hydroxyalkanoyloxy) alkanoic acids (HAAs), a precursor of rhamnolipid synthesis, are formed from (R) 3-hydroxyacyl-ACP, an intermediate of fatty acid *de novo* biosynthesis. Rhamnosyl transferase I (rhlB) and rhamnosyl transferase II (rhlC) activities transformed HAAs further into mono- and di-rhamnolipids, resulting in the total synthesis of rhamnolipids [33].

In addition to accumulating intercellular biopolymers including polyhydroxyalkanoates (PHA) and polyhydroxybutyrate (PHB), *Burkholderia thailandensis* pathogenic and nonpathogenic strains utilized cooking oil as a carbon source to create Rhamnolipids. According to studies, to lower surface tension to 37.7 mN/m and interfacial tension to 4.2 and 1.5 mN/m, respectively, against benzene and oleic acid, *B. thailandensis* may produce up to 2.2 g/L of di-rhamnolipid Rha-Rha-C14-C14 [34]. *B. thailandensis* has a second, as-yet-unidentified PHA production route. Recombinant *B. thailandensis* with altered polyhydroxyalkanoate (PHA) synthesis genes exhibit higher Rhamnolipid production. To increase the production of Rhamnolipids, three knockout strains (ΔphbA1, ΔphbB1, and ΔphbC1) were developed. The wild-type strain produced the least amount of purified Rhamnolipids (1.28 g/L), while phbB1 produced the most (3.78 g/L), with a higher concentration of mono-Rhamnolipids [35].

USING OMICS TO IDENTIFY BIOSURFACTANT PRODUCERS AS AN ADVANCED TECHNIQUE METHOD

Metabolomics

The study of an organism's entire metabolic makeup is called metabolomics. The cell's molecules and metabolites are photographed in their entirety representing the physiological state and metabolic activity, and it delivers astonishingly important information [36]. In a study project, Floros and colleagues used an untargeted metabolomics technique based on LC-MS/MS to examine the chemistries of different marine organisms [37]. Valinomycin, desferrioxamine E, and actinomycin D, three chemicals that are crucial to industry, have been discovered as a result of this. Interestingly, along with these metabolites, this investigation also detected surfactin biosurfactants [37]. It was therefore hypothesized that untargeted metabolic profiling is a useful method for locating specialist chemistries, especially for identifying biosurfactant producers amid a large range of other non-producers [38, 39].

Using metabolomics, it was determined if *B. velezensis* FZB42 was able to kill the phytopathogen Xanthomonas campestris. It was discovered that the killing mechanism involved bacillibactin, siderophore, and lipopeptides. The lipopeptides bacillomycin, surfactin, and fengycin, as well as bacillibactin, were contained in *B. velezensis* mono- and co-cultures. The 24 distinct forms found in the co-culture supernatant stood out among the four chemicals according to mass spectrophotometric analysis. Moreover, it was discovered that the side chains of fengycin and surfactin differed [40, 41]. A metabolomics analysis was carried out of more than 250 *Pseudomonas* strains. The strains were environmentally varied, and a mass spectrometric analysis of the molecular networking revealed structural linkages. There are peptide families known as massetolides, putisolvins, tolaasins, orfamides, and cyclic lipopeptides that have been found to have biosurfactant characteristics [41]. Furthermore, lipopeptides with antibacterial properties have been discovered as a result of *S. marcescens'* molecular networking.

A fatty acid chain, butyric acid, valine, glucose, and serratamolide—a lipopeptide with two L-serine residues linked to two fatty acid chains—were all present in these lipopeptides (with C10 to C12 chain length) [42]. During a media-dependent metabolomic analysis, it was shown that different solvent extracts of the yeast *Rhodotorula mucilaginosa* 50-3-19/20B contained anticancer and antibiotic substances. According to the manual de-replication against SciFinder and the Dictionary of Natural Products, glycolipids were the biggest molecular family in the molecular network. These glycolipids, which were hypothesized to be biosurfactants, belonged to a family of polyol esters of fatty acids and had various degrees of acetylation [43]. These research works showed that metabolomics methods may be used to find microbial strains that make biosurfactants or manufacture biosurfactants on their own.

Sequencing Based

Microorganisms, which are the most numerous, varied, adaptable, and evolutionary forms of life, have thrived in the environment of the world. The use of these bacteria has evolved into a crucial instrument for the well-being of society because they are crucial for maintaining our ecology [44]. For the formation and research of microbial communities, well-established and well-known culture and identification procedures were traditionally used, leaving a significant portion of the unknown and unidentified [45]. The ability to identify and fully comprehend a microbial strain's propensity to make biosurfactants led to the development of omics technology, which consists of metagenomics (whole DNA), metaproteomics (whole protein), and metatranscriptomics (whole RNA) [46].

Metagenomics

The term "metagenomics" refers to the omics technique utilized to find unidentified microorganisms. Without using any enrichment or culture techniques beforehand, the genetic material for this investigation was directly extracted from the ambient samples. DNA extraction, metagenome library construction, 16S rRNA sequencing, and data processing are all parts of the metagenomics technique [45]. The two screening techniques used in metagenomics are function-based screening, which does not require an existing metagenomics library, and sequence-based screening, which uses primers based on known coding genes. This method relied on cloning DNA to obtain the needed function sequences, which could have led to the discovery of novel sequences [47]. It simplifies the process of analyzing and comprehending the genetic makeup of untamed bacteria. the discovery of novel microbes and their byproducts, as well as their biochemical connections, processes, and roles [46]. One benefit of the metagenomics technique is the identification of novel biosurfactant-producing bacteria.

The use of the metagenomics technique has allowed scientists to investigate and analyse microorganisms with a deeper understanding of their genetic makeup. Researchers can learn more about the discovered microorganisms, their roles, and connections between various microbial communities thanks to the information gathered in metagenomic libraries [45]. Nevertheless, this technique has certain limitations, such as the difficulty of sequence-based metagenomics to locate sparsely populated bacteria and the incapacity of metagenomics to discriminate between expressed and undetected genes [48]. To determine gene expression, metagenomics and metatranscriptomics are typically combined.

Metatranscriptomics

Studying mRNA transcripts of the microbiome is known as metatranscriptomics. This method explains the genes that are overexpressed and under expressed as a result of environmental changes [46]. Direct RNA isolation from ambient materials, mRNA enrichment, mRNA to cDNA conversion, and next-generation sequencing are the three main stages of metatranscriptomics [49]. Brazil's Jundiai River soil samples were used to extract DNA by da Silva Araujo and his team [50]. The creation of a metagenome library after the functional screening, a clone with the name 3C6 which included the novel protein MBSP1 with biosurfactant activity, was found. Haloferax lucentense, a hydrocarbon degrader, was most similar to the sequence of MBSP1 [50]. In South Africa's Bufflespruit Lake, Williams *et al*. (2019) collected soil samples for a study. *Escherichia coli*, *Pseudomonas putida*, and *Streptomyces lividans* fosmids were utilized to create a metagenomic library. Using the paraffin spray test, the *P. putida* clone was discovered to exhibit biosurfactant activity, which was attributed to ornithine acyl-ACP N-acyltransferase (olsB). Although the fosmid proved inefficient in *E. coli*, the T7 promoter allowed the olsB gene to be overexpressed, resulting in the creation of a biosurfactant-active lyso-ornithine lipid [51].

The impact of L and D-Leucine on *B. velezensis* BS-37, which produced 1000 mg/L of surfactin biosurfactant in the presence of glycerol, was investigated by Zhou and team [52] using a transcriptome technique. Production increased to 2000 mg/L with the addition of 10 mM L-Leu; however, when 10 mM D-Leu was added, the output fell to 250 mg/L. Genes for surfactin synthase, the branched-chain amino acid synthesis pathway, and the glycerol utilization pathway had higher levels of expression in the strain of BS-37 transcriptome, and L-Leu was used as a precursor while D-Leu served as a competitive inhibitor. Metatranscriptomics has proven incredibly useful for studying and creating high surfactin-producing strains [52]. According to the research, strain MT45 had improved carbon metabolism, fatty acid biosynthesis, and enrichment of the surfactin production pathway. Also, it was discovered that MT45 strain's surfactin synthase was upregulated from 9 to 49 times more than DSM7T, leading to the high surfactin synthesis of MT45 strain [53]. Although the analysis of gene activities and gene regulators in the microbial community is aided by metatranscriptomics, the usefulness of the technique is constrained by mRNA's short half-life, instability, and susceptibility to degradation [49, 54]. Hence, the alternative omics strategy known as metaproteomics is used to get around these drawbacks [46].

Metaproteomics

Metaproteomics is the study of proteins produced by bacteria in the environment. It looks into the microbial population's functional genes and offers details about them [46]. Metaproteomics links the genetic and functional data of microbial communities through the analysis of data [55].

Pitocchi and team identified two fungal strains from an oil-contaminated maritime location in order to research new biosurfactant proteins: *Aspergillus terreus* MUT271 and *Trichoderma harzianum* MUT290 [56]. A common protein that belongs to the cerato-platanins (CP) family has been discovered thanks to proteomics. They found that both strains' CPs functioned as a biosurfactant and a bioemulsifier. On the other hand, *T. Harzianum* CP showed improved surfactant activity and can be used in industrial applications [56]. Adding glycerol to *Burkholderia* sp. C3 enhanced DBT breakdown and increased rhamnolipid (RL) synthesis, according to a study. Degradation increased by 25%-30% in tandem with an increase in RL biosynthesis. Proteomics was utilized to analyse the rise of enzymes involved in RL biosynthesis and DBT breakdown [57].

Researchers can discover novel functional genes and metabolic pathways using the metaproteomics approach, as well as proteins that are expressed in response to changes in substrate or environmental conditions [15]. Despite these restrictions, metaproteomics outperforms metagenomics and metatranscriptomics in terms of information output [46]. Although the omics technique is helpful, there are still many issues to be resolved. To obtain improved data and in-depth insights into genes, however, the multi-metaomics technique, which integrates two or more approaches, must be taken into account [46].

IMPROVEMENTS TO THE PRODUCTION PLAN

Metabolic Engineering and Tailoring Strategies

The commercial production of surface active compounds using chemical or enzymatic catalysis, as well as microbial fermentation (biosurfactants), necessitates further investigation [58]. There is evidence that biosurfactants can withstand difficult environmental conditions. In-depth gene-level research is required to improve the generation of microbial surfactants, which can replace synthetic surfactants and do so at a substantially lower cost. A large variety of different strains might be transformed by using metabolic engineering to develop high-yielding variants [59]. According to study hotspots (Fig. **2**), tailoring the genome and changing metabolic pathways by metabolic engineering can greatly increase the predicted yield of biosurfactants (0.47g biosurfactants/substrates). According to studies, the alternative surfactin route can increase biosurfactants' output by minimising the imbalance between cofactors and ATP [58, 60]. Genome reduction has emerged as a critical way to mimic a compelling functional framework in microbial cell factories. Genome-reduced *B. amyloliquefaciens* strain LL3 was created by slicing/tailoring 4.18% barren genomic sections of strain LL3 in order to boost growth rate, the ability for heterologous protein production, transformation efficiency, intracellular reduction power, and surfactin synthesis. Strain GR167ID was created through additional metabolic engineering, specifically by chopping the iturin and fengycin biosynthetic gene clusters from strain GR167's genome. After introducing the PRsuc and PRtpxi promoters from strain LL3 into strain GR167ID, respectively, strains GR167IDS and GR167IDT were created. The best mutant was identified to be strain GR167IDS, which demonstrated a 10.4-fold increase in surfactant production (11.35 mg/L) as compared to strain GR167. GR167ID and a 678-fold improvement at the transcriptional level in the srfA operon [60].

In order to construct modified Mannosylerythritol Lipids (MELs), Becker and the team combined the genes Mac1 and Mac2, which code for acyl-transferases from several fungal species. This resulted in fungal variations with specific physical and antibacterial properties. MELs are glycolipids having sugar moiety side chains that have acetyl- or acyl-group attachments that vary. Ustilago maydis can take part in the production of altered MEL variants since it contains a Mac2 paralog with variable substrate specificity [61]. Maydis demonstrates how the gene repository for Mac enzymes may be tailored to produce MELs that are specifically designed to meet the needs of the biotechnological or pharmaceutical industries. Outperforming plasmid-based expression systems' genomes were merged with Rhamnolipid synthesis genes to create an expression cassette that improved the strain's genetic stability by adding an inducible promoter [62].

Further elimination of carbon/energy-consuming characteristics boosted the synthesis of rhamnolipids. By recycling the foam in a 1 L bioreactor without applying any antifoaming agents or mechanical foam destructors, a maximum of 1.5 g/L of Rhamnolipid titer was attained.

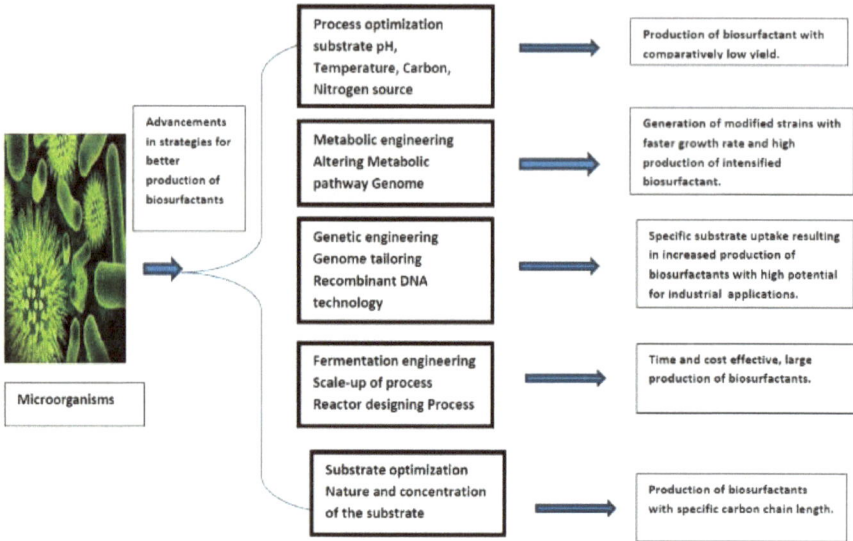

Fig. (2). Methods used to produce biosurfactants.

Surfactin biosynthetic activity was restored and a surfactin titer of 0.4 g/L was obtained by introducing a full sfp gene into the genome of a *B. subtilis* strain 168 that was not producing surfactants. The surfactin titer increased by 3.3-fold when 3.8% of the genome was removed, which includes genes for nonribosomal peptide synthetase, polyketide synthase pathways, and biofilm formation [63]. By overexpressing self-resistance-associated proteins, the surfactin titer may be boosted by 8.5-fold, enhancing cellular tolerance to surfactins. The surfactin titer has increased to 8.5 g/L as a result of modifying the branched-chain fatty acid biosynthesis procedure and increasing the precursor supply of branched-chain fatty acids (20.3-fold increase) [63].

The transposon-mediated chromosome integration of the rhlAB operon in *P. aeruginosa* PAO1, an engineered strain, produced 1.819 g/L of rhamnolipids at a constant production rate of 18.95 mg/L/h. When glucose was the only carbon source, the highest yield of 0.784 g/L was obtained, and the productivity rate was 8.19 mg/L/h. The modified strain *P. aeruginosa* PG201 generated 2.2 g/L of rhamnolipids using glycerol (2 g/L) as a substrate at a productivity rate of 13.1 mg/L/h [59].

Developments in the methods used to produce biosurfactants are shown in Fig. (**2**).

CASE STUDIES

Terrestrial

Several chemical compounds, petroleum hydrocarbons, and poisonous heavy metals are collected in the soil as a result of natural disasters and human activity, increasing soil pollution and affecting both human and environmental health [64]. Several techniques, such as physio-chemical and thermal processes, were created to clean up these polluted areas, but they were expensive and inefficient [65]. Due to their low toxicity, ease of biodegradation, cheap cost of manufacture, and environmentally friendly composition, the use of microbial biosurfactants for the restoration of a polluted site is a successful technique in this situation [66]. This has drawn researchers' interest throughout the world, and various studies with a particular focus on adding biosurfactants to contaminated locations have been published. The process of the addition of the biosurfactant to the soil is known as surfactant-assisted or surfactant-enhanced bioremediation, and it decreases surface and interfacial tension by facilitating the mass transfer of pollutants and permitting the admission of hydrocarbons to the microorganisms. It was suggested that biosurfactants be used carefully chosen in accordance with the qualities and characteristics of the pollutants for effective cleanup [67].

Two distinct procedures for removing petroleum hydrocarbons from polluted environments were introduced by Ambust and team [68]. The first is an increase in the substrate's bioavailability, and the second is an interaction between the bacterial cell surface and hydrophobic substrates by enhancing hydrophobicity and reducing the cell wall's lipopolysaccharides index without causing damage to the cell membrane, consequently, preventing the formation of hydrogen bonds, increasing hydrophobic-hydrophilic interaction and decreasing surface tension of water. Biosurfactants have a significant effect on improving bioavailability and biodegradability [68 - 70]. A test was conducted to investigate the impact of a *P. aeruginosa* derived rhamnolipid biosurfactant on the total petroleum hydrocarbon component of an artificially oil-contaminated soil. In two concentrations, 4 g and 8 g per kg of soil, rhamnolipid were sprayed. At 4 grams per kilogram, 8 grams per kilogram, and untreated dirt, respectively, the soil's total petroleum hydrocarbon percentage decreased from 6% to 1.3%, 2%, and 2.75%. This suggested that the elimination of petroleum components is aided by the rhamnolipid biosurfactant [70]. A biosurfactant that resembles surfactin was discovered in a *Bacillus nealsonii* S2MT environmental isolate that was tested for its capacity to lessen pollution in soil contaminated with 10% heavy motor oil.

According to calculations, salt and biosurfactant remediate the same quantity of oil at 10 and 40 mg/L, whereas biosurfactant remediates 43.6 and 46.7% of oil, respectively.

Moreover, the sophorolipid from *C. bombicola* ATCC 22214 accelerated the biodegradation of crude oil and synthetic hydrocarbons in the soil. On 6, 6, and 2 days, the biodegradation of virgin, hexadecane, and 2-methylnaphthalene was enhanced to 85, 97, and 95% by the addition of sophorolipid at a concentration of 10 g/L, respectively. Moreover, 56 d after sophorolipid addition, aromatics and saturates degraded by 72% and 80%, respectively [71]. Regar and his team found that rhamnolipid biosurfactant was helpful in accelerating the breakdown of chlorinated pesticides present in the soil during a metagenomic study for the comparative microbial dynamics in pesticide-contaminated locations [72]. The degradation of -hexachlorocyclohexane by *Rhodococcus sp.*, *Sphingomonas sp.*, and *Pseudomonas sp.* was found to be 58%; however, the addition of rhamnolipid JBR 425 (a commercial biosurfactant generated from *P. aeruginosa*) at 20-100 M increased the degradation to 64% compared to Triton X (synthetic surfactant). Endosulfan was reported to degrade at a maximum of 61% in the presence of rhamnolipid, at 50% when Triton X-100 was used, and at 50% when no surfactant was present. Moreover, for DDT, 38% for Triton X-100 and 34% by enhanced bacteria alone were reported to degrade at a rate of 57% in the presence of rhamnolipid. This suggested that the degrader strain's capacity to degrade was significantly impacted by the addition of biosurfactants and that microbial biosurfactants were more effective than manufactured ones [72].

Marine/Aquatic

Accidental hydrocarbon spills into aquatic environments, such as lakes, ponds, bays, or seas, and have a significant impact on the aquatic food chain, resulting in the extinction of aquatic species, *etc.* Because of the continued devastation that oil spills have caused to the marine ecosystem, oil is one of the most abundant organic contaminants in the sea [74]. Polyaromatic hydrocarbons, monoaromatic hydrocarbons, and heavy metals are the main pollutants that the petroleum industry produces and releases into the environment. These pollutants have a negative impact on humans, plants, animals, and microbes [73]. Due to their surface-active nature, biosurfactants improve the effectiveness of biodegradation, assisting in the cleanup of these contaminants [75]. Instead, chemical dispersants are harmful to aquatic life; hence they are recommended to use a bio-based, non-toxic solution [74, 76]. One possible use for the crude biosurfactant produced by *B. subtilis* and *P. aeruginosa* is microbial-improved oil recovery [74]. When the surface properties of the biosurfactant improve, so do the hydrocarbon pollutants' solubility and mobility. A biosurfactant's effectiveness in marine remediation may be determined by looking at its CMC value; one with a low CMC of less than 1 to 1.5 is particularly successful in accelerating biodegradation [74]. 80% hydrocarbon breakdown was shown by two biosurfactant-producing bacteria, *Ralstonia pickettii* and *Alcaligenes piechaudii*. The strains produced biosurfactant and digested hydrocarbons concurrently. The production of biosurfactants was linked to better hydrocarbon degradation through higher hydrocarbon transport and absorption [77].

According to reports, biosurfactant was essential in cleaning up 5.73 million tonnes of oil spills in water globally between 1970 and 2016. To encourage microbial-assisted degradation, the dispersion agent biosurfactant was created [78]. In order to effectively clean up oil spills in marine environments, it has been suggested that surface active chemicals produced by marine microorganisms such as the *Halomonas* species can be employed as biosurfactants. of hydrocarbons and start the bioavailability of those substances for processes of oil recovery and degradation [78]. Nievas and co-workers undertook a decomposition attempt on the hazardous bilge waste, a fuel oil-like residue made up of a combination of seawater, n-alkanes, which resolved total hydrocarbons, and unsolvable complicated mixes using a microbial community that creates emulsifiers (branched, cyclic, and aromatic hydrocarbons). N-alkane, resolvent hydrocarbon, and insolvent reduction combinations in fewer than 6 days were about 91%, 78%, and 61%, respectively [79]. Studies claim that marine bacteria's production of a biosurfactant decreased oil slicks from the water's surface by promoting the dispersion of oil in water and the creation of a stable emulsion [80]. It is advised to employ biosurfactants to clean up oil spills on shorelines and in the water as a result. The marine bacterial isolate reported to be capable of digesting the hydrocarbon-polluted marine environment is named hydrocarbonoclatic bacteria [80]. Rhamnolipid biosurfactant enabled the discovery of more than 15 carbon chain alkanes to biodegrade faster. *Alcaligenes, Alcanivorax, Rhodococcus, Bacillus, Acinetobacter, Pseudomonas,* and *Corynebacterium* are biosurfactant-producing bacterial species that have been used in numerous recent studies to remove hydrocarbons from oil-polluted environments with better remediation rates [81].

In addition, heavy metal removal from aquatic settings was claimed to be possible with biosurfactants. Rhamnolipid from *Pseudomonas aeruginosa* has been found to be more efficient (than sodium dodecyl sulphate) at removing heavy metals from intertidal silt. At a critical micelle concentration of 43.73 mg/L, it was revealed that rhamnolipid was capable of removing 50.20% mercury (Hg) and 62.50% lead (Pb). The initial concentrations of Hg were also analysed [82]. Also, it was discovered that the rhamnolipid improved the breakdown of heavy metals from river sediments. Rhamnolipid greatly improved the washing efficiency when added, and at a concentration of 3%, it removed 47.85, 86.87, 80.21, and 63.54% of Cr, Cd, Cu, and Pd, respectively. High pH and extended washing times were discovered to be ideal for effective washing [83].

PERSPECTIVES AND BOTTLENECKS

Despite having remarkable environmental applications and showing a great potential for replacing synthetic, petroleum-derived surfactants, the market for biosurfactants is inadequate (s). They incur hefty production expenses [84]. The price of the substrates, the subsequent processing, and the yield of the biosurfactant are a few aspects contributing to the high cost of production [85]. Co-production may be used to create additional value-added goods simultaneously, which might boost the manufacturing process' profitability [86, 87]. As it is costly and difficult to purify goods with identical polarity against organic solvents, the product recovery process is the main constraint here. Existing methods have been modified to address this issue such that changing the substrate might lead to the production of one product while preventing the expression of another, similar to how Roelants and colleagues did the modification when high glucose levels were present. Although suppressing the oxidation pathway, which lowers or stops the formation of polyhydroxyalkanoates, Candida species boost the production of sophorolipid biosurfactants [88]. It will be much easier to remove oil when microbial strains have the ability to grow actively and produce biosurfactants in anaerobic settings. Using genetic engineering techniques, it is possible to create modified strains that release biosurfactants that can be injected into reservoirs to recover the leftover oils [89]. Also, by applying omics methodologies that may lead to the discovery of novel nature biosurfactants, the tedious and time-consuming biosurfactant screening procedures can be replaced. A key drawback of the majority of investigated glycolipid biosurfactants has been recognized as the lack of structural variation, which might be increased using any of these techniques [88]. Future studies may thus focus on using a cheap, renewable substrate for biosurfactant synthesis by modified strains that will produce molecules that are specifically adapted for different purposes.

CONCLUSION

The current review gives details on advanced techniques and methods for locating biosurfactants and their makers. Co-production may increase the industrial usability of biosurfactants by making them more cost-effective in production. Backspace identification of new biosurfactants has been made possible because of the use of metabolomic and metagenomic technologies, which have also enhanced and expedited the identification process. As a result, we now have a better grasp of the dynamics and structure of microbial communities, and the technology is now more dependable and secure. The pollution load in terrestrial and aquatic habitats was significantly reduced by the use of biosurfactants. This research included a thorough examination of the engineering approaches that might increase the production of a specific biosurfactant and increase the cost-effectiveness of the process as a whole. Future development in this field of study could be the metabolic engineering of microbes for product specificity.

REFERENCES

[1] Mohanty, S.S.; Koul, Y.; Varjani, S.; Pandey, A.; Ngo, H.H.; Chang, J.S.; Wong, J.W.C.; Bui, X.T. A critical review on various feedstocks as sustainable substrates for biosurfactants production: A way towards cleaner production. *Microb. Cell Fact.,* **2021**, *20*(1), 120.
[http://dx.doi.org/10.1186/s12934-021-01613-3] [PMID: 34174898]

[2] Varjani, S.J.; Upasani, V.N. Critical review on biosurfactant analysis, purification and characterization using rhamnolipid as a model biosurfactant. *Bioresour. Technol.,* **2017**, *232*, 389-397.
[http://dx.doi.org/10.1016/j.biortech.2017.02.047] [PMID: 28238638]

[3] Gaur, V.K.; Tripathi, V.; Gupta, P.; Dhiman, N.; Regar, R.K.; Gautam, K.; Srivastava, J.K.; Patnaik, S.; Patel, D.K.; Manickam, N. Rhamnolipids from *Planococcus* spp. and their mechanism of action against pathogenic bacteria. *Bioresour. Technol.,* **2020**, *307*, 123206.

[4] Liu, G.; Zhong, H.; Yang, X.; Liu, Y.; Shao, B.; Liu, Z. Advances in applications of rhamnolipids biosurfactant in environmental remediation: A review. *Biotechnol. Bioeng.,* **2018**, *115*(4), 796-814.
[http://dx.doi.org/10.1002/bit.26517] [PMID: 29240227]

[5] Devda, V.; Chaudhary, K.; Varjani, S.; Pathak, B.; Patel, A.K.; Singhania, R.R.; Taherzadeh, M.J.; Ngo, H.H.; Wong, J.W.C.; Guo, W.; Chaturvedi, P. **2021**.
[http://dx.doi.org/10.1080/21655979.2021.1946631]

[6] Sharma, P.; Pandey, A.K.; Kim, S.-H.; Singh, S.P.; Chaturvedi, P.; Varjani, S. Critical review on microbial community during *in-situ* bioremediation of heavy metals from industrial wastewater. *Environ. Technol. Innov.,* **2021**, 101826.
[http://dx.doi.org/10.1016/j.eti.2021.101826]

[7] Varjani, S.; Pandey, A.; Upasani, V.N. Petroleum sludge polluted soil remediation: Integrated approach involving novel bacterial consortium and nutrient application. *Sci. Total Environ.,* **2021**, *763*, 142934.
[http://dx.doi.org/10.1016/j.scitotenv.2020.142934] [PMID: 33268247]

[8] Gaur, V.K.; Manickam, N. Microbial biosurfactants: Production and applications in circular bioeconomy. In: *Biomass, Biofuels, Biochemicals*; Elsevier, **2021**; pp. 353-378.
[http://dx.doi.org/10.1016/B978-0-12-821878-5.00011-8]

[9] Gaur, V.K.; Sharma, P.; Sirohi, R.; Varjani, S.; Taherzadeh, M.J.; Chang, J.S.; Yong Ng, H.; Wong, J.W.C.; Kim, S.H. Production of biosurfactants from agro-industrial waste and waste cooking oil in a circular bioeconomy: An overview. *Bioresour. Technol.,* **2022**, *343*, 126059.
[http://dx.doi.org/10.1016/j.biortech.2021.126059] [PMID: 34606921]

[10] Wisjnuprapto, N.A.; Helmy, Q.; Kardena, E.; Funamizu, N. Strategies toward commercial scale of biosurfactant production as potential substitute for it's chemically counterparts. *Int. J. Biotechnol.,* **2011**, *12*(1/2), 66-86.
[http://dx.doi.org/10.1504/IJBT.2011.042682]

[11] Jackson, S.A.; Borchert, E.; O'Gara, F.; Dobson, A.D.W. Metagenomics for the discovery of novel biosurfactants of environmental interest from marine ecosystems. *Curr. Opin. Biotechnol.,* **2015**, *33*, 176-182.

[http://dx.doi.org/10.1016/j.copbio.2015.03.004] [PMID: 25812477]

[12]   Markande, A.R.; Patel, D.; Varjani, S. A review on biosurfactants: Properties, applications and current developments. *Bioresour. Technol.,* **2021,** *330,* 124963.
[http://dx.doi.org/10.1016/j.biortech.2021.124963] [PMID: 33744735]

[13]   Varjani, S.J.; Upasani, V.N. Carbon spectrum utilization by an indigenous strain of *Pseudomonas aeruginosa* NCIM 5514: Production, characterization and surface active properties of biosurfactant. *Bioresour. Technol.,* **2016,** *221,* 510-516.
[http://dx.doi.org/10.1016/j.biortech.2016.09.080] [PMID: 27677153]

[14]   Datta, S.; Rajnish, K.N.; Samuel, M.S.; Pugazlendhi, A.; Selvarajan, E. Metagenomic applications in microbial diversity, bioremediation, pollution monitoring, enzyme and drug discovery. A review. *Environ. Chem. Lett.,* **2020,** *18*(4), 1229-1241.
[http://dx.doi.org/10.1007/s10311-020-01010-z]

[15]   Maron, P.A.; Ranjard, L.; Mougel, C.; Lemanceau, P. Metaproteomics: a new approach for studying functional microbial ecology. *Microb. Ecol.,* **2007,** *53*(3), 486-493.
[http://dx.doi.org/10.1007/s00248-006-9196-8] [PMID: 17431707]

[16]   Das, A.J.; Ambust, S.; Singh, T.; Kumar, R. Biosurfactant assisted design treatments for remediation of petroleum contaminated soil and metabolomics based interactive study with Brassica nigra L. *Environ. Challen.,* **2021,** *4,* 100080.
[http://dx.doi.org/10.1016/j.envc.2021.100080]

[17]   Palma Esposito, F.; Giugliano, R.; Della Sala, G.; Vitale, G.A.; Buonocore, C.; Ausuri, J.; Galasso, C.; Coppola, D.; Franci, G.; Galdiero, M. **2021.**
[http://dx.doi.org/10.3390/ijms22169055]

[18]   Williams, W.; Trindade, M. Metagenomics for the discovery of novel biosurfactants. In: *Functional Metagenomics: Tools and Applications*; Springer, **2017**; pp. 95-117.
[http://dx.doi.org/10.1007/978-3-319-61510-3_6]

[19]   Patel, G.B.; Shah, K.R.; Shindhal, T.; Rakholiya, P.; Varjani, S. Process parameter studies by central composite design of response surface methodology for lipase activity of newly obtained Actinomycete. *Environm. Technol. Innov.,* **2021,** *23,* 101724.
[http://dx.doi.org/10.1016/j.eti.2021.101724]

[20]   Gerc, A.J.; Stanley-Wall, N.R.; Coulthurst, S.J. Role of the phosphopantetheinyltransferase enzyme, PswP, in the biosynthesis of antimicrobial secondary metabolites by Serratia marcescens Db10. *Microbiology,* **2014,** *160*(8), 1609-1617.
[http://dx.doi.org/10.1099/mic.0.078576-0] [PMID: 24847000]

[21]   Grosso-Becerra, M.V.; González-Valdez, A.; Granados-Martínez, M.J.; Morales, E.; Servín-González, L.; Méndez, J.L.; Delgado, G.; Morales-Espinosa, R.; Ponce-Soto, G.Y.; Cocotl-Yañez, M.; Soberón-Chávez, G. *Pseudomonas aeruginosa* ATCC 9027 is a non-virulent strain suitable for mono-rhamnolipids production. *Appl. Microbiol. Biotechnol.,* **2016,** *100*(23), 9995-10004.
[http://dx.doi.org/10.1007/s00253-016-7789-9] [PMID: 27566690]

[22]   Roelants, S.L.K.W.; Saerens, K.M.J.; Derycke, T.; Li, B.; Lin, Y.C.; Van de Peer, Y.; De Maeseneire, S.L.; Van Bogaert, I.N.A.; Soetaert, W. *Candida bombicola* as a platform organism for the production of tailor-made biomolecules. *Biotechnol. Bioeng.,* **2013,** *110*(9), 2494-2503.
[http://dx.doi.org/10.1002/bit.24895] [PMID: 23475585]

[23]   Das, A.J.; Kumar, R. Production of biosurfactant from agro-industrial waste by *Bacillus safensis* J2 and exploring its oil recovery efficiency and role in restoration of diesel contaminated soil. *Environ. Technol. Innov.,* **2019,** *16,* 100450.
[http://dx.doi.org/10.1016/j.eti.2019.100450]

[24]   Feng, S.; Hao Ngo, H.; Guo, W.; Woong Chang, S.; Duc Nguyen, D.; Cheng, D.; Varjani, S.; Lei, Z.; Liu, Y. Roles and applications of enzymes for resistant pollutants removal in wastewater treatment. *Bioresour. Technol.,* **2021,** *335,* 125278.

[http://dx.doi.org/10.1016/j.biortech.2021.125278] [PMID: 34015565]

[25] Shah, A.V.; Srivastava, V.K.; Mohanty, S.S.; Varjani, S. Municipal solid waste as a sustainable resource for energy production: State-of-the-art review. *J. Environ. Chem. Eng.,* **2021**, *9*(4), 105717.
[http://dx.doi.org/10.1016/j.jece.2021.105717]

[26] Kavuthodi, B.; Thomas, S.K.; Sebastian, D. Co-production of pectinase and biosurfactant by the newly isolated strainBacillus subtilis BKDS1. *Microbiol. Res. J. Int.,* **2015**, 1-12.

[27] Hmidet, N.; Jemil, N.; Nasri, M. Simultaneous production of alkaline amylase and biosurfactant by Bacillus methylotrophicus DCS1: application as detergent additive. *Biodegradation,* **2019**, *30*(4), 247-258.
[http://dx.doi.org/10.1007/s10532-018-9847-8] [PMID: 30097752]

[28] Guzmán-López, O.; Cuevas-Díaz, M. del C. Martínez Toledo, A., Contreras-Morales, M.E., Ruiz-Reyes, C.I., Ortega Martínez. A. del C. **2021**.
[http://dx.doi.org/10.1080/02757540.2021.1909003]

[29] Shekhar, S.; Sundaramanickam, A.; Balasubramanian, T. Biosurfactant producing microbes and their potential applications: A review. *Crit. Rev. Environ. Sci. Technol.,* **2015**, *45*(14), 1522-1554.
[http://dx.doi.org/10.1080/10643389.2014.955631]

[30] Li, T.; Elhadi, D.; Chen, G.Q. Co-production of microbial polyhydroxyalkanoates with other chemicals. *Metab. Eng.,* **2017**, *43*(Pt A), 29-36.
[http://dx.doi.org/10.1016/j.ymben.2017.07.007] [PMID: 28782693]

[31] Pantazaki, A.A.; Papaneophytou, C.P.; Lambropoulou, D.A. Simultaneous polyhydroxyalkanoates and rhamnolipids production by Thermus thermophilus HB8. *AMB Express,* **2011**, *1*(1), 17.
[http://dx.doi.org/10.1186/2191-0855-1-17] [PMID: 21906373]

[32] Varjani, S.; Upasani, V.N. Evaluation of rhamnolipid production by a halotolerant novel strain of *Pseudomonas aeruginosa. Bioresour. Technol.,* **2019**, *288*, 121577.
[http://dx.doi.org/10.1016/j.biortech.2019.121577] [PMID: 31174086]

[33] Hori, K.; Ichinohe, R.; Unno, H.; Marsudi, S. Simultaneous syntheses of polyhydroxyalkanoates and rhamnolipids by *Pseudomonas aeruginosa* IFO3924 at various temperatures and from various fatty acids. *Biochem. Eng. J.,* **2011**, *53*(2), 196-202.
[http://dx.doi.org/10.1016/j.bej.2010.10.011]

[34] Kourmentza, C.; Freitas, F.; Alves, V.; Reis, M.A.M. Microbial conversion of waste and surplus materials into high-value added products: The case of biosurfactants. In: *Microbial Applications*; Springer, **2017**; 1, pp. 29-77.
[http://dx.doi.org/10.1007/978-3-319-52666-9_2]

[35] Funston, S.J.; Tsaousi, K.; Smyth, T.J.; Twigg, M.S.; Marchant, R.; Banat, I.M. Enhanced rhamnolipid production in Burkholderia thailandensis transposon knockout strains deficient in polyhydroxyalkanoate (PHA) synthesis. *Appl. Microbiol. Biotechnol.,* **2017**, *101*(23-24), 8443-8454.
[http://dx.doi.org/10.1007/s00253-017-8540-x] [PMID: 29043376]

[36] Worley, B.; Powers, R. Multivariate analysis in metabolomics. *Curr. Metabolomics,* **2013**, *1*(1), 92-107.
[PMID: 26078916]

[37] Floros, D.J.; Jensen, P.R.; Dorrestein, P.C.; Koyama, N. A metabolomics guided exploration of marine natural product chemical space. *Metabolomics,* **2016**, *12*(9), 145.
[http://dx.doi.org/10.1007/s11306-016-1087-5] [PMID: 28819353]

[38] Gaur, V.K.; Manickam, N. Microbial production of rhamnolipid: Synthesis and potential application in bioremediation of hydrophobic pollutants. In: *Microbial and Natural Macromolecules*; Elsevier, **2021**; pp. 143-176.
[http://dx.doi.org/10.1016/B978-0-12-820084-1.00007-7]

[39] Adetunji, C.O.; Jeevanandam, J.; Anani, O.A.; Inobeme, A.; Thangadurai, D.; Islam, S.; Olaniyan,

O.T. Strain improvement methodology and genetic engineering that could lead to an increase in the production of biosurfactants. In: *Green Sustainable Process for Chemical and Environmental Engineering and Science*; Elsevier, **2021**; pp. 299-315.
[http://dx.doi.org/10.1016/B978-0-12-823380-1.00002-2]

[40] Mácha, H.; Marešová, H.; Juříková, T.; Švecová, M.; Benada, O.; Škríba, A.; Baránek, M.; Novotný, Č.; Palyzová, A. *Killing effect of Bacillus velezensis,* **2021**.

[41] Nguyen, D.D.; Melnik, A.V.; Koyama, N.; Lu, X.; Schorn, M.; Fang, J.; Aguinaldo, K.; Lincecum, T.L., Jr; Ghequire, M.G.K.; Carrion, V.J.; Cheng, T.L.; Duggan, B.M.; Malone, J.G.; Mauchline, T.H.; Sanchez, L.M.; Kilpatrick, A.M.; Raaijmakers, J.M.; De Mot, R.; Moore, B.S.; Medema, M.H.; Dorrestein, P.C. Indexing the *Pseudomonas* specialized metabolome enabled the discovery of poaeamide B and the bananamides. *Nat. Microbiol.,* **2016**, *2*(1), 16197.
[http://dx.doi.org/10.1038/nmicrobiol.2016.197] [PMID: 27798598]

[42] Clements, T.; Rautenbach, M.; Ndlovu, T.; Khan, S.; Khan, W. A metabolomics and molecular networking approach to elucidate the structures of secondary metabolites produced by Serratia marcescens strains. *Front Chem.,* **2021**, *9*, 633870.
[http://dx.doi.org/10.3389/fchem.2021.633870] [PMID: 33796505]

[43] Buedenbender, L.; Kumar, A.; Blümel, M.; Kempken, F.; Tasdemir, D. Genomics-and metabolomics-based investigation of the deep-sea sediment-derived yeast, Rhodotorula mucilaginosa 50-3-19/20B. *Mar. Drugs,* **2020**, *19*(1), 14.
[http://dx.doi.org/10.3390/md19010014] [PMID: 33396687]

[44] Varjani, S.J. Remediation processes for petroleum oil polluted soil. *Indian J. Biotechnol.,* **2017**, *16*, 157-163.

[45] Dhanjal, D.S.; Sharma, D. Microbial metagenomics for industrial and environmental bioprospecting: The unknown envoy. In: *Microbial Bioprospecting for Sustainable Development*; Springer, **2018**; pp. 327-352.
[http://dx.doi.org/10.1007/978-981-13-0053-0_18]

[46] Malik, G.; Arora, R.; Chaturvedi, R.; Paul, M.S. Implementation of genetic engineering and novel omics approaches to enhance bioremediation: A focused review. *Bull. Environ. Contam. Toxicol.,* **2021**, 1-8.
[PMID: 33837794]

[47] Ngara, T.R.; Zhang, H. Recent advances in function-based metagenomic screening. *Genomics Proteomics Bioinformatics,* **2018**, *16*(6), 405-415.
[http://dx.doi.org/10.1016/j.gpb.2018.01.002] [PMID: 30597257]

[48] Hazen, T.C.; Rocha, A.M.; Techtmann, S.M. Advances in monitoring environmental microbes. *Curr. Opin. Biotechnol.,* **2013**, *24*(3), 526-533.
[http://dx.doi.org/10.1016/j.copbio.2012.10.020] [PMID: 23183250]

[49] Ranjan, R.; Rani, A.; Kumar, R. Exploration of microbial cells: The storehouse of bio-wealth through metagenomics and metatranscriptomics. *Microb. Factories,* **2015**, 7-27.

[50] Araújo, S.C.S.; Silva-Portela, R.C.B.; de Lima, D.C.; da Fonsêca, M.M.B.; Araújo, W.J.; da Silva, U.B.; Napp, A.P.; Pereira, E.; Vainstein, M.H.; Agnez-Lima, L.F. MBSP1: a biosurfactant protein derived from a metagenomic library with activity in oil degradation. *Sci. Rep.,* **2020**, *10*(1), 1340.
[http://dx.doi.org/10.1038/s41598-020-58330-x] [PMID: 31992807]

[51] Williams, W.; Kunorozva, L.; Klaiber, I.; Henkel, M.; Pfannstiel, J.; Van Zyl, L.J.; Hausmann, R.; Burger, A.; Trindade, M. Novel metagenome-derived ornithine lipids identified by functional screening for biosurfactants. *Appl. Microbiol. Biotechnol.,* **2019**, *103*(11), 4429-4441.
[http://dx.doi.org/10.1007/s00253-019-09768-1] [PMID: 30972461]

[52] Zhou, D.; Hu, F.; Lin, J.; Wang, W.; Li, S. Genome and transcriptome analysis of *Bacillus velezensis* BS-37, an efficient surfactin producer from glycerol, in response to d-/l-leucine. *Microbiology Open,* **2019**, *8*(8), e00794.

[http://dx.doi.org/10.1002/mbo3.794] [PMID: 30793535]

[53]   Zhi, Y.; Wu, Q.; Xu, Y. Genome and transcriptome analysis of surfactin biosynthesis in *Bacillus amyloliquefaciens* MT45. *Sci. Rep.,* **2017**, *7*(1), 40976.
[http://dx.doi.org/10.1038/srep40976] [PMID: 28112210]

[54]   Zhang, Y.; Thompson, K.N.; Branck, T.; Yan Yan, ; Nguyen, L.H.; Franzosa, E.A.; Huttenhower, C. Metatranscriptomics for the human microbiome and microbial community functional profiling. *Annu. Rev. Biomed. Data Sci.,* **2021**, *4*(1), 279-311.
[http://dx.doi.org/10.1146/annurev-biodatasci-031121-103035] [PMID: 34465175]

[55]   Gaur, V.K.; Gupta, S.; Pandey, A. Evolution in mitigation approaches for petroleum oil-polluted environment: Recent advances and future directions. *Environ. Sci. Pollut. Res. Int.,* **2021**, 1-17.
[PMID: 34420173]

[56]   Pitocchi, R.; Cicatiello, P.; Birolo, L.; Piscitelli, A.; Bovio, E.; Varese, G.C.; Giardina, P. Cerato-platanins from marine fungi as effective protein biosurfactants and bioemulsifiers. *Int. J. Mol. Sci.,* **2020**, *21*(8), 2913.
[http://dx.doi.org/10.3390/ijms21082913] [PMID: 32326352]

[57]   Ortega Ramirez, C.A.; Kwan, A.; Li, Q.X. Rhamnolipids induced by glycerol enhance dibenzothiophene biodegradation in *Burkholderia* sp. C3. *Engineering,* **2020**, *6*(5), 533-540.
[http://dx.doi.org/10.1016/j.eng.2020.01.006]

[58]   Moutinho, L.F.; Moura, F.R.; Silvestre, R.C.; Romão-Dumaresq, A.S. Microbial biosurfactants: A broad analysis of properties, applications, biosynthesis, and techno-economical assessment of rhamnolipid production. *Biotechnol. Prog.,* **2021**, *37*(2), e3093.
[http://dx.doi.org/10.1002/btpr.3093] [PMID: 33067929]

[59]   Dobler, L.; Vilela, L.F.; Almeida, R.V.; Neves, B.C. Rhamnolipids in perspective: Gene regulatory pathways, metabolic engineering, production and technological forecasting. *N. Biotechnol.,* **2016**, *33*(1), 123-135.
[http://dx.doi.org/10.1016/j.nbt.2015.09.005] [PMID: 26409933]

[60]   Zhang, F.; Huo, K.; Song, X.; Quan, Y.; Gao, W.; Wang, S.; Yang, C. Engineering modification of genome-reduced strain *Bacillus amyloliquefaciens* for enhancing surfactin production. *Res. Sq.,* **2020**.
[http://dx.doi.org/10.21203/rs.3.rs-41198/v3]

[61]   Becker, F.; Stehlik, T.; Linne, U.; Bölker, M.; Freitag, J.; Sandrock, B. Engineering *Ustilago maydis* for production of tailor-made mannosylerythritol lipids. *Metab. Eng. Commun.,* **2021**, *12*, e00165.
[http://dx.doi.org/10.1016/j.mec.2021.e00165] [PMID: 33659181]

[62]   Tiso, T.; Ihling, N.; Kubicki, S.; Biselli, A.; Schonhoff, A.; Bator, I.; Thies, S.; Karmainski, T.; Kruth, S.; Willenbrink, A.L.; Loeschcke, A.; Zapp, P.; Jupke, A.; Jaeger, K.E.; Büchs, J.; Blank, L.M. Integration of genetic and process engineering for optimized rhamnolipid production using *Pseudomonas putida. Front. Bioeng. Biotechnol.,* **2020**, *8*(976), 976.
[http://dx.doi.org/10.3389/fbioe.2020.00976] [PMID: 32974309]

[63]   Wu, Q.; Zhi, Y.; Xu, Y. Systematically engineering the biosynthesis of a green biosurfactant surfactin by *Bacillus subtilis* 168. *Metab. Eng.,* **2019**, *52*, 87-97.
[http://dx.doi.org/10.1016/j.ymben.2018.11.004] [PMID: 30453038]

[64]   Rathankumar, A.K.; Saikia, K.; Kumar, P.S.; Varjani, S.; Kalita, S.; Bharadwaj, N.; George, J.; Kumar, V.V. Surfactant aided mycoremediation of soil contaminated with polycyclic aromatic hydrocarbon (PAHs): Progress, limitation and countermeasures. *J. Chem. Technol. Biotechnol.,* **2021**, 1-18.
[http://dx.doi.org/10.1002/jctb.6721]

[65]   Bustamante, M.; Durán, N.; Diez, M.C. Biosurfactants are useful tools for the bioremediation of contaminated soil: a review. *J. Soil Sci. Plant Nutr.,* **2012**, *12*(667), 687.
[http://dx.doi.org/10.4067/S0718-95162012005000024]

[66]   Mulligan, C.N. Sustainable remediation of contaminated soil using biosurfactants. *Front. Bioeng. Biotechnol.,* **2021**, *9*(195), 635196.
[http://dx.doi.org/10.3389/fbioe.2021.635196] [PMID: 33791286]

[67]   Mohanty, S.; Jasmine, J.; Mukherji, S. Practical considerations and challenges involved in surfactant enhanced bioremediation of oil. *BioMed Res. Int.,* **2013**, *2013*, 1-16.
[http://dx.doi.org/10.1155/2013/328608] [PMID: 24350261]

[68]   Ambust, S.; Das, A.J.; Kumar, R. Bioremediation of petroleum contaminated soil through biosurfactant and *Pseudomonas sp.* SA3 amended design treatments. *Current Research in Microbial Sciences,* **2021**, *2*, 100031.
[http://dx.doi.org/10.1016/j.crmicr.2021.100031] [PMID: 34841322]

[69]   Souza, E.C.; Vessoni-Penna, T.C.; de Souza Oliveira, R.P. Biosurfactant-enhanced hydrocarbon bioremediation: An overview. *Int. Biodeterior. Biodegradation,* **2014**, *89*, 88-94.
[http://dx.doi.org/10.1016/j.ibiod.2014.01.007]

[70]   Pradeep, N.V.; Anupama, S.; Anitha, G.; Renukamma, A.S.; Afreen, S.S. Bioremediation of oil contaminated soil using biosurfactant produced by *Pseudomonas aeruginosa. J. Res. Biol.,* **2012**, *2*, 281-286.

[71]   Kang, S.W.; Kim, Y.B.; Shin, J.D.; Kim, E.K. Enhanced biodegradation of hydrocarbons in soil by microbial biosurfactant, sophorolipid. *Appl. Biochem. Biotechnol.,* **2010**, *160*(3), 780-790.
[http://dx.doi.org/10.1007/s12010-009-8580-5] [PMID: 19253005]

[72]   Regar, R.K.; Gaur, V.K.; Bajaj, A.; Tambat, S.; Manickam, N. Comparative microbiome analysis of two different long-term pesticide contaminated soils revealed the anthropogenic influence on functional potential of microbial communities. *Sci. Total Environ.,* **2019**, *681*, 413-423.
[http://dx.doi.org/10.1016/j.scitotenv.2019.05.090] [PMID: 31108361]

[73]   Varjani, S.J.; Gnansounou, E.; Pandey, A. Comprehensive review on toxicity of persistent organic pollutants from petroleum refinery waste and their degradation by microorganisms. *Chemosphere,* **2017**, *188*, 280-291.
[http://dx.doi.org/10.1016/j.chemosphere.2017.09.005] [PMID: 28888116]

[74]   Shindhal, T.; Rakholiya, P.; Varjani, S.; Pandey, A.; Ngo, H.H.; Guo, W.; Ng, H.Y.; Taherzadeh, M.J. A critical review on advances in the practices and perspectives for the treatment of dye industry wastewater. *Bioengineered,* **2021**, *12*(1), 70-87.
[http://dx.doi.org/10.1080/21655979.2020.1863034] [PMID: 33356799]

[75]   Fenibo, E.O.; Ijoma, G.N.; Selvarajan, R.; Chikere, C.B. Microbial surfactants: The next generation multifunctional biomolecules for applications in the petroleum industry and its associated environmental remediation. *Microorganisms,* **2019**, *7*(11), 581.
[http://dx.doi.org/10.3390/microorganisms7110581] [PMID: 31752381]

[76]   Dang, H.; Klotz, M.G.; Lovell, C.R.; Sievert, S.M. The responses of marine microorganisms, communities and ecofunctions to environmental gradients. *Front. Microbiol.,* **2019**, *10*, 115.
[http://dx.doi.org/10.3389/fmicb.2019.00115] [PMID: 30800101]

[77]   Płaza, G.A.; Łukasik, K.; Wypych, J.; Nałęcz-Jawecki, G.; Berry, C.; Brigmon, R.L. Biodegradation of crude oil and distillation products by biosurfactant-producing bacteria. *Pol. J. Environ. Stud.,* **2008**, 17.

[78]   Tripathi, V.; Gaur, V.K.; Dhiman, N.; Gautam, K.; Manickam, N. Characterization and properties of the biosurfactant produced by PAH-degrading bacteria isolated from contaminated oily sludge environment. *Environ. Sci. Pollut. Res. Int.,* **2020**, *27*(22), 27268-27278.
[http://dx.doi.org/10.1007/s11356-019-05591-3] [PMID: 31190304]

[79]   Nievas, M.L.; Commendatore, M.G.; Esteves, J.L.; Bucalá, V. Biodegradation pattern of hydrocarbons from a fuel oil-type complex residue by an emulsifier-producing microbial consortium. *J. Hazard. Mater.,* **2008**, *154*(1-3), 96-104.

[http://dx.doi.org/10.1016/j.jhazmat.2007.09.112] [PMID: 17997031]

[80]   Karlapudi, A.P.; Venkateswarulu, T.C.; Tammineedi, J.; Kanumuri, L.; Ravuru, B.K.; Dirisala, V.; Kodali, V.P. Role of biosurfactants in bioremediation of oil pollution : A review. *Petroleum,* **2018**, *4*(3), 241-249.
[http://dx.doi.org/10.1016/j.petlm.2018.03.007]

[81]   Lee, D.W.; Lee, H.; Kwon, B.O.; Khim, J.S.; Yim, U.H.; Kim, B.S.; Kim, J.J. Biosurfactant-assisted bioremediation of crude oil by indigenous bacteria isolated from Taean beach sediment. *Environ. Pollut.,* **2018**, *241*, 254-264.
[http://dx.doi.org/10.1016/j.envpol.2018.05.070] [PMID: 29807284]

[82]   Chen, Q.; Li, Y.; Liu, M.; Zhu, B.; Mu, J.; Chen, Z. Removal of Pb and Hg from marine intertidal sediment by using rhamnolipid biosurfactant produced by a *Pseudomonas aeruginosa* strain. *Environ. Technol. Innov.,* **2021**, *22*, 101456.
[http://dx.doi.org/10.1016/j.eti.2021.101456]

[83]   Chen, W.; Qu, Y.; Xu, Z.; He, F.; Chen, Z.; Huang, S.; Li, Y. Heavy metal (Cu, Cd, Pb, Cr) washing from river sediment using biosurfactant rhamnolipid. *Environ. Sci. Pollut. Res. Int.,* **2017**, *24*(19), 16344-16350.
[http://dx.doi.org/10.1007/s11356-017-9272-2] [PMID: 28547372]

[84]   Mishra, B.; Varjani, S.; Agrawal, D.C.; Mandal, S.K.; Ngo, H.H.; Taherzadeh, M.J.; Chang, J.S.; You, S.; Guo, W. Engineering biocatalytic material for the remediation of pollutants: A comprehensive review. *Environ. Technol. Innov.,* **2020**, *20*, 101063.
[http://dx.doi.org/10.1016/j.eti.2020.101063]

[85]   Prajapati, P.; Varjani, S.; Singhania, R.R.; Patel, A.K.; Awasthi, M.K.; Sindhu, R.; Zhang, Z.; Binod, P.; Awasthi, S.K.; Chaturvedi, P. Critical review on technological advancements for effective waste management of municipal solid waste : Updates and way forward. *Environ. Technol. Innov.,* **2021**, *23*, 101749.
[http://dx.doi.org/10.1016/j.eti.2021.101749]

[86]   Adesra, A.; Srivastava, V.K.; Varjani, S. Valorization of dairy wastes: Integrative approaches for value added products. *Indian J. Microbiol.,* **2021**, *61*(3), 270-278.
[http://dx.doi.org/10.1007/s12088-021-00943-5] [PMID: 34294992]

[87]   R, J.; Gurunathan, B.; K, S.; Varjani, S.; Ngo, H.H.; Gnansounou, E. Advancements in heavy metals removal from effluents employing nano-adsorbents: Way towards cleaner production. *Environ. Res.,* **2022**, *203*, 111815.
[http://dx.doi.org/10.1016/j.envres.2021.111815] [PMID: 34352231]

[88]   Roelants, S.L.K.W.; Ciesielska, K.; De Maeseneire, S.L.; Moens, H.; Everaert, B.; Verweire, S.; Denon, Q.; Vanlerberghe, B.; Van Bogaert, I.N.A.; Van der Meeren, P.; Devreese, B.; Soetaert, W. Towards the industrialization of new biosurfactants: Biotechnological opportunities for the lactone esterase gene from *Starmerella bombicola. Biotechnol. Bioeng.,* **2016**, *113*(3), 550-559.
[http://dx.doi.org/10.1002/bit.25815] [PMID: 26301720]

[89]   Quraishi, M.; Bhatia, S.K.; Pandit, S.; Gupta, P.K.; Rangarajan, V.; Lahiri, D.; Varjani, S.; Mehariya, S.; Yang, Y.H. Exploiting microbes in the petroleum field: Analyzing the credibility of microbial enhanced oil recovery (MEOR). *Energies,* **2021**, *14*(15), 4684.
[http://dx.doi.org/10.3390/en14154684]

SUBJECT INDEX

A

www.ingramcontent.com/pod-product-compliance
Lightning Source LLC
Chambersburg PA
CBHW050830220326
41598CB00006B/343